Distributed Energy Resources Management

Distributed Energy Resources Management

Special Issue Editor

Pedro Faria

MDPI • Basel • Beijing • Wuhan • Barcelona • Belgrade

MDPI

Special Issue Editor
Pedro Faria
GECAD—Research Group on Intelligent Engineering and Computing
for Advanced Innovation and Development
Portugal

Editorial Office
MDPI
St. Alban-Anlage 66
4052 Basel, Switzerland

This is a reprint of articles from the Special Issue published online in the open access journal *Energies* (ISSN 1996-1073) from 2017 to 2019 (available at: https://www.mdpi.com/journal/energies/special_issues/distributed_energy_resources_management)

For citation purposes, cite each article independently as indicated on the article page online and as indicated below:

LastName, A.A.; LastName, B.B.; LastName, C.C. Article Title. *Journal Name* **Year**, *Article Number*, Page Range.

ISBN 978-3-03897-718-6 (Pbk)
ISBN 978-3-03897-719-3 (PDF)

Contents

About the Special Issue Editor

Pedro Faria, who completed his Ph.D in 2016, works in the field of power systems with a focus on energy markets, smart grids, and demand response and has published 1 patent and over 150 papers. His current work includes renewable-based distributed generation, energy storage, and electric vehicles. In these fields, optimization, clustering, and classification methods have been applied to real and simulated environment problems. He has been developing business models for modeling, aggregation, and remuneration of consumers participating in electricity markets and in demand response programs. He has also worked in the real-time simulation of power and energy systems. Pedro Faria participated in a number of relevant national and international research projects, contributing with models and their implementation, testing, demonstration, and piloting.

energies

MDPI

Editorial

Distributed Energy Resources Management

Pedro Faria

GECAD-Research Group on Intelligent Engineering and Computing for Advanced Innovation and Development, Polytechnic of Porto, Rua Dr. Antonio Bernardino de Almeida, 431, 4200-072 Porto, Portugal; pnf@isep.ipp.pt; Tel.: +351-228-340-511; Fax: +351-228-321-159

Received: 29 November 2018; Accepted: 24 January 2019; Published: 11 February 2019

1. Introduction

The impact of distributed energy resources in the operation of power and energy systems is nowadays unquestionable at the distribution level but also at the whole power system management level. Increased flexibility is required to accommodate intermittent distributed generation and electric vehicle charging. Demand response has already been proven to have great potential to contribute to increased system efficiency while bringing additional benefits, especially to consumers. Distributed storage is also promising, particularly when used jointly with photovoltaic (PV) panels.

This Special Issue addresses the management of distributed energy resources, which is increasingly important to ensure sustainable and efficient power and energy systems. The issue focuses on methods and techniques to achieve optimized operation, aggregate the resources by means of virtual power players, and remunerate them. The integration of distributed resources in electricity markets is also addressed as a main path for the efficient use of resources.

The topics of this Special Issue include the following:

- Demand response
- Distributed energy resources
- Distributed generation
- Electric vehicles
- Energy resource optimization
- Energy storage
- Intelligent resource management
- Renewable energy sources
- Smart grids

Thirteen research papers have been published in this Special Issue. The following statistics apply:

- Submissions: 23, published: 13, rejected: 10
- Average article processing time: 58.76 days
- Authors' geographical distribution:

  - Spain (3), Portugal (3), China (3)
  - Korea (2), Denmark (2)
  - Italy (1), USA (1), Japan (1), India (1), Brazil (1)

2. Contributions

This paper provides a summary of the *Energies* Special Issue covering the published articles [1–13], which address several topics related to distributed energy resources management. Table 1 identifies the most relevant topics in each publication; most of them cover three or more topics.

Table 1. Topics covered in each publication.

Topic	References												
	[1]	[2]	[3]	[4]	[5]	[6]	[7]	[8]	[9]	[10]	[11]	[12]	[13]
Demand response		x	x	x			x	x		x		x	x
Distributed generation	x		x		x	x	x	x	x				x
Operation and control	x			x	x	x				x	x		
Electricity markets and aggregation		x	x					x	x			x	x
Energy storage		x			x	x							
Intelligent resource management				x	x	x	x		x				x
Renewable energy sources		x	x			x	x		x			x	
Laboratory simulation	x			x									
Total	3	4	4	4	4	5	4	3	4	2	1	3	4

One can see that, regarding the type of resources, most of the publications focus on demand response and distributed generation. Energy storage is also included in four papers. Looking at the proposed methods and/or addressed problems, most of the papers are dedicated to operation and control aspects and intelligent resource management. Electricity markets and resource aggregation are addressed in five papers. Specific challenges of integrating renewable energy sources are addressed in five papers. Finally, two papers make relevant contributions regarding laboratory simulation with some hardware for emulating power system components.

Reference [1] proposed a coordinated distributed control strategy for a hybrid AC/DC microgrid, taking into consideration several resource characteristics. A two-level control structure was developed, with local controllers linked to a central controller and a central controller that performs the energy management.

With a deep focus on demand response and aggregation, the authors of [2] developed a method of producing optimal bidding curves for an aggregator participating in day-ahead and intraday markets, with the objective of minimizing the costs of purchasing energy. The three-step approach involves optimal bidding to the day-ahead market, after the day-ahead market clearing when rescheduling is fulfilled, and new optimal bidding to the intraday market, taking advantage of the lower marginal prices.

Another perspective on energy trading and pricing is provided in [3], which formulates a hierarchical game between the energy provider as the leader and consumers as the followers. The uncertainty of the energy supply is also considered.

As seen in Table 1, one relevant topic for this Special Issue is simulation, which was addressed in [1,4]. Reference [4] presented a platform with real-time simulation skills adequate for demand response and distributed generation. The integration of centralized and distributed control approaches is discussed and validated through the emulation of power system components for a more realistic simulation of the microgrid and the validation of the computational models. A virtual power player manages the resources, aiming at minimizing operational costs.

A microgrid operation methodology was proposed in [5]. The economic operation strategy is devoted to both normal and emergency operation modes. Without a central controller, the proposed methodology is able to minimize the global operation cost. Looking more specifically at combined heat and power (CHP) generation, the microgrid operation costs were minimized in [6] by using the Lyapunov approach. Fault location detection is addressed in [11].

A multiagent-based approach is used in [9], supporting a decentralized method for microgrid restoration. In the proposed approach, local controllers are assigned to specific agents. The available information on generation and consumption is used to establish the best sequence for the restoration.

In [7], a predictive dispatch model was used for home energy management, and the uncertainty of PV generation is modeled by the InterStoch hybrid method. In the first stage of the method, day-ahead energy management is performed. The second stage runs in real time.

Energies **2019**, *12*, 550

From a different perspective, the discomfort costs associated with demand response and the generation costs are minimized in [10]. The discomfort costs are formulated based on Fanger thermal comfort.

Moving to large-size consumption and generation, reference [8] applied the cat swarm optimization technique to a demand–response-based unit commitment, including a real-time-based demand response program that is used during peak hours. The developed approach makes it possible to maximize the profit of both generation companies and demand response providers.

A case study of the Nordic electricity market was presented in [12]. It includes a strengths, weaknesses, opportunities, and threats (SWOT) analysis of four business models devoted to building participation in demand response programs. There is also a focus on aggregation aspects.

Finally, reference [13] presented a methodology addressing the rescheduling of resources in a sequence of the definition of a new aggregation and remuneration process. A representative tariff for each group of distributed energy resources is obtained.

Conflicts of Interest: The authors declare no conflict of interest.

References

1. Baek, J.; Choi, W.; Chae, S. Distributed Control Strategy for Autonomous Operation of Hybrid AC/DC Microgrid. *Energies* **2017**, *10*, 373. [CrossRef]
2. Ayón, X.; Moreno, M.; Usaola, J. Aggregators' Optimal Bidding Strategy in Sequential Day-Ahead and Intraday Electricity Spot Markets. *Energies* **2017**, *10*, 450. [CrossRef]
3. Ma, K.; Hu, S.; Yang, J.; Dou, C.; Guerrero, J. Energy Trading and Pricing in Microgrids with Uncertain Energy Supply: A Three-Stage Hierarchical Game Approach. *Energies* **2017**, *10*, 670. [CrossRef]
4. Abrishambaf, O.; Faria, P.; Gomes, L.; Spínola, J.; Vale, Z.; Corchado, J. Implementation of a Real-Time Microgrid Simulation Platform Based on Centralized and Distributed Management. *Energies* **2017**, *10*, 806. [CrossRef]
5. Bui, V.; Hussain, A.; Kim, H. Diffusion Strategy-Based Distributed Operation of Microgrids Using Multiagent System. *Energies* **2017**, *10*, 903. [CrossRef]
6. Zhang, G.; Cao, Y.; Cao, Y.; Li, D.; Wang, L. Optimal Energy Management for Microgrids with Combined Heat and Power (CHP) Generation, Energy Storages, and Renewable Energy Sources. *Energies* **2017**, *10*, 1288. [CrossRef]
7. Gazafroudi, A.; Prieto-Castrillo, F.; Pinto, T.; Prieto, J.; Corchado, J.; Bajo, J. Energy Flexibility Management Based on Predictive Dispatch Model of Domestic Energy Management System. *Energies* **2017**, *10*, 1397. [CrossRef]
8. Selvakumar, K.; Vijayakumar, K.; Boopathi, C. Demand Response Unit Commitment Problem Solution for Maximizing Generating Companies' Profit. *Energies* **2017**, *10*, 1465. [CrossRef]
9. Rokrok, E.; Shafie-khah, M.; Siano, P.; Catalão, J. A Decentralized Multi-Agent-Based Approach for Low Voltage Microgrid Restoration. *Energies* **2017**, *10*, 1491. [CrossRef]
10. Ma, K.; Bai, Y.; Yang, J.; Yu, Y.; Yang, Q. Demand-Side Energy Management Based on Nonconvex Optimization in Smart Grid. *Energies* **2017**, *10*, 1538. [CrossRef]
11. Pinto Moreira de Souza, D.; da Silva Christo, E.; Rocha Almeida, A. Location of Faults in Power Transmission Lines Using the ARIMA Method. *Energies* **2017**, *10*, 1596. [CrossRef]
12. Ma, Z.; Billanes, J.; Jørgensen, B. Aggregation Potentials for Buildings—Business Models of Demand Response and Virtual Power Plants. *Energies* **2017**, *10*, 1646. [CrossRef]
13. Faria, P.; Spínola, J.; Vale, Z. Reschedule of Distributed Energy Resources by an Aggregator for Market Participation. *Energies* **2018**, *11*, 713. [CrossRef]

energies

MDPI

Article

Distributed Control Strategy for Autonomous Operation of Hybrid AC/DC Microgrid

Jongbok Baek [1,*], Wooin Choi [2] and Suyong Chae [1]

[1]   Energy Efficiency Research Division, Korea Institute of Energy Research, Daejeon 34129, Korea; sychae@kier.re.kr

[2]   Samsung Electronics, Suwon 16677, Korea; wooin.choi@gmail.com

*   Correspondence: jonngbok.baek@kier.re.kr; Tel.: +82-042-860-3575

Academic Editor: Pedro Faria

Received: 6 December 2016; Accepted: 10 March 2017; Published: 16 March 2017

Abstract: This paper proposes a distributed control strategy that considers several source characteristics to achieve reliable and efficient operation of a hybrid ac/dc microgrid. The proposed control strategy has a two-level structure. The primary control layer is based on an adaptive droop method, which allows local controllers to operate autonomously and flexibly during disturbances such as fault, load variation, and environmental changes. For efficient distribution of power, a higher control layer adjusts voltage reference points based on optimized energy scheduling decisions. The proposed hybrid ac/dc microgrid is composed of converters and distributed generation units that include renewable energy sources (RESs) and energy storage systems (ESSs). The proposed control strategy is verified in various scenarios experimentally and by simulation.

Keywords: ac/dc hybrid microgrid; adaptive droop control; autonomous operation; distributed generation; energy management system

1. Introduction

To reduce carbon emissions, increased penetration of renewable energy sources (RESs) in power systems is desirable. This adoption of distributed energy resources can enhance energy security for local regions [1,2]. However, the effective utilization of intermittent RES generation and the integration of multiple distributed energy resources remain significant challenges. Furthermore, power quality and system reliability requirements are also increasing. Therefore, microgrids are attracting interest as alternative systems that could enable an intelligent power grid in the future, owing to the capability of microgrids to strengthen grid resilience and to enable the integration of distributed energy resources such as RESs, diesel generation, and energy storage systems (ESSs) [2–5].

A microgrid is a localized small grid that can operate in both grid-connected and off-grid modes to enhance energy security. Depending on the type of bus voltage, microgrids are categorized into ac, dc, and hybrid systems [6–9]. Comparing ac and dc systems, dc microgrid systems feature improved efficiency, requiring fewer conversion stages for RESs than ac systems. In addition, dc systems substantially reduce the impacts of synchronization and harmonic distortion, resulting in improved power quality compared to ac systems. However, low-voltage dc distribution systems require consideration of technical issues such as protection and grounding, as well as practical issues such as the limited number of commercially available dc components [10–12]. For these reasons, hybrid ac/dc microgrid systems are often investigated as alternative distribution networks. In hybrid ac/dc systems, there are separate ac and dc voltage buses for ac and dc loads, respectively, and the buses are interfaced through power electronics devices [7,13].

A microgrid contains multiple power electronics blocks connected to the system in parallel operation. These converters must be controlled to satisfy several essential microgrid requirements, including reliability, voltage regulation, and power sharing [14–17]. To address the aforementioned challenges, a number of control approaches have been proposed in microgrid applications. The control approaches can be divided into two classes based on their architectures: centralized and decentralized [17–22]. The centralized strategy increases efficient energy management through high-level communications, but is inadequate for microgrids requiring high reliability and scalability. The decentralized strategy, which is usually based on a droop scheme in a local controller, has improved reliability and facilitated power sharing without the need for communication between the components, although mode transition flexibility and optimized energy management are restricted [8,23–26].

This paper proposes a distributed control strategy for autonomous operation of a hybrid ac/dc microgrid. A hybrid ac/dc microgrid is considered in which distributed generation units and ESSs are connected to the dc bus as shown in Figure 1. The overall control structure is formulated with low-speed communication between two layers of controllers: the primary decentralized local controllers and the higher central controller. This hybrid control strategy enables autonomous operating mode transitions including in a fault situation; a supervised controller is not required because operating modes are based on events and bus voltage levels. The central controller executes an energy management system (EMS) to optimize the energy utilization of the system. Optimal energy scheduling is derived based on a dynamic programming method, using the information measured by the local controllers. To minimize energy costs, both the state of charge (SOC) and energy fluctuation trends are considered, and the optimal power dispatch is performed by adjusting the offset voltage level. The control architectures of the converters are discussed in more detail in Sections 2 and 3.

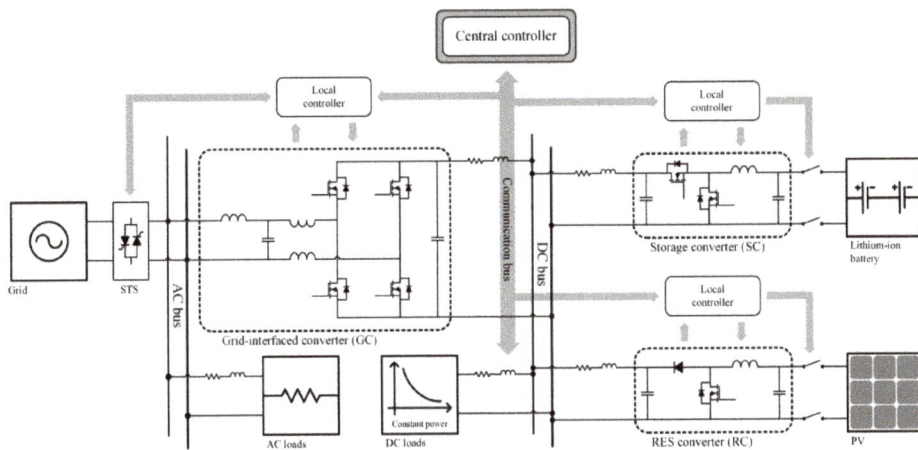

Figure 1. System diagram of hybrid ac/dc microgrid with communication links.

The local controllers are operated following the droop control method and are designed to inspect the operation conditions of each power electronics block: (1) the grid-interfaced converter (GC) manages islanding and reconnection to the grid; (2) the storage converter (SC) is used to implement the energy management strategy for energy optimization; and (3) the RES converter (RC) maximizes the RES output power.

The paper is organized as follows. In Section 2, the overall system structure of the proposed hybrid ac/dc microgrid is described, and the fundamental control philosophy of the proposed strategy is introduced, with descriptions of the converters' operation modes. In Section 3, the design method of the central control is discussed, with a mathematical formulation of the EMS strategy and its brief results. Section 4 presents primary control designs for different power sources with different control objectives. In Section 5, the proposed control strategy is experimentally verified in various scenarios. Finally, Section 6 presents the conclusions.

2. Configuration and Control Strategy of a Hybrid AC/DC Microgrid

2.1. System Description

Figure 1 diagrams the entire system, including the electric network and communication network. The proposed microgrid consists of a photovoltaic (PV) RES, ESS, and utility grid, all of which are coupled to the bus using converters. Ac and dc loads are connected to each bus. The loads are either a resistive load or a constant power load. Connection of the distributed generation units to the dc bus improves the system efficiency by reducing the number of conversion stages if the combined generated power is consumed in the dc network. Moreover, connection to the dc bus eliminates the control issues associated with synchronization and reactive power. The static transfer switch can connect and disconnect to the utility grid by fault signals or by a supervisory control strategy. The dc bus is interfaced to the ac bus through an ac/dc converter. The GC located between the ac bus and the dc bus works as a rectifier to regulate dc bus voltage during grid-connected operation, and as an inverter to form the ac bus and feed the ac load during off-grid operation. The topology of the GC is a single-phase voltage-source converter with an LCL filter. A lithium-ion battery set as an ESS is connected through a bidirectional synchronous buck converter. The PV source is the RES and is connected to the dc bus through a boost converter. The RC performs the maximum power point tracking (MPPT).

The local controllers of each converter share a single communication bus. Each local controller measures local voltage and current, and controls the dedicated converter and the switch of the nearby source. The specific designs of these controllers will be detailed in the Sections 3 and 4.

2.2. Control Strategy

The overall control structure is formulated with two layers. To retain reliability, primary local control is based on an adaptive droop method. Considering the source characteristics and operating mode, local controllers regulate bus voltage or perform MPPT. Because bus voltage is shared, each local controller can realize seamless mode transitions. To operate the microgrid efficiently, a central controller optimizes the EMS using a dynamic programming algorithm to optimize the battery usage schedule. The resulting commands are implemented by a droop curve compensator in the SC's outer controller. In this manner, in which the droop-based local controllers are coordinated with the central controller, the system reliability and efficiency are greatly enhanced. The objectives of the proposed control design are listed as follows.

- *Reliable and Autonomous Control*

To avoid a single point of failure due to device or communication malfunction, the converters are controlled in a decentralized manner using a droop-based method. In addition, the operating modes of converters transition autonomously during unpredictable situations to improve the power system's resilience.

- *DC Bus Voltage Regulation*

Regulation of the dc bus voltage (e.g., at 380 V), is one of the power quality criteria required of a dc microgrid. To overcome the poor voltage regulation of the typical droop method, the GC adjusts dc voltage offset.

- *Energy Optimization*

Energy optimization is performed to maximize the benefits of RESs and the ESS. An EMS module in the central controller obtains energy scheduling for optimization solutions and communicates the derived scheduling to the SC.

Based on the operation requirements above, Table 1 classifies the operating modes of the converters, including failure cases. In this classification, states of the entire system are characterized by combining the states of each converter. For example, State 121 represents the operating condition in which GC regulates V_{dc} under grid-connected conditions, SC regulates P_{ESS} for the EMS, and RC performs MPPT.

The shaded cells in Table 1 can be implemented using the adaptive droop-based method. The droop curves of each converter are shown in Figure 2, in which Figure 2a–c show the GC, SC, and RC curves, respectively. The GC curve shifts vertically to compensate for the dc voltage deviation. The SC curve can be expanded within the shaded region to achieve the required power control and SOC compensation. The RC performs an autonomous mode transition between MPPT and off-MPPT without any curve manipulation. According to the grid condition, the GC performs a seamless transition from the grid-connected mode to the off-grid mode, in which case, from the perspective of the dc bus, only the SC and RC regulate the dc bus voltage in droop control, while the GC appears as a load.

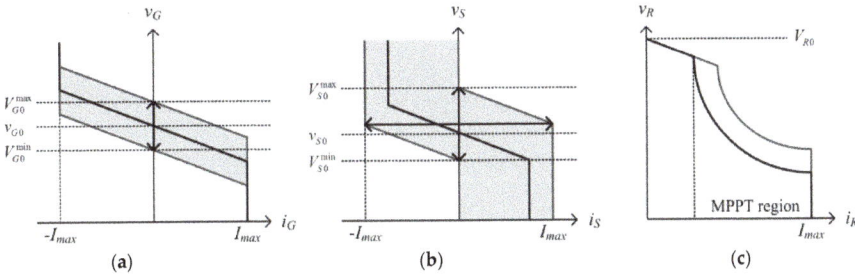

Figure 2. Droop characteristics in V–I curves of (**a**) grid-interfaced converter (GC); (**b**) storage converter (SC); and (**c**) renewable energy source (RES) converter (RC).

Table 1. Operating modes of converters.

State	Grid-Interfaced Converter (GC)	Storage Converter (SC)	RES Converter (RC)
1	Grid-connected: V_{dc}	Idle: V_{dc}	MPPT: P_{PV}
2	Off-grid: V_{ac}	EMS: P_{ESS}	Off-MPPT: V_{dc}
3	Fail	Fail	Fail

2.3. Operation Description

Figure 2 shows the V–I curves of the converters, where i_G, v_G, i_S, v_S, i_R, and v_R represent the currents and voltages of the GC, SC, and RC, respectively. From these curves, in the ideal case, the steady-state operating points of the dc bus under the droop control are determined by

$$v_{dc} = v_G = v_S = v_R \tag{1}$$

$$i_L = i_G + i_S + i_R \tag{2}$$

in which v_{dc} is the dc bus voltage, and i_L is the total dc load current. In this subsection, several examples of system operation will be described to highlight the features of the proposed control scheme. This series of examples shows operational transitions, in which IG, IS, and IR are the steady-state currents of the GC, SC, and RC, respectively, and v_{dc} is the steady-state dc bus voltage. In the following examples, shown in Figure 3, the steady-state value v_{dc1} moves to v_{dc2} after the relevant transitions, and the other values shift accordingly. Assuming constant load consumption, the following relationship is satisfied.

$$I_L = I_{G1} + I_{S1} + I_{R1} = I_{G2} + I_{S2} + I_{R2} \tag{3}$$

- *DC Bus Voltage Compensation at State 111*

According to Table 1, this state represents the condition in which the GC and SC regulate the dc bus voltage and the RC performs MPPT of the PV RES. Because the dc bus voltage v_{dc1} is less than the nominal voltage of 380 V, an additional outer loop of the GC compensates for the voltage deviation, as shown in Figure 3a. Consequently, the GC curve shifts upward until the steady-state voltage v_{dc2} is regulated to 380 V. The operating points of the other converters also change: the RC remains in MPPT, and the SC's output power returns to zero in steady-state.

- EMS at State 121

State 121 is identical to State 111, except that the SC operates in the EMS mode. The objective of the SC's local controller is to regulate the output power to the reference given by the central controller. Before the transition, the reference from the central module is I_{S1}. When the reference increases to I_{S2}, the SC curve shifts upward until the output current reaches the reference as shown in Figure 3b.

- Reliability under Failure from State 311 to State 332

At State 311, the GC is not involved in the droop control of the dc bus. At least one of two sources, the SC and/or RC, should operate in the dc bus voltage regulation mode. After a transition in which the SC fails, the RC may regulate the dc bus's voltage level. If total load power is less than the maximum PV power, the dc bus voltage is regulated by the RC as shown in Figure 3c. Even if the irradiation changes, the RC tracks the new maximum power point while maintaining the dc bus voltage as in Figure 3d.

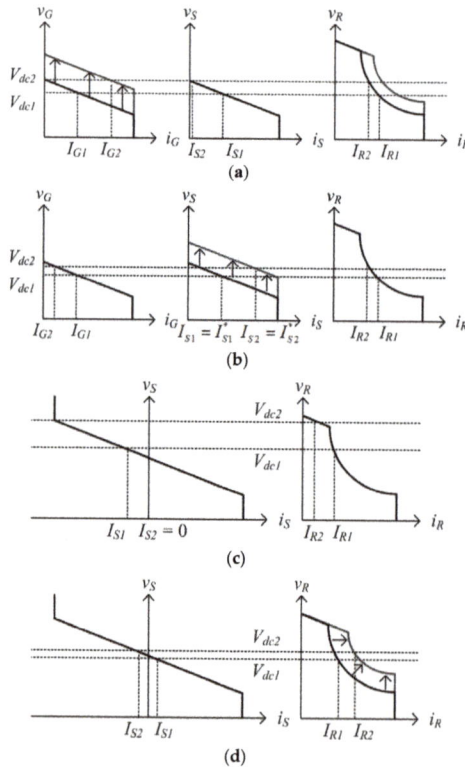

Figure 3. *V–I* curves of the converters for various operation examples. (**a**) Voltage reference change of GC after relevant transition; (**b**) voltage reference change of SC by energy management system (EMS); (**c**) failure of GC; and (**d**) change of photovoltaic (PV) generating power.

3. Control Design: Central Controller

As shown in Figure 1, the central controller shares the communication bus with the local controllers. Table 2 shows the information that the central controller processes for each local controller. In this section, the EMS feature of the central controller is highlighted. Inputs from the local controllers for this energy scheduling optimization stage include the source and load power information and the SOC of the battery; additional inputs include meteorological and pricing information from a higher-level operator, such as a distribution system operator as shown in Figure 4. The EMS scheme is implemented using a dynamic programming method. After an optimal solution is derived by the EMS module, the central controller dispatches the EMS power reference and operation mode to the SC.

Table 2. Communication of the central controller. SOC: state of charge.

Target	Transmit	Receive
GC	Protection	Measurements, V_{dc} restoration
SC	Protection, EMS, Mode selection	Measurements, SOC, V_{ocv}
RC	Protection	Measurements

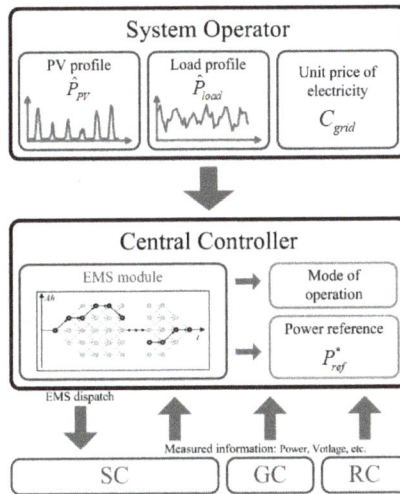

Figure 4. *V–I* curves and the operating points at State 111.

3.1. EMS Optimization

Determining the optimal energy dispatch solution for the battery's charge and discharge profile is accomplished by a shortest-path problem in which the path length represents the operator-defined cost. Using the previous and estimated RES generation profile and the load consumption profile, an optimal energy scheduling solution is derived to minimize the objective function under a set of constraints associated with the problem.

The EMS optimization is solved by a dynamic programming method. With hourly profiles of the RES and load power, the cost of the objective function is calculated for every hour t. Scheduling 1 day ahead, the path with the lowest cost from 1 h to 24 h is determined.

(1) *Objective Function*

The objective function is defined as in (4), where T is the total time of a day. $J_1[t]$ is the grid electricity consumption, which is computed by multiplying the grid power P_{grid} and the unit electricity cost C_{grid}. P_{grid} is the net energy consumed by the utility during 1 h. Electricity cost is based on time-of-use pricing, which is set for a specific time period in advance of the calculation. $J_2[t]$ is the equivalent cost of battery usage at time t, where α is a weighting factor and $Ah[t]$ is the state variable. The weighting factor is calculated to reflect the battery's cost and life cycle. $J_2[t]$ is proportional to energy transferred to and from the battery, which includes both charging and discharging energy; therefore, this term can restrict indiscriminate battery usage.

$$J = \sum_{t=1}^{T} (J_1[t] + J_2[t])$$

where

$$J_1[t] = P_{grid}[t] \cdot C_{grid}[t]$$
$$J_2[t] = \alpha \cdot \Delta Ah[t]$$

(4)

(2) *State Variable*

The state variable is defined as the energy flow of the ESS, as determined by the integration of the battery current over time, following (5).

$$Ah[t] = \sum_{k=t-1}^{t} i_{bat}[k]$$

(5)

(3) *Input*

Estimated PV generation P_{PV}, load consumption P_{load}, and electricity pricing information C_{grid} are given from the distribution system operator.

(4) *Constraint 1*

Power processed by the GC is calculated as:

$$P_G[t] = P_{load}[t] - P_S[t] - P_R[t]$$

(6)

where $P_G[t]$, $P_S[t]$, and $P_R[t]$ are the power delivered to the dc bus by GC, SC, and RC, respectively; $P_{load}[t]$ is given as an input. Using η_G, η_S, and η_R as the conversion efficiencies of the GC, SC, and RC, respectively, $P_S[t]$ is computed as

$$P_S[t] = \begin{cases} \frac{1}{\eta_S} \cdot i_{bat}[t] \cdot v_{bat}[t], & (\text{charge}: i_{bat}[t] \leq 0) \\ \eta_S \cdot i_{bat}[t] \cdot v_{bat}[t], & (\text{discharge}: i_{bat}[t] > 0) \end{cases}$$

(7)

where i_{bat} and v_{bat} are the current and voltage of the battery terminal, and the conversion efficiency is applied according to the direction of power flow. $P_R[t]$ is obtained as:

$$P_R[t] = \eta_R \cdot P_{PV}[t]$$

(8)

and the inflow grid power P_{grid} is:

$$P_{grid}[t] = \begin{cases} \frac{1}{\eta_G} \cdot P_G[t], & (\text{import}: P_G[t] \geq 0) \\ \eta_G \cdot P_G[t], & (\text{export}: P_G[t] < 0) \end{cases}$$

(9)

Estimated PV generation P_{PV}, load consumption P_{load}, and electricity pricing information C_{grid} are given from the distribution system operator.

(5) *Constraint 2*

To maintain a constant SOC level of the battery at the beginning and the end of EMS cycle, the net stored energy during a day is maintained at zero:

$$Ah[t = T] = Ah[t = 0] = 0 \qquad (10)$$

3.2. Energy Scheduling Results

Figure 5 shows the simulated results of the EMS formulated above where PS, PR, and P_{load} are one day's SC, RC, and load consumption power profiles, respectively. Figure 5a,b show the optimization results of the proposed EMS with fixed pricing. It is seen that the SC tends to charge the battery during the day when the PV generation is larger than the peak load. Figure 5c,d show the optimized profiles with variable pricing. Because the price during the night is lower than during the day, the SC charges the battery during the night and during the peak generation time, and discharges the stored energy during the peak load at early morning and late evening. In both cases, the net stored energy at the beginning and the end of the day is zero to satisfy the constraint. Figure 5e shows the optimization result of scheduling 6 days ahead. The calculated results are dispatched to the local controller. Even in the case of a communication failure, the dc bus voltage can be maintained by adopting the adaptive droop method; thus, the proposed method does not require high-bandwidth communication.

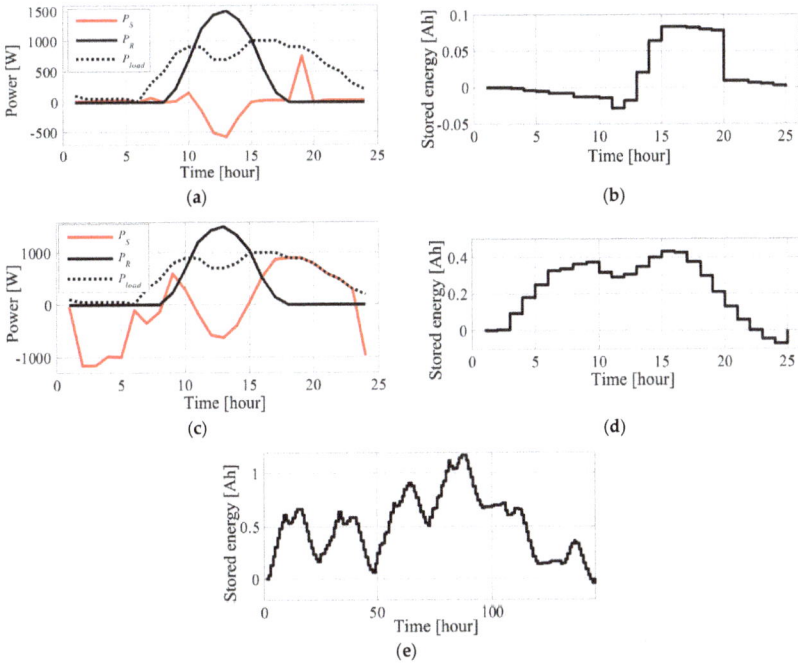

Figure 5. EMS optimization results with various conditions. (**a**) Power profile with fixed price; (**b**) energy profile with fixed price; (**c**) power profile with variable price; (**d**) energy profile with variable price; and (**e**) optimization result of a calculation executed 6 days ahead of time, with variable price.

4. Control Design: Local Controllers

4.1. GC Local Controller

Figure 6 shows a block diagram of the GC's local controller. The measurement variables are i_{dc}, v_{dc}, i_{ac}, and v_{ac}, which are the currents and voltages at the dc and ac terminals, respectively.

As described in Table 1, the GC is controlled in two different modes: the grid-connected mode and the off-grid mode. For both modes, the single-phase d-q current-loop is used to effect seamless mode transition. The current references, $i_{q,ref}$ and $i_{d,ref}$, are selected according to the operation mode. In the grid-connected mode, the voltage-loop is composed of two sub-blocks. The voltage reference, $v_{ref}[k]$, is computed using two additional terms:

$$v_{ref}[k] = V_{dc,ref} - v_{d,ref}[k] + v_{o,ref}[k] \tag{11}$$

where $V_{dc,ref}$ is the nominal dc bus voltage (i.e., 380 V), and $v_{d,ref}[k]$ is the droop voltage given as

$$v_{d,ref}[k] = K_d \cdot LPF(i_{dc}[k]) \tag{12}$$

where K_d is the droop gain, and LPF($\cdot$) is a low-pass filtering function. $v_{o,ref}[k]$ is the offset for the dc bus voltage restoration, which is given as

$$v_{o,ref}(z) = H_o(z)\varepsilon_o(z)$$
$$\text{where } \varepsilon_o(z) = V_{dc,ref} - N(z)v_{dc}(z) \tag{13}$$

$H_o(z)$ is a low-band-width PI-controller for the offset loop, and $N(z)$ is a 120 Hz notch filter to eliminate the 120 Hz ripple in the v_{dc}. The error between the nominal reference and $v_{dc}[k]$ is compensated by a slow PI controller to restore the deviation induced by the droop control. The relationship of the terms in (11) is shown in Figure 7. At a certain operating point, $v_{d,ref}[k]$ is determined by the droop gain and the output current, and then the error, $\varepsilon_o[k]$, is computed to restore the voltage to the nominal level (i.e., 380 V). The voltage restoration information $v_{o,ref}[k]$ is transmitted to the central controller. $v_{d,ref}[k]$ and $v_{o,ref}[k]$ in (11) are processed through two limiters, both of which are designed to consider the droop gain and the maximum current rating of the converter, as in:

$$v_{d,ref}^{max} = V_{dc,ref} + K_d I_{max}$$
$$v_{d,ref}^{min} = V_{dc,ref} - K_d I_{max} \tag{14}$$

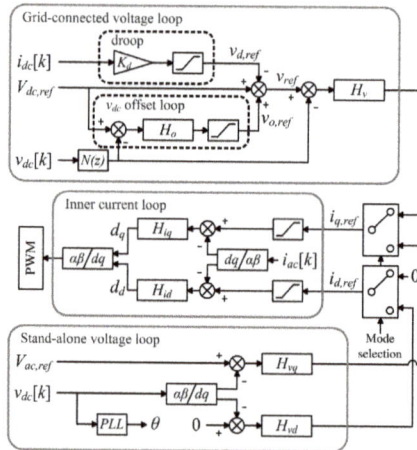

Figure 6. Control diagram of GC local controller.

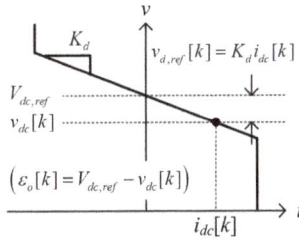

Figure 7. Voltage-loop reference on *V–I* curve of GC local controller.

4.2. SC Local Controller

The block diagram of the SC's local controller is shown in Figure 8. Although the converter is controlled by a two-loop controller, the detailed diagrams of the current- and voltage-loops are not depicted. The voltage reference is given as:

$$v_{ref}[k] = \begin{cases} V_{dc,ref} - v_{d,ref}[k] + v_{SOC,ref}[k], & (\text{Mode} = 1) \\ V_{dc,ref} - v_{d,ref}[k] + v_{SOC,ref}[k] + v_{EMS,ref}[k], & (\text{Mode} = 2) \end{cases}, \quad (15)$$

where $v_{d,ref}$ is the droop voltage, and $v_{SOC,ref}$ is the droop offset to reflect the SOC of battery, computed as:

$$v_{SOC,ref}[k] = (SOC[k] - 0.5) \cdot K_{SOC} \quad (16)$$

where K_{SOC} is a weighting factor. $v_{SOC,ref}[k]$ is zero when $SOC[k]$ is 0.5. When $SOC[k]$ is less than 0.5, $v_{SOC,ref}[k]$ becomes positive, and the converter will tend to lower the output of the battery. When $SOC[k]$ is greater than 0.5, $v_{SOC,ref}[k]$ becomes negative, and the converter will tend to increase the output of the battery. When the mode of the controller is switched to Mode 2 by the signal from the central controller, the additional term $v_{EMS,ref}[k]$ becomes effective, which is given as:

$$v_{EMS,ref}(z) = H_{EMS}(z)\varepsilon_{EMS}(z) \\ \text{where } \varepsilon_{EMS}(z) = Z\{i_{EMS}[k] - i_L[k]\} \quad (17)$$

where $i_L[k]$ is the inductor current, and $i_{EMS}[k]$ is the current reference delivered from the central controller.

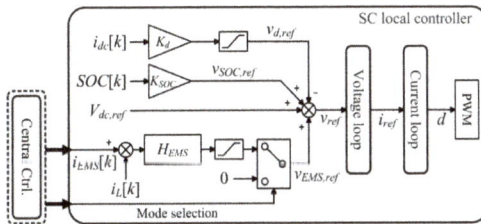

Figure 8. Control diagram of SC local controller.

4.3. RC Local Controller

The RC's local controller performs MPPT under normal conditions, and performs droop control of the dc voltage if off-MPPT is unavoidable. In Figure 9, the voltage-loops for both modes are operating continuously, and the current-loop references are generated. As seen in the *V–I* curve in Figure 2c, the reference with the lower value is selected, which is expressed by min block in the diagram. The voltage-loop compensators for the two modes are designed based on two different models in which the control objectives are the regulation of v_{dc} and v_{PV}, respectively.

13

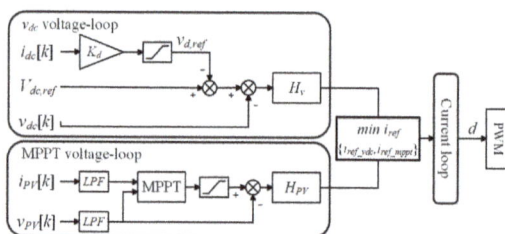

Figure 9. Control diagram of RC local controller.

5. Experimental Results

To validate the proposed control strategy, a hybrid ac/dc microgrid was constructed in the laboratory as shown in Figure 10. Table 3 shows the specifications for this experimental setup, following the electric diagram shown in Figure 1. The ac and PV sources were replaced by a 1 kVA grid simulator and 1.8 kW PV simulator, respectively. The central controller was designed using Matlab/Simulink (MathWorks, Natick, MA, USA), and microcontrollers were used for the local controllers. The power dispatch command is transferred through controller area network (CAN) communication at 100 bps. In the following figures, v_{dc} is the dc bus voltage, i_{load} is the load current, and i_G, i_R, and i_S are the output currents of GC, RC, and SC, respectively.

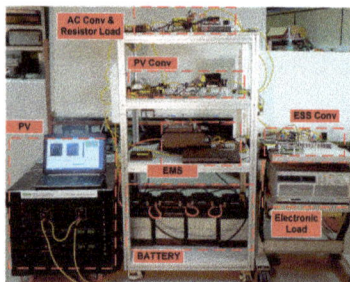

Figure 10. Experimental setup.

Table 3. Specifications of experiment set-up.

Component	Rating
AC source	Grid simulator 1 kVA
PV source	OCV 150 V/SCC 14 A/MPP 1800 W
Battery	4.2–2.7 V/31 Ah/Li-polymer cell/56S1P
Grid converter	1 kW/18 kHz
Storage converter	2 kW/50 kHz
RES converter	2 kW/50 kHz
DC load	Electric load/1 kW
AC load	Resistive load/108 Ω
Central controller	Matlab/Simulink
Local controller	Texas Instruments TMS320F28335 (Dallas, TX, USA)

5.1. Single-Mode Operation

Figure 11a,b shows experimental results of load step change on State 111 with its operational description in Figure 3b. GC, SC, and RC are in the grid-connected mode, idle mode, and MPPT mode, respectively. During the load step changes from 400 W to 1 kW and from 1 kW to 600 W, with the

RC output power remaining constant by performing MPPT. After each transition, the dc bus voltage fluctuates and then restores to the nominal voltage, 380 V. The dc bus voltage restoration is achieved by the slow PI controller in the GC's local controller. It is seen that the SC's output power remains zero, except for the transient period required to buffer the GC's slow voltage regulation characteristic.

Figure 12 shows the experimental results of load step changes in State 211, in which the GC, SC, and RC are operated in the off-grid, idle, and MPPT modes, respectively. The dc load is changed from 0.4 kW to 1 kW, and from 1 kW to 0.6 kW, and the ac load is maintained at 450 W. In the absence of the grid power, the GC regulates the ac bus to the nominal voltage 220 Vrms, feeding the ac load. During the transitions, the dc bus voltage is regulated within 374–378 V because the dc bus voltage is not restored by the GC. The RC consistently performs MPPT, while the power balance condition is satisfied by the SC. Figure 12c,d show the detailed waveforms of the transition.

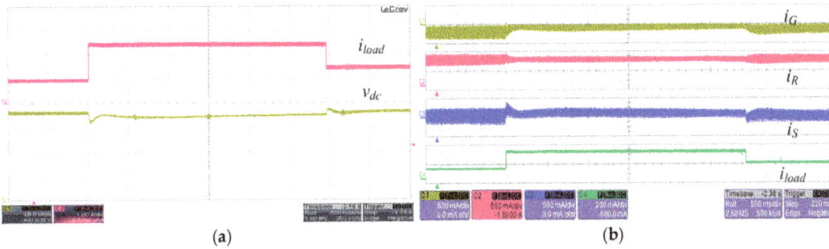

Figure 11. Experimental waveforms of load step change on State 111 (0.4 kW to 1 kW to 0.6 kW). (a) Load current and dc bus voltage; and (b) output current.

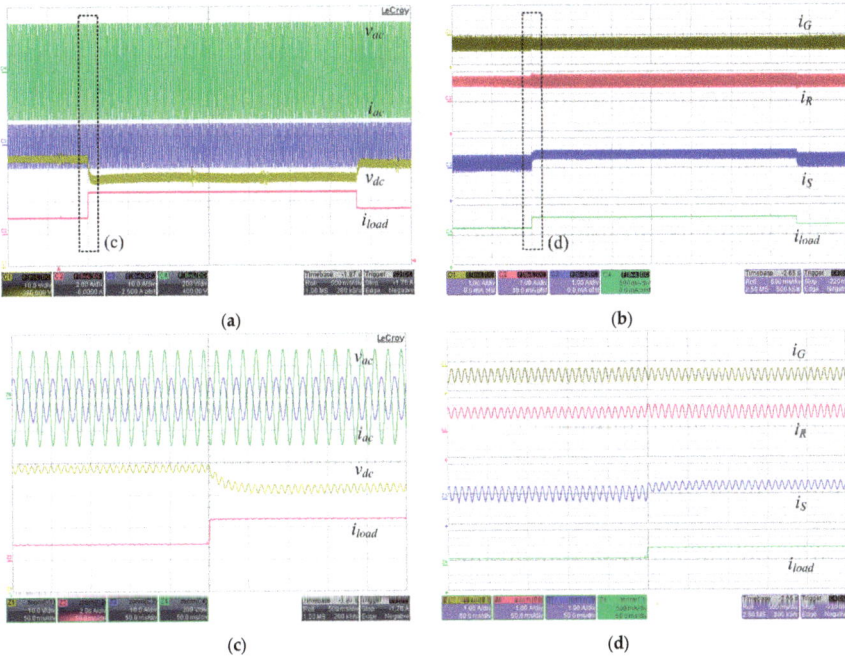

Figure 12. Experimental waveforms of load step change on State 211. (a) Bus voltage and current; (b) output current; (c) zoomed-in waveform of (a); and (d) zoomed-in waveform of (b).

5.2. EMS Dispatch

Figure 13 depicts the reference dispatch of the EMS module in the central controller in fixed load consumption and PV generation. The system is operated in State 121, described in Figure 3c, in which the GC, SC, and RC are in the grid-connected, EMS, and MPPT modes, respectively. The central controller dispatches the power reference to SC, 300 W of charge, 300 W of discharge, and 0 W of output power. While the RC is in the MPPT mode, the SC regulates its output power to the given references.

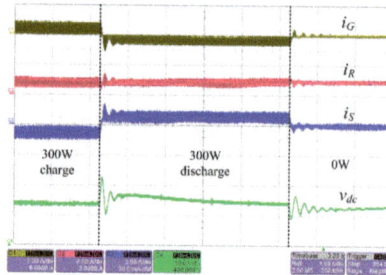

Figure 13. Experimental waveforms of EMS dispatch on State 121.

5.3. Mode Transitions

Figure 14 shows the experimentally determined waveforms for the mode transitions in the proposed control scheme. The GC's state transitions according to the grid condition are shown, where v_{util} and i_{util} are the voltage and current of the utility grid power, respectively. When the grid voltage is interrupted, the GC's mode is changed from the grid-connected mode to the off-grid mode. The GC feeds the ac loads with a sinusoidal ac voltage of 220 Vrms. After a few seconds, the grid voltage is restored, and the GC returns to the grid-connected mode.

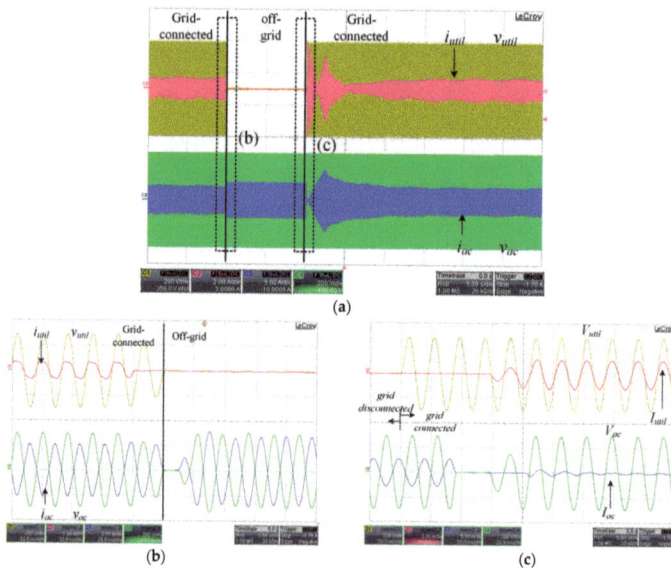

Figure 14. State transition of GC according to the grid condition. (**a**) Transition of GC; (**b**) zoomed-in waveforms during transition of grid-connected to off-grid; and (**c**) zoomed-in waveforms during transition of off-grid to grid-connected.

In Figure 15, the SC's operating mode switches from the idle mode to the EMS mode, and, accordingly, the state of the system is changed from State 111 to State 121. Before the transition, the SC participates in the droop control, with 0 W of steady-state output power. After the transition, the SC actively controls the output power of the ESS and charges the battery at the power reference, 500 W of charge, delivered from the central controller.

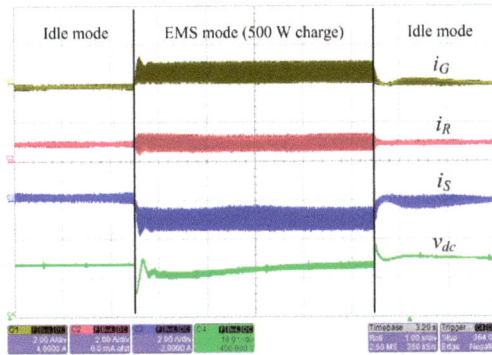

Figure 15. State transition of SC between idle mode and EMS mode.

In Figure 16, the waveforms of the converters' output currents and the dc voltage are depicted in the cases of failure of a unit. Figure 16a describes a GC failure event, and it is shown that the bus condition is maintained by the SC and RC. Whereas before the fault, the excessive power of the RC is exported to the grid through the GC, after the fault, the power charges the battery through the SC. The steady-state value of the dc voltage is increased because the dc voltage is not restored when the GC fails. In Figure 16b, whereas before the RC fault, the RC performs MPPT to supply a large portion of the total load, after the fault, the output power of the RC decreases to zero, and consequently, the GC and SC autonomously feed the loads without disrupting the operation of the system.

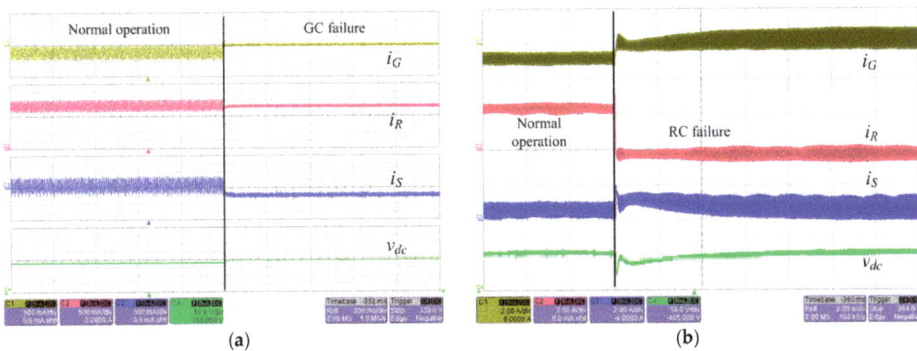

Figure 16. Failure modes of (a) GC; and (b) RC.

These experimental results verify that the proposed control strategy satisfies the control objectives discussed previously. The system maintains its bus quality under various fault conditions. The dc voltage is maintained within a limited range with the voltage restoration implemented through the GC. Moreover, efficient energy utilization is achieved through the low-speed communication. Furthermore, other control features, including the RC's autonomous transition between MPPT and off-MPPT and the SC's charge/discharge current limitation, are performed, although these results are not delivered in this paper.

6. Conclusions

This paper proposes a coordinated distributed control strategy for a hybrid ac/dc microgrid, considering several source characteristics. To achieve reliable operation and efficient management of energy, a two-level control structure is developed. Local controllers for the various sources are designed based on the droop method to optimally utilize the sources with high reliability. In the proposed scheme, the local controllers are linked to a central controller through a low-bandwidth communication device. The central controller executes EMS to optimally utilize the energy produced in the system. The proposed distributed control strategy is experimentally verified to demonstrate enhanced reliability and efficient operation.

Acknowledgments: This work was supported by the International Collaborative Energy Technology R&D Program of the Korea Institute of Energy Technology Evaluation and Planning (KETEP) granted financial resource from the Ministry of Trade, Industry & Energy, Korea (No. 20158530050130).

Author Contributions: The research presented in this paper was a collaborative effort among all authors. All authors developed the methodology, conducted the experiment, discussed results and wrote the paper.

Conflicts of Interest: The authors declare no conflict of interest.

References

1. Yan, Y.; Qian, Y.; Sharif, H.; Tipper, D. A survey on smart grid communication infrastructures: Motivations, requirements and challenges. *IEEE Commun. Surv. Tutor.* **2013**, *15*, 5–20. [CrossRef]
2. Xu, Z.; Yang, P.; Zeng, Z.; Peng, J.; Zhao, Z. Black start strategy for PV-ESS multi-microgrids with three-phase/single-phase architecture. *Energies* **2016**, *9*, 372. [CrossRef]
3. Carrasco, J.M.; Franquelo, L.G.; Bialasiewicz, J.T.; Galván, E.; PortilloGuisado, R.C.; Prats, M.M.; León, J.I.; Moreno-Alfonso, N. Power-electronic systems for the grid integration of renewable energy sources: A survey. *IEEE Trans. Ind. Electron.* **2006**, *53*, 1002–1016. [CrossRef]
4. Lasseter, R. Smart distribution: Coupled microgrids. *Proc. IEEE* **2011**, *99*, 1074–1082. [CrossRef]
5. Arefifar, S.A.; Mohamed, Y.A.R.I. DG Mix, Reactive sources and energy storage units for optimizing microgrid reliability and supply security. *IEEE Trans. Smart Grid* **2014**, *5*, 1835–1844. [CrossRef]
6. Loh, P.C.; Li, D.; Chai, Y.K.; Blaabjerg, F. Autonomous operation of hybrid microgrid with AC and DC subgrids. *IEEE Trans. Power Electron.* **2013**, *28*, 2214–2223. [CrossRef]
7. Cho, B.H.; Choi, W.; Baek, J.B. Control of the dc distribution microgrid system. In Proceedings of the IEEE International Power Electronics Conference (IPEC), Hiroshima, Japan, 13–21 May 2014.
8. Solanki, A.; Nasiri, A.; Bhavaraju, V.; Familiant, Y.L.; Fu, Q. A new framework for microgrid management: Virtual droop control. *IEEE Trans. Smart Grid* **2016**, *7*, 554–566. [CrossRef]
9. Tan, K.T.; So, P.L.; Chu, Y.C.; Chen, M.Z.Q. A flexible AC distribution system device for a microgrid. *IEEE Trans. Energy Convers.* **2013**, *28*, 601–610. [CrossRef]
10. Justo, J.J.; Mwasilu, F.; Ju, L.; Jung, J.W. AC-microgrids versus DC-microgrids with distributed energy resources: A review. *Renew. Sustain. Energy Rev.* **2013**, *24*, 387–405. [CrossRef]
11. Yuan, C.; Haj-Ahmed, M.A.; Mahesh, S.I. Protection strategies for medium-voltage direct-current microgrid at a remote area mine site. *IEEE Trans. Ind. Appl.* **2015**, *51*, 2846–2853. [CrossRef]
12. Bui, D.M.; Chen, S.L.; Wu, C.H.; Lien, K.Y.; Huang, C.H.; Jen, K.K. Review on protection coordination strategies and development of an effective protection coordination system for DC microgrid. In Proceedings of the 2014 IEEE Asia-Pacific Power and Energy Engineering Conference (APPEEC), Hong Kong, China, 7–10 December 2014; pp. 1–10.
13. Wang, C.; Yang, X.; Wu, Z.; Che, Y.; Guo, L.; Zhang, S.; Liu, Y. A Highly integrated and reconfigurable microgrid testbed with hybrid distributed energy sources. *IEEE Trans. Smart Grid* **2016**, *7*, 451–459. [CrossRef]
14. Hong, M.; Yu, X.; Yu, N.P.; Loparo, K.A. An energy scheduling algorithm supporting power quality management in commercial building microgrids. *IEEE Trans. Smart Grid* **2016**, *7*, 1044–1056. [CrossRef]

15. Zhu, Y.; Zhuo, F.; Wang, F.; Liu, B.; Gou, R.; Zhao, Y. A virtual impedance optimization method for reactive power sharing in networked microgrid. *IEEE Trans. Power Electron.* **2016**, *31*, 2890–2904. [CrossRef]

16. Vidyanandan, K.V.; Senroy, N. Frequency regulation in microgrid using wind—Fuel cell—Diesel generator. In Proceedings of the 2012 IEEE Power and Energy Society General Meeting (PESGM), San Diego, CA, USA, 22–26 July 2012; pp. 1–8.

17. He, M.; Giesselmann, M. Reliability-constrained self-organization and energy management towards a resilient microgrid cluster. In Proceedings of the IEEE Innovative Smart Grid Technologies Conference (ISGT), Washington, DC, USA, 18–20 February 2015; pp. 1–5.

18. Guerrero, J.M.; Chandorkar, M.; Lee, T.L.; Loh, P.C. Advanced control architectures for intelligent microgrids—Part I: Decentralized and hierarchical control. *IEEE Trans. Ind. Electron.* **2013**, *60*, 1254–1262. [CrossRef]

19. Karlsson, P.; Svensson, J. DC bus voltage control for a distributed power system. *IEEE Trans. Power Electron.* **2003**, *18*, 1405–1412. [CrossRef]

20. Rocabert, J.; Luna, A.; Blaabjerg, F.; Rodri, X.; Guez, P. Control of power converters in AC microgrids. *IEEE Trans. Power Electron.* **2012**, *27*, 4734–4749. [CrossRef]

21. Guerrero, J.M.; Vasquez, J.C.; Matas, J.; Vicuña, L.G.; Castilla, M. Hierarchical control of droop-controlled AC and DC microgrids—A general approach toward standardization. *IEEE Trans. Ind. Electron.* **2011**, *58*, 158–172. [CrossRef]

22. Choi, W.; Baek, J.B.; Cho, B.H. Control design of coordinated droop control for hybrid AC/DC microgrid considering distributed generation characteristics. In Proceedings of the IEEE Energy Conversion Congress and Exposition (ECCE), Pittsburgh, PA, USA, 14–18 September 2014; pp. 4276–4281.

23. Farhadi, M.; Mohammed, O. Adaptive energy management in redundant hybrid DC microgrid for pulse load mitigation. *IEEE Trans. Smart Grid* **2015**, *6*, 54–62. [CrossRef]

24. Dragičević, T.; Guerrero, J.M.; Vasquez, J.C.; Škrlec, D. Supervisory control of an adaptive-droop regulated DC microgrid with battery management capability. *IEEE Trans. Power Electron.* **2014**, *29*, 695–706. [CrossRef]

25. Aymen, C.; Rashad, M.K.; Ridha, A.; Ken, N. Multiobjective intelligent energy management for a microgrid. *IEEE Trans. Ind. Electron.* **2013**, *60*, 1688–1699.

26. Li, F.; Xie, K.; Yang, J. Optimization and analysis of a hybrid energy storage system in a small-scale standalone microgrid for remote area power supply (RAPS). *Energies* **2015**, *8*, 4802–4826. [CrossRef]

energies

MDPI

Article

Aggregators' Optimal Bidding Strategy in Sequential Day-Ahead and Intraday Electricity Spot Markets

Xiaolin Ayón *, María Ángeles Moreno and Julio Usaola

Department of Electrical Engineering, Universidad Carlos III de Madrid, Avenida de la Universidad 30, 28911 Leganés, Madrid, Spain; amoreno@ing.uc3m.es (M.Á.M.); jusaola@ing.uc3m.es (J.U.)
* Correspondence: xayon@pa.uc3m.es; Tel.: +34-916-278-853

Academic Editor: Pedro Faria
Received: 2 February 2017; Accepted: 22 March 2017; Published: 1 April 2017

Abstract: This paper proposes a probabilistic optimization method that produces optimal bidding curves to be submitted by an aggregator to the day-ahead electricity market and the intraday market, considering the flexible demand of his customers (based in time dependent resources such as batteries and shiftable demand) and taking into account the possible imbalance costs as well as the uncertainty of forecasts (market prices, demand, and renewable energy sources (RES) generation). The optimization strategy aims to minimize the total cost of the traded energy over a whole day, taking into account the intertemporal constraints. The proposed formulation leads to the solution of different linear optimization problems, following the natural temporal sequence of electricity spot markets. Intertemporal constraints regarding time dependent resources are fulfilled through a scheduling process performed after the day-ahead market clearing. Each of the different problems is of moderate dimension and requires short computation times. The benefits of the proposed strategy are assessed comparing the payments done by an aggregator over a sample period of one year following different deterministic and probabilistic strategies. Results show that probabilistic strategy reports better benefits for aggregators participating in power markets.

Keywords: aggregator; optimal bidding; electricity markets; probabilistic programming

1. Introduction

The smart grid will be the future standard at the distribution level, after generalization of active demand and distributed generation, mainly from renewable energy sources. The spread of automation and control is currently a major challenge for regulators and grid operators and it also opens a large field of opportunities to make a better use of all of the available resources in the grid, in order to achieve a safer, cheaper, and more sustainable electric supply [1]. In this context and from the demand side, a new player emerges: the aggregator, which could encompass the role of a retailer, a flexibility manager, and a balanced responsible party or market agent [2–5]. The participation of the aggregator in the power markets is relatively new since it exploits the flexibility of customers, as well as the optimal management of distributed generation resources. The aggregator needs to solve optimal scheduling and bidding problems to manage their prosumers' resources and participate in the power markets in an efficient way. However, the approach from the aggregator point of view is new and different from traditional producers and retailers regarding the supply-demand balance, the bounds of the possible imbalance incurred by the aggregator, and the uncertainties involved in the problem. Considering these differences, the optimal participation of an aggregator in sequential electricity spot markets (only day-ahead and intraday markets are considered), with the objective of minimizing the cost of the traded energy is addressed in this paper.

Optimal scheduling problem is addressed in the literature in different ways, for instance the objectives functions including maximization of profits, social welfare, utility for the demands and minimization of energy capacity, and cost of imbalances and operational costs are considered in References [6–10]. The focus on the optimal bidding problem involves different objectives, some of which are the minimization of negative returns and the cost related to emissions in [11], evaluation of different prosumer risk tolerance in [12], as well as jointly minimizing the risk and maximizing profits in [13–15]. The bids to submit to the electricity market depends on the energy market rules and many solutions of the bidding problem result in one optimal price-energy pair [16] for a certain time period, but a more realistic bid consists of a curve composed of multiple price-energy pairs, as that presented in references [17,18]. There are few documents in the literature concerning both scheduling and bidding problems; for instance, in [19], a two-stage stochastic mixed integer linear program where the bidding decision is made in the first stage and the scheduling in the second is solved for aggregators that sell electricity to prosumers and buy back surplus electricity in the spot market.

In practice, the aggregator must face some difficulties, mainly related to the uncertainties of the intermittent and non-dispatchable nature of renewable energy sources (RES) generation, but also because the demand and the energy prices cannot be accurately predicted in advance. Moreover, the forecast errors derive unpredictable imbalances between the real-time production/consumption and the energy previously scheduled in the electricity market. Imbalance penalties also depend on the energy market rules. Some works in the literature such as [20] envisage the likely imbalances leading to unpredicted imbalance costs. In [7], penalties for failure to supply the market and customers are considered. Penalties due to over-production or under-production status are penalized with different values of a weight coefficient introduced into the model [12]. In [19], since they penalize imbalances heavily, the case becomes not realistic because avoiding imbalance is forced. In this formulation, it is assumed that market prices, loads, and generation are known with certainty before the optimal schedule is decided, and they then run a deterministic optimization for the scheduling process, but only market results are revealed. In our work, realistic imbalance prices and day-ahead market results are considered. The penalties due to deviations of a wind power producer are formulated in references [17,21], but in our case we are considering an aggregator, and then the problem is different. In [22], the formulation of deviations and its penalizations is like our proposal, but they assume that only unidirectional bids are allowed for the aggregator in the electricity market, i.e., it can only buy energy but cannot sell excessive energy back to the wholesale energy market.

Participating in intraday markets is a way for any market participant to reduce the forecast error costs, updating previous day-ahead scheduling as in [23]. The cost of the purchased energy can also be reduced in this market as in [24]. In addition, the flexibility offered by shiftable demand and battery storage also contributes to an increase of the expected benefit of the aggregator as shown in [25]. This flexibility allows the aggregator to shift the consumption from peak hours to valley hours, buying energy at low-cost hours, selling it at high-cost hours, and reducing the energy imbalance caused by bid deviation.

The optimal bidding in the day-ahead market taking into account all subsequent markets as intraday or balancing markets is assessed commonly in the literature. However, profit comparisons of taking into account or not the intraday markets when preparing the bidding for the day-ahead market (coordinated or separate bidding) are questioned in previous works, such as [26,27]. Additionally, several optimization techniques are widely used to solve the optimal bidding problem under uncertainty, such as modified particle swarm, stochastic, robust, fuzzy, model predictive control in [7,9,28,29] and other metaheuristic techniques, whose effectiveness and efficiencies considering different initial solution algorithms are compared in [30]. In [31,32], a comparison between the stochastic and robust optimization with the perfect information case is carried out and the results show that stochastic programming provides better solutions. However, new optimization techniques and improvements over the existent techniques are still under research.

The proposed approach solves separate probabilistic optimization problems, which considers the uncertainty of market prices (day-ahead and intraday), RES generation and fixed demand, and takes into account the possible imbalance costs the aggregator may incur. We assume a neutral risk aggregator, because reducing risk through changing operating decisions can be costly compared to financial operations [33]. The imbalances are regulated by a dual pricing mechanism that implies penalizations for those incurring energy deviations against the system [34]. The proposed probabilistic approach is based on the formulation presented in [35] and the work presented in this paper extends the method presented in [36] by considering the intraday market, and improves the mathematical formulation of the problem.

Our approach differs from other works in the literature and the main difference with related previous works is that we consider an independent bidding strategy in each electricity spot market, which allows taking advantage of both the gain in certainty of forecasts and the knowledge of previous market results. We also take into consideration time dependent constraints. The optimization problems proposed here are set for a whole day aggregator's portfolio that includes shiftable demand, RES generation, and batteries. A simple modelling of shiftable demand is used with the purpose of testing the method. A thorough modelling of this demand is out of the scope of the paper. Furthermore, the method of aggregating and coordinating the flexibility of customers is out of the scope of this paper, since it requires much information on customers' behavior and preferences that is not available. The intended contributions of this paper are listed as follows:

(1)   To propose a simple and effective optimization model that provides hourly optimal bidding curves for an aggregator who manages fixed and shiftable demand, RES generation, and storage devices when participating in the electricity markets (daily and intraday markets), aiming to minimize the daily energy cost.
(2)   To include in the optimization model in (1) the different uncertainties faced by the aggregator, namely fixed demand, RES generation and market prices, and the possible imbalance costs in which the aggregator may incur.
(3)   To assess the benefits of the optimization model in (1) over a whole year comparing the yearly payments performed by the aggregator under different strategies using realistic data taken from publicly available sources; case studies based on a whole year with realistic data are not widely assessed in the literature.

The paper continues with some previous considerations regarding the market framework and the uncertainty of the random variables involved in Section 2. Then, the main assumptions and constraints are described in Section 3. The formulation of each decision-making problem follows in Section 4. Next, Section 5 describes the case study and in Section 6, the participation of an aggregator in the Spanish electricity market is simulated over a year, considering different strategies in order to assess the benefits of the proposed approach. Conclusions and future work are given at the end of the paper.

2. Previous Considerations

Consider an aggregator that represents a cluster of prosumers with RES generators (wind and photovoltaic) and storage devices (batteries). If this aggregator wants to participate in the electricity markets (day-ahead and intraday markets), he has to solve different decision-making problems that involve uncertainties (demand, RES generation, and market energy prices), with the aim of minimizing the cost of the daily traded energy. The aggregator is considered as a price taker, because its participation does not affect the resulting market prices. The participation in the reserves market is not considered. The approach for the aggregator decision-making process follows the sequence of the markets considered and therefore optimization problems may be solved independently, deriving the optimal bidding curves to be submitted first to the day-ahead market and next to the intraday markets. The market framework is based in the Iberian electricity market [37] because it has a more liquid intraday market compared with other European intraday markets (most of them with continuous trading implementing a pay-as-bid matching algorithm). According to [38], the Spanish intraday market has effectively contributed to RES generation balancing and, intermittent energy sources have more flexibility to bid in this market aimed at the maximization of their economic profits.

The optimization problems solved by the aggregator and their mathematical formulations are detailed in Sections 3 and 4.

2.1. Market Framework

The energy traded in the Iberian electricity spot market is managed by OMIE (the Spanish division of the Iberian Energy Market Operator), which is in charge of collecting orders, clearing the markets, and publishing results, available in [39,40]. Most of the energy is negotiated in the day-ahead market (or daily market) where purchase and sale bids for day D must be sent to OMIE before the gate closure at 12 a.m. of day D-1. Once the daily market has ended, and until 12:45 p.m. of the following day, six sessions of the intraday market are held, which allow participants to adjust their generation and consumption schedules to their best forecasts for their real-time needs. The agents who have participated in the day-ahead market have there an opportunity to change their energy bids to reduce their imbalances. A review of different European market designs and the importance of sufficient liquidity in intraday markets can be found in [41], which concludes that the Spanish mechanism auctions is considered the most attractive market design for systems with a high share of non-dispatchable generation. Finally, the Transmission System Operator (TSO) is in charge of ensuring a balanced and secure system operation. The net energy system imbalance between generation and demand is corrected through the balancing services, whose costs are covered by those incurring imbalances. In the Iberian market, a dual imbalance pricing mechanism is followed, where the imbalance prices depend on the sign of the net system imbalance. Thus, if an aggregator incurs a positive imbalance (higher production or lower consumption than scheduled) in any period of time, then the energy surplus is paid at the sell imbalance price (π_t^+), lower than or equal to the day-ahead market price (π_t^d). On the contrary, if the imbalance is negative (lower production or higher consumption than scheduled) at any time, the energy deficit must be bought at the buy imbalance price (π_t^-), higher than or equal to the day-ahead market price. The relation between the imbalance prices and the market price at any time period t may be written as follows:

$$r_t^+ = \frac{\pi_t^+}{\pi_t^d} \leq 1 \tag{1}$$

$$r_t^- = \frac{\pi_t^-}{\pi_t^d} \geq 1 \tag{2}$$

2.2. Predictions and Uncertainties

In the decision-making problem faced by the aggregator, there are three main sources of uncertainty: demand, RES production, and market prices. This uncertainty increases with the horizon of the forecast; thus, it is greater in the day-ahead market than in the intraday market. The uncertainty can be obtained by some prediction programs [42] and the aggregator can use this information to produce optimal bids to the market. However in this work, given that no prediction programs were available, the uncertainty is modelled through scenarios, i.e., we try to reproduce the results of prediction programs by creating scenarios from a given time series.

Scenarios are created in two steps:

(1) First, the basic trajectories of forecasts are generated from historical data or synthesized production series.
(2) From the basic trajectory and using an autoregressive time series AR(1) for modelling the forecast error, the desired number of equiprobable scenarios for each time period t are created.

Note that uncertainty characterization or scenario generation is not a goal of this work because the uncertainty quantification would come, in reality, from advanced forecasting tools, as said before. A brief description of the random variables considered in this work is given. An example of basic trajectories and scenarios generated is given in Section 5.2.

2.2.1. Demand

Demand is divided into two kinds, namely fixed and shiftable demand. Fixed demand is supplied to the user at any time without restrictions (within the limit of the contracted power), but its value is not known beforehand, and the aggregator must forecast it. It might correspond to certain manually operated appliances or systems such as lighting, computers, etc. Shiftable demand can be shifted along a given time period but it is assumed that the daily amount of the energy required by this shiftable demand is known and previously agreed on between the aggregator and his customers through a contract, which reflects the will of the aggregator's customers in shifting their consumption along the day. It could correspond to electric vehicles, or noncritical devices such as washing machines, dishwashers, etc. Thus, only fixed demand uncertainty is considered in this paper through scenario generation.

2.2.2. RES Generation

Only renewable energy (solar and wind generation) is considered in the study. Hence, the aggregator must forecast the power supplied by RES generators for the considered time interval. For the basic pattern of solar production, a site is chosen in order to fit real conditions, and the basic trajectory is obtained as in [43]. For the wind production, wind speed scenarios are generated from a basic trajectory of wind speed. Once the wind speed scenarios are generated, they are transformed into power scenarios through the power curve associated to the turbine model of the wind farm.

2.2.3. Fixed Demand Minus RES Generation

Once scenarios of RES generation (photovoltaic and wind production) and fixed demand are obtained, a new random variable ($P^s_{rnd,t}$) can be defined as the fixed demand minus photovoltaic and wind productions, for any period of time t and scenario s.

$$P^s_{rnd,t} = P^s_{fix,t} - P^s_{pv,t} - P^s_{w,t}, \quad \forall t, \forall s \tag{3}$$

These random variables are not independent, but their predictions errors are.

2.2.4. Energy and Imbalance Prices

The aggregator must also perform forecasts of energy prices. Day-ahead market price scenarios and intraday market price scenarios can be created from historic prices of a given period. Imbalance prices are modelled in a simpler way: since they are extremely volatile, an hourly constant ratio between the imbalance price and the daily energy price is taken, as in [20]. Thus, the uncertainty of imbalance prices is not modelled directly.

2.2.5. Global Uncertainty

Thus, the uncertainty characterizing the aggregator bidding problem in day-ahead/intraday market is modeled through a symmetric scenario tree that is specifically built as follows:

(1) Generate N_{dp}/N_{ip} price scenarios for the day-ahead/intraday market.
(2) From the basic trajectory and using an autoregressive time series AR(1) for modelling the forecast error, the desired number of equiprobable scenarios for each time period t is created. For each realization of the day-ahead/intraday market prices, generate N_{dr}/N_{ir} wind power realizations,

N_{dr}/N_{ir} photovoltaic power realizations, and N_{dr}/N_{ir} fix demand realizations, and calculate the N_{dr}/N_{ir} realizations of the new variable fix demand minus RES generation with (3).

Hence, the total number of scenarios composing the tree is $N_w = N_{dp} \cdot N_{dr}$ for the day-ahead market and $N_{iw} = N_{ip} \cdot N_{ir}$ for the intraday market.

3. Optimization Problem Assumptions and Modelling Details

This section presents an overview of the optimization process as well as the constraints that should be considered. The mathematical formulation of the problem is given in the next section.

The aggregator participates in the day-ahead market in order to purchase the net energy for his customers' portfolio. With this purpose, the aggregator solves a probabilistic optimization problem resulting in the optimal quantity of energy to be purchased (or sold) in each period of time, depending on the market price. Once the day-ahead market is cleared, and the scheduled energy for each period of time is known, it may happen that the constraints related to the flexible demand (shiftable demand and batteries) are not fulfilled, and the aggregator performs an adjustment process of scheduling in order to ensure the fulfillment of constraints within the time horizon. For the intraday market, the aggregator can update his previous market position aimed at minimizing the total cost of the energy, using fresh and more accurate predictions and the knowledge on day-ahead market prices and flexible demand schedule. The complete process of the aggregator's participation in sequential electricity markets is illustrated in Figure 1. Note that the day-ahead (DA) and intraday (ID) market clearing processes are performed by the Market Operator (MO), and thus they are external to the aggregator.

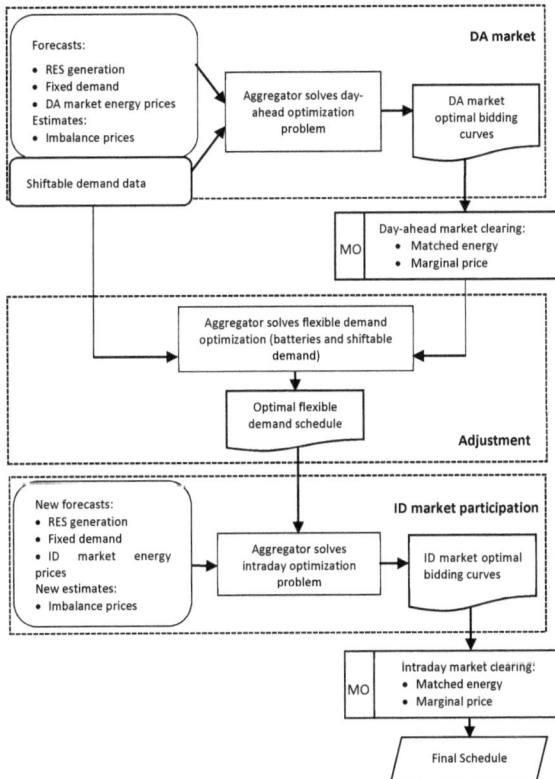

Figure 1. Aggregator participation in the day-ahead and intraday markets.

In the problem formulation, it has been assumed that:

(1) The energy prices in the market are not affected by the aggregator bids, because the market is large enough.
(2) The aggregator buys and purchases energy at the same price, i.e., grid access tariffs have not been included. Losses are included, according to the Spanish regulation, as a fixed percentage of the demand, added to the forecasted consumption.

The aggregator must solve two different decision-making problems involving uncertainties, one for the day-ahead market and another for the intraday market. Furthermore, the aggregator must solve two additional optimization problems in order to schedule the flexible demand after the day-ahead market clearing. When solving those problems, the aggregator has to take into account several constraints related to shiftable demand, batteries, RES generation, and energy imbalances, as described next. In the following subsections, all the decision variables are denoted with a superscript s representing a generic scenario s.

3.1. Shiftable Demand

This demand can be shifted over a given period of time but the amount of the daily energy to be consumed is known and previously agreed upon between the aggregator and his consumers through a contract. It could correspond to electric vehicles, or noncritical devices such as washing machines, dishwashers, etc. The optimization process will tend to shift this demand to lower price hours. Regarding this type of demand, (4) defines the total energy consumed by the shiftable demand, for a certain scenario s over a planning horizon of N_h periods of time. This equation could be reformulated if the periods of time for consumption are limited to a given set (e.g., tariff charging of electric vehicles during the night). Equation (5) models the bounds of the hourly shiftable demand for any period of time t and scenario s.

$$\sum_{t=1}^{N_h} P^s_{shift,t} \cdot d_t = E_{shift}, \ \forall s \tag{4}$$

$$0 \leq P^s_{shift,t} \leq P^{max}_{shift}, \ \forall t, \forall s \tag{5}$$

3.2. Batteries

Batteries are modelled in a simple way (rated power, maximum/minimum capacity) as in [24], but in this paper losses are also included, which are considered as constant. The following constraints must be satisfied for any period of time t and any scenario s:

$$0 \leq P^{+s}_{B,t} \leq y^s_t P^{+max}_B; \ 0 \leq P^{-s}_{B,t} \leq (1 - y^s_t) P^{-max}_B, \ \forall t, \forall s \tag{6}$$

$$E^{min}_B \leq E^s_{B,t} \leq E^{max}_B; \ E^s_{B,N_h} = E^s_{B,1}, \ \forall t, \forall s \tag{7}$$

$$E^s_{B,t} = E^s_{B,t-1} - \left(\frac{1}{\eta^+}\right) P^{+s}_{B,t} d_t + \eta^- P^{-s}_{B,t} d_t - \Delta E, \ \forall t, \forall s \tag{8}$$

The constraint (6) set the bounds of the rated power. The binary variable y^s_t avoids battery charge and discharge at the same time step; it is equal to 1 if batteries are discharging in period t and 0 otherwise. Constraint (7) set the limits of the storage energy, respectively. Here, the level of battery storage at the end of the scheduling horizon is equal to its initial energy level. It is assumed that $E^s_{B,1} > E^{min}_B$ to exploit the flexibility of the batteries during the first periods of time of the horizon. Constraint (8) represents the energy balance in the batteries. Note that no battery degradation costs are considered in this work because we have compared the results from the participation of the aggregator in the power markets for the next day or few hours before the energy delivery time.

3.3. Energy Imbalances

The imbalance which the aggregator could incur is defined as the gap between the energy traded in the electricity market (day-ahead or intraday) and the actual consumption/production. If the actual energy of the aggregator is greater than the scheduled energy in the market, the aggregator's imbalance is positive, otherwise it is negative. Note that this is different from the point of view of a producer.

As the aggregator's imbalance makes the problem nonlinear, in order to keep the linearity, Equation (9) decomposes the energy imbalance into positive and negative imbalances, as in [35,44]. Constraint (10) set the limits of the imbalances, which could reach the sum of the contracted power (maximum buying bid) and the installed generating power (maximum selling offer) in both senses, for any period of time t and scenario s.

$$\Delta_t^s = \Delta_t^{+s} - \Delta_t^{-s}, \ \ \forall t, \forall s \tag{9}$$

$$0 \le \Delta_t^{+s} \le Lim \cdot d_t; \ 0 \le \Delta_t^{-s} \le Lim \cdot d_t, \ \left(Lim = \sum P_{gen} + \sum P_{cont} \right), \ \forall t, \forall s \tag{10}$$

4. Problem Formulation

In this section the probabilistic formulations of the proposed method are presented.

4.1. Day-Ahead Market

The aim of the aggregator is to buy energy in the day-ahead market at the minimum cost, taking into account the likely imbalance cost. This problem is formulated as a mixed integer linear probabilistic programming of one stage. The number of scenarios considered is $N_w = N_{dp} \cdot N_{dr}$, with N_{dp} being the number of day-ahead market prices scenarios, each one with a probability of occurrence p^{dp}, and N_{dr} as the number of fixed demand minus RES production scenarios, each one with a probability of occurrence p^{dr}. It is assumed that both random variables are independent because the size of the aggregator is small compared to the market. Thus, the probability of occurrence of scenario w is $p^w = p^{dp} \cdot p^{dr}$. Then the optimization problem may be formulated as follows:

$$\underset{P_{n,t}^{dp}, \forall t; \ \Delta_t^{+w}, \ \forall t, \forall w; \ \Delta_t^{-w}, \forall t, \forall w}{\text{Min}} \sum_{t=1}^{N_h} \left[\sum_{dp=1}^{N_{dp}} p^{dp} \left(\pi_t^{dp} P_{n,t}^{dp} \right) \cdot d_t + \sum_{w=1}^{N_w} p^w \left(\pi_t^w r_t^{-est} \Delta_t^{-w} - \pi_t^w r_t^{+est} \Delta_t^{+w} \right) \right] \tag{11}$$

subject to constraints (3)–(10),

$$\Delta_t^w = \left[P_{n,t}^w - \left(P_{rnd,t}^w + P_{shift,t}^w - P_{B,t}^{+w} + P_{B,t}^{-w} \right) \right] \cdot d_t, \ \ \forall t, \forall w \tag{12}$$

$$P_{n,t}^{dp} - P_{n,t}^{dp'} \le 0 \ : \ \pi_t^{dp} \ge \pi_t^{dp'}, \ \ \forall t, \forall dp \tag{13}$$

$$-\sum P_{gen} < P_{n,t}^{dp} < \sum P_{cont}, \ \ \forall t, \forall dp \tag{14}$$

The objective function in (11) is the expected cost of the energy traded by the aggregator in the day-ahead market considering the possible imbalance cost. The first term corresponds to the cost from the purchase of energy and the second term to the cost due to the imbalance. It must be remarked that only one direction of the net system imbalance is possible at a given period of time. The first term in (11) is affected by the daily price probability p^{dp} because it depends only on the scenarios of energy prices at this market whereas the second term is affected by the global probability p^w. Constraints (3)–(10) have already been explained in Section 3. Equation (12) defines the deviation of the aggregator in each period of time t and scenario w of the day-ahead market. Constraint (13) forces the bidding curves to be monotonically decreasing, which is a requirement in most markets. Constraint (14) limits the amount of power that can be sold or purchased in the day-ahead market in any period of time t and scenario dp.

The optimization problems (3)–(14) derive the N_h bidding curves, one for each period of time t of the time horizon, with N_{dp} pairs of possible values of energy-price, which correspond to the scenarios of the energy prices. It must be remarked that constraints related to shiftable demand (4) and batteries (6) and (7) are fulfilled for every scenario, but it may happens that after the market clearing, the committed energy for each period of time corresponds to different scenarios, and thus, those constraints will not be satisfied. A possible solution is to perform an adjustment process for rescheduling the flexible power, as explained below, but other solutions could be followed.

4.2. Flexible Power Scheduling

Once the day-ahead market is cleared and the daily prices and the committed energy are already known, the flexible power, i.e., the power from/to batteries and the shiftable demand, must be scheduled to ensure the satisfaction of constraints (4), (6), and (7). This flexible power is optimized along the time horizon according to the daily market price π_t^d. The batteries are adjusted with the objective function (15) and constraints (6)–(8), whereas the shiftable demand is optimized with the objective function (16) and constraints (4) and (5), in which market results are used. It should be remarked that only one scenario is considered in this process.

$$\underset{P_{B,t}^+, \forall t;\, P_{B,t}^-, \forall t}{Min} \sum_{t=1}^{N_h} \pi_t^d \cdot \left(-P_{B,t}^+ + P_{B,t}^- \right) \cdot d_t \tag{15}$$

$$\underset{P_{shift,t}, \forall t}{Min} \sum_{t=1}^{N_h} \pi_t^d \cdot P_{shift,t} \cdot d_t, \tag{16}$$

4.3. Intraday Market

The aggregator participates in the intraday market in order to correct the previous position taken in the day-ahead market, and it is assumed that the aggregator corrects the position only once for each period of time, participating in the ID market session closer to that period.

At this market level, daily market prices are already known, the flexible power (power from/to batteries and shiftable power) has been scheduled and new forecasts of intraday energy prices, fixed demand, and RES production are available, thus the global uncertainty decreases. The number of scenarios considered in this problem is $N_{iw} = N_{ip} \cdot N_{ir}$, with N_{ip} being the number of intraday market price scenarios, each one with a probability of occurrence p^{ip}, and N_{ir} as the number of fixed demand minus RES production scenarios, each one with a probability of occurrence p^{ir}. Again, it is assumed that both random variables are independent and therefore the probability of occurrence of scenario iw is $p^{iw} = p^{ip} \cdot p^{ir}$. The optimization problem in the intraday market may be formulated as follows:

$$\underset{P_{n,t}^{ip}, \forall t, \forall ip;\, \Delta_t^{+iw}, \forall t, \forall iw;\, \Delta_t^{-iw}, \forall t, \forall iw}{Min} \sum_{t=1}^{N_h} \left[\sum_{ip=1}^{N_{ip}} p^{ip} \left(\pi_t^{ip} P_{n,t}^{ip} \right) \cdot d_t + \sum_{iw=1}^{N_{iw}} p^{iw} \left(\pi_t^d r_t^{-est} \Delta_t^{-iw} - \pi_t^d r_t^{+est} \Delta_t^{+iw} \right) \right] \tag{17}$$

subject to constraints (3), (9), and (10),

$$\Delta_t^{iw} = \left[P_{n,t}^{iw} + P_{n,t}^d - \left(P_{rnd,t}^{iw} + P_{shift,t}^d - P_{B,t}^{+d} + P_{B,t}^{-d} \right) \right] \cdot d_t, \quad \forall t, \forall iw \tag{18}$$

$$P_{n,t}^{ip} - P_{n,t}^{ip'} \leq 0 : \pi_t^{ip} \geq \pi_t^{ip'}, \quad \forall t, \forall ip \tag{19}$$

$$-\sum P_{gen} \leq P_{n,t}^{ip} + P_{n,t}^d \leq \sum P_{cont}, \quad \forall t, \forall ip \tag{20}$$

the objective function (17) is the expected cost of the traded energy by the aggregator in this market, envisaging the likely imbalance cost. The first term in (17) corresponds to the cost from the purchase/sale of the energy in the intraday market and it is affected by the intraday energy price probability p^{ip}, since this term only depends on the intraday market price scenarios. The second term in (17) represents the final imbalance cost which the aggregator can incur and it is affected by the global probability p^{iw}. The intraday market bids can be positive (purchase of energy) or negative (sale of energy) and this has an effect over the total traded energy and the final imbalance at the end of both markets, defined in (18), where the terms with an upper index d correspond to results from the day-ahead market or the scheduling process for the flexible demand carried out after the day-ahead market clearing. Equation (19) forces the bidding curves to be monotonically decreasing and constraint (20) set the limits of the total offered power in both markets.

The optimization problem (3), (9)–(11), and (17)–(20) derive N_h bidding curves, one for each period of time t of the time horizon, with N_{ip} pairs of possible values of energy-price, which correspond to the scenarios of intraday market energy prices.

5. Case Study

The performance of the proposed method has been assessed through a case study where the aggregator participates in the Spanish electricity market [45] over a period of one year. Following the rules of this market, the schedule for updating bids in the intraday market for day D is depicted in Table 1. For example, the bids for hourly periods 1 to 4 of day D are updated in the second ID market session. In this way, the lead time for the six existing intraday markets spread throughout the day varies from 3 to 4 h until 5 to 8 h.

Table 1. Rules for updating energy bids for day D in intraday markets.

Hourly time period of day D	1–4	5–7	8–11	12–15	16–21	22–24
ID market session for updating bids	2	3	4	5	6	1 (D + 1)

Input data used in this realistic example is explained below. The entire problem was modelled in Matlab R2015a (version 8.5.0.197613, The Mathworks, Inc., Natick, MA, USA).

5.1. Grid Data

The data of consumers and generators are taken from the Conseil International des Grands Réseaux Électriques (CIGRE) medium voltage European benchmark grid [46]. Dispatchable units have not been included because they do not add uncertainty to the problem. The total installed generating power capacity considered is 1710 kW, with 1500 kW of wind power and 210 kW of photovoltaic power. The total load is 5291.6 kW, with 3843.6 kW and 1448 kW of fixed and shiftable demand, respectively. The storage consists of two batteries of 600 kW and 200 kW with 90% charge and discharge efficiency. The total capacity is 1600 kWh and the initial/final capacity in the batteries is taken as 80 kWh.

5.2. Forecast Data Scenarios

The set of scenarios for the probabilistic problems was created from available real data from the year 2013. Energy prices were taken from the web of the Spanish Market Operator [39]. Twenty-four ratios r_t^+ and r_t^- for the imbalance prices were estimated for each hour of the day as an hourly average of the 365 values of each hour of the year 2013. The average energy prices analyzed with their standard deviation during the whole year are shown in Table 2.

Table 2. Real data of energy prices for the whole year analyzed.

Energy Prices	Average (c€/kWh)	Standard Deviation (c€/kWh)
Day-ahead	4.43	2.07
Intraday	4.3	2.13
Buy Imbalance	5.2	2.06
Sell Imbalance	3.63	2.23

Scenarios of photovoltaic power were created from a synthesized production pattern of a solar plant located in Girona, Spain. The wind power scenarios were produced from data of a real wind farm that was used as a basic trajectory. A typical demand profile for residential customers from the Spanish System Operator Red Eléctrica de España (REE) [47] was used to create the probabilistic scenarios of demand. An example of the basic trajectories of demand and RES generation for a winter day in January is depicted in Figure 2.

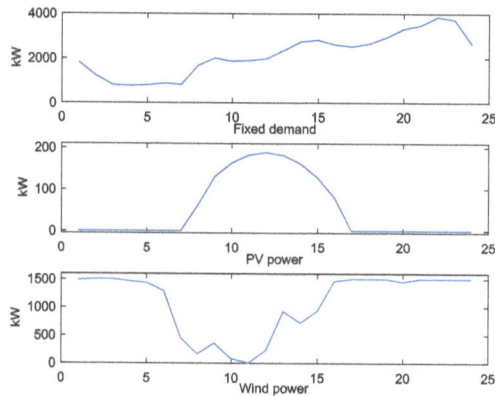

Figure 2. Basic trajectories of fixed demand, photovoltaics (PV), and wind power.

The number of total scenarios created for each one of the probabilistic optimization problems (day-ahead and intraday markets) was 10^6, 1000 for energy prices and 1000 for fixed demand minus RES production. Given the large number of scenarios created, a backward scenario reduction algorithm [48] was used to decrease the computational complexity and time, while preserving the most representative scenarios. Thus, the final number of scenarios was 200 (10 for energy prices and 20 for fixed demand minus RES production). From the basic trajectories shown in Figure 2, the reduced scenarios for the random variable $P_{rnd,t}^s$ for the day-ahead and intraday markets are depicted in Figure 3.

The normalized mean squared errors (NMSE) for each set of forecasts are shown in Table 3. Note that the forecast error for fixed demand is adequate, taking into account that the level of aggregation is low (the more aggregation, the lower forecast error); a NMSE of 13.6% corresponds to a maximum root mean squared error of 98 kW. The reduction of the PV forecast error from 24% to 15% has been considered analogous to the improvement of the new statistical learning methods over the numerical weather prediction model output, which is 10–15% [49]. We have considered that this reduction of the error is reasonable and it is enough for the purposes of the paper.

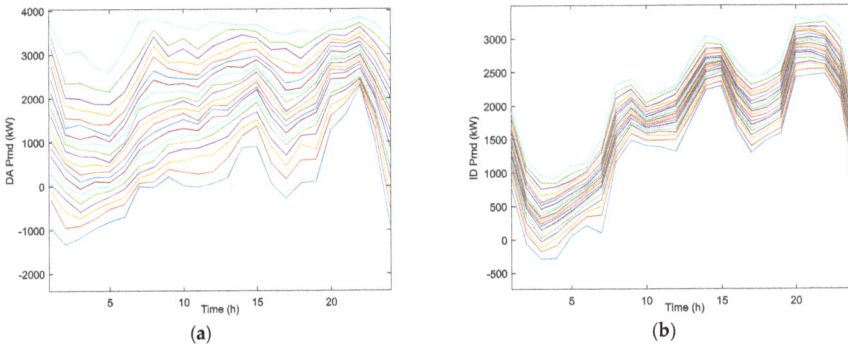

Figure 3. Reduced scenarios of fixed demand minus renewable energy sources (RES) generation to participate in (**a**) Day-ahead market and (**b**) Intraday markets.

Table 3. Normalized mean squared errors of random variables.

Random Variables	Forecast Errors (%)	
	Day-Ahead	Intraday
Photovoltaic Power	24.0	15.0
Wind Power	9.8	3.7
Fixed Demand	13.6	3.1
Energy Prices	0.2	0.08

5.3. Assessment of the Possible Aggregator Strategies

Different strategies are compared to assess the benefits of the proposed method along a year (8760 periods of one hour). These strategies refer to the participation or not in the intraday market and the application of the proposed probabilistic optimization compared to a more conventional deterministic one, as in [36]. The comparison among them is based on the annual payment performed by the aggregator, calculated with (21), where the first summation represents the payment in day-ahead and intraday markets for the purchase of energy over the period, and the second one is the payment due to the true real-time energy imbalances. If the aggregator only participates in the day-ahead market, the second term of the first brackets is null.

$$Payment = \sum_{t=1}^{8760} \left(\pi_t^d P_{n,t}^d + \pi_t^{id} P_{n,t}^{id} \right) \cdot d_t + \sum_{t=1}^{8760} \left(\pi_t^d r_t^{-true} \Delta_t^{-true} - \pi_t^d r_t^{+true} \Delta_t^{+true} \right) \tag{21}$$

6. Results

Given the uncertainty involved, the results of each market optimization problem are the bidding curves that would minimize the cost of the traded energy in the corresponding market. Each curve consists of 10 pairs of values of energy and price (one for each price scenario considered). Figure 4 shows the day-ahead optimal bidding curves for some hours of a sample day. For instance, the bidding curve for $t = 9$ h means that the aggregator would buy 4082 kWh if the market price is less than 6.37 c€/kWh at this hour, but he would buy 3717 kWh if the price were higher.

Figure 5 shows the optimal bidding curves for different hours and different intraday market sessions. Some curves change from positive (buy) to negative (sell) energy values, reflecting the aggregator's opportunity of submitting buying or selling bids for each hour aiming to update the committed energy at the day-ahead market framework.

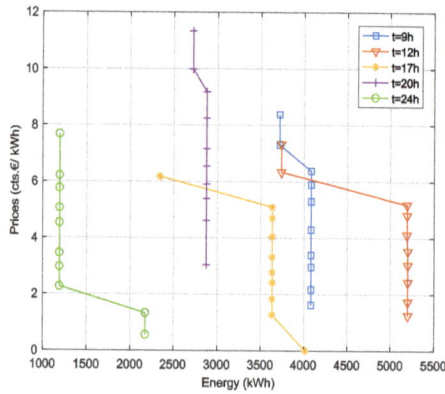

Figure 4. Optimal bidding curves to buy energy in the day-ahead market.

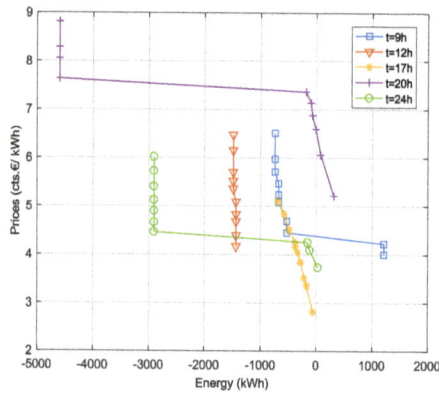

Figure 5. Optimal bidding curves to buy energy in the intraday market.

6.1. Benefit Comparisons

6.1.1. Case 1: Participation Only in the Day-Ahead Market

In this case, the aggregator decides to only participate in the day-ahead (DA) market representing a cluster of RES generators, batteries, and fixed and shiftable demand. In Table 4, the annual payment made by the aggregator following a deterministic and a probabilistic strategy are shown. This payment has been disaggregated in both terms of (21). The deterministic strategy does not consider the uncertainty and the likely imbalances, and is formulated as in [36]. The results show a reduction in the payment made with a probabilistic strategy over a deterministic one of 1.7%. Note that the probabilistic strategy leads to a higher purchase of energy in the DA market. In Table 4, the negative sign of the payment for the probabilistic net imbalance means that the imbalance term represents an income, i.e., there were more hours in which the consumption scheduled in the day-ahead market was greater than the actual consumption of the aggregator's customers, and this energy difference was paid at the positive imbalance price.

Table 4. Aggregator annual payment from the participation only in the Day-ahead (DA) market.

Payment (k€)	Deterministic	Probabilistic
Day-ahead market (k€)	999	1308
Net Imbalance (k€)	22	−304
Total (k€)	1021	1004
Gain (%)	-	1.7

6.1.2. Case 2: Participation in the Day-Ahead and Intraday Markets

If the aggregator also decides to participate in the intraday (ID) markets, several strategies could be followed, depending on the type of the optimization problem followed in each market (probabilistic (Prob) or deterministic (Det)); the strategies considered in this work are depicted in Table 5. The annual payments derived from each strategy are compared in Table 6. As in Table 4, the payment is disaggregated in payments in the markets and payments due to the imbalances; the last row shows the gain in payment obtained by following the different strategies over the pure deterministic one (strategy S1). It may be thought that, given the high uncertainty at the time of the day-ahead market regarding RES generation random variables, strategy S2 could be good enough (a gain of 2.84% over strategy S1 is achieved), but the results show a better performance of the strategy S3 proposed in this work, which leads to a benefit of 4.06% over strategy S1.

Table 5. Possible aggregator strategies to participate in the DA plus intraday (ID) markets.

Market	Strategy S1	Strategy S2	Strategy S3
DA + ID	Det + Det	Det + Prob	Prob + Prob

Table 6. Aggregator annual payment from the participation in the DA plus ID markets.

Payment (k€)	Strategy S1	Strategy S2	Strategy S3
DA + ID markets (k€)	977	1108	1111
Net Imbalance (k€)	8	−151	−166
Total (k€)	985	957	945
Gain (%)	-	2.84	4.06

From Table 6, it is seen that payments done in the markets increase from strategies S1 to S2 and S3, but the total payments decrease. This is due to the behavior of the imbalance term. With strategy S1, the aggregator pays 8 k€ because of the deviations, but with strategies S2 and S3, the aggregator is paid because of the deviations (negative payment), meaning that positive deviations have been more frequent than negative deviations.

An in-depth analysis of each type of payment in Table 6 is performed next. First, the comparison between the energy traded in each market for the best and worse strategies (i.e., strategies S3 and S1, respectively) is shown in Table 7, distinguishing buying and selling energy. Second, the comparison between the total energy deviations incurred by the aggregator after the participation in the markets is included in Table 8.

From Table 7, it is derived that the consideration of the uncertainties leads to a greater quantity of energy traded in the markets (with the probabilistic strategy S3, the aggregator trades more energy than with the deterministic one S1), and thus the net payment in the markets is also higher, as shown in Table 6. As the net position in the ID market is selling energy, the aggregator tends to purchase an excess of energy in the DA market with both strategies, but this effect is more accentuated with the strategy S3. The reason may be that strategy S3 takes into account the possible payments due to imbalances, and while sell imbalance prices are limited ($0 \leq \pi_t^+ \leq \pi_t^d$), buy imbalance prices could reach very high values. Then, it is seen that the aggregator tends to trade more energy in the markets with a knowledge improvement. From Table 8, it can be said that: energy deviations are reduced with the participation in the ID market, as expected; and the probabilistic strategy leads to higher energy deviations than the deterministic one.

Table 7. Energy traded in DA and ID markets.

Strategy	Annual Quantity	DA Market	ID Market	Total Traded/Paid
S3 Probabilistic Energy (MWh)	Total buying	30,402	4122	34,524
	Total selling	0	6586	6586
	Net (buying)	30,402	−2464	27,938
S1 Deterministic Energy (MWh)	Total buying	21,867	2941	24,808
	Total selling	73	3197	3270
	Net (buying)	21,794	−256	21,538

Table 8. Total energy deviation.

Term	Strategy	DA Market	ID Market
Total Energy Deviation (MWh)	S3 Probabilistic	9013	6529
	S1 Deterministic	406	144

For a better understanding of the payments made due to energy deviations (Net Imbalance term in Table 6), approximate probability density functions (PDF) of the annual energy deviations and the payment/revenue derived from the deviations are shown in Figures 6 and 7 for strategies S3 and S1. Figure 6 shows the results after the participation only in the DA market, whereas Figure 7 shows the results after the participation in the ID market (having participated in the DA market). After the participation in the DA market, it can be remarked in Figure 6a that the energy deviations derived from the deterministic strategy S1 are symmetrically distributed with a slightly displacement to the negative side. By contrast, energy deviations derived from the probabilistic strategy S3 have a right-skewed bi-modal distribution, showing higher positive imbalances (i.e., real-time demand of energy lower than scheduled in the DA market), both in number and value, although most of the imbalances are associated with the left mode. If the aggregator participates in the ID market after the DA market, the PDFs of final energy deviations are shown in Figure 7a. In this case, the imbalances from the deterministic strategy S1 do not present any significant variation but most of the imbalances from the probabilistic strategy S3 are now located around the mean. Also, the tails are longer than in Figure 6a, which means that there are few values of deviated energy higher than in the DA market.

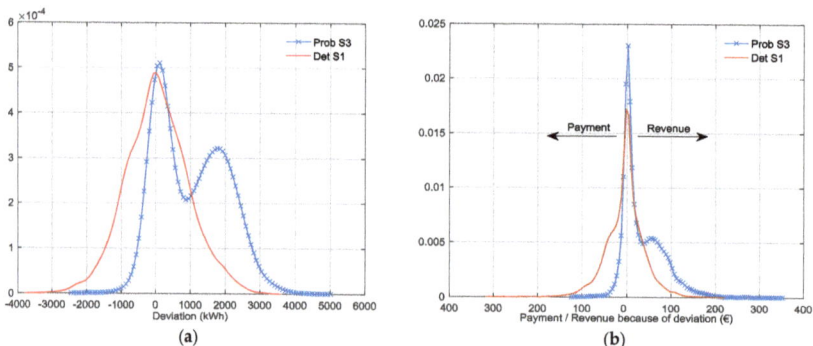

Figure 6. (a) Approximate probability density functions (PDF) of the energy deviation after the day-ahead market; (b) PDF of the payment/revenue due to deviation after the day-ahead market.

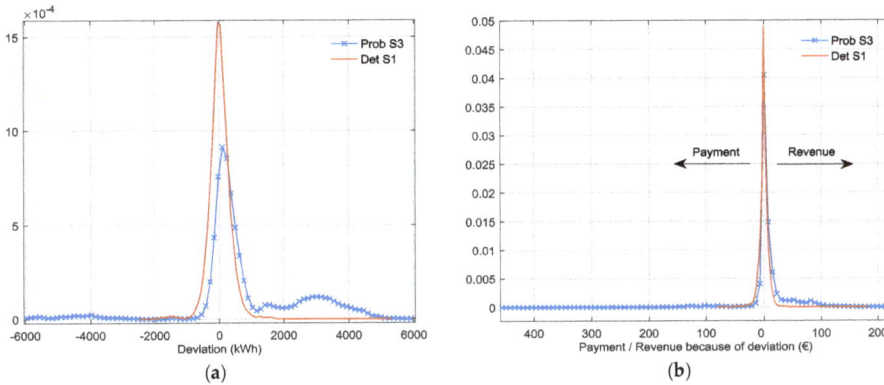

Figure 7. (**a**) PDF of the energy deviation after the intraday market; (**b**) PDF of the payment/revenue due to deviation after the intraday market.

Regarding the payment/revenue derived from the energy deviations, it can be seen in Figures 6b and 7b that:

- Using the deterministic strategy S1, its variability after the ID market decreases compared to the variability after the DA market, but with the probabilistic strategy S3, the payment/revenue variability increases after the ID market.
- With the probabilistic strategy S3, the PDFs present a right-skewed distribution compared to the deterministic S1, which means that the probability of the aggregator to be remunerated for a positive imbalance is higher with S3 than with S1.

Thus, the probabilistic approach tends to maximize the revenue due to positive deviations and minimize the payment due to the negative deviations. The deterministic strategy does not consider this, leading to a higher payment. The probabilistic strategy may also benefit from arbitrage between the ID markets and the imbalance prices, selling or buying in the ID market, when the expected imbalance price is, respectively, lower or higher than the forecasted intraday market price. Of course, this strategy might lead to occasional losses, but the overall results are favorable. Risk-averse strategies to limit these losses are less profitable in the long term [20].

Overall, the proposed method has a good performance and the computation time is reduced (the computation time of both problems (day-ahead plus intraday) for a planning period of 24 h is around 15 s).

7. Conclusions and Future Work

A method for producing optimal bidding curves for an aggregator participating in day-ahead and intraday markets has been presented. The objective is minimizing the payment done by the aggregator for the energy purchase for his customers. The method consists of different optimization problems which considers flexible consumption through shiftable demand and the use of batteries, and takes into account the uncertainty of the forecasts (RES generation, market prices, and fixed demand) and the likely imbalance costs. The overall process is performed in three steps: First, the optimal bidding curves are produced and submitted to the day-ahead market; Second, after the day-ahead market clearing, intertemporal constraints related to the flexible consumption are fulfilled through a rescheduling process; and finally, new optimal bidding curves are produced and submitted to the intraday market, trying to take advantage of the lower lead time and the knowledge gained with the day-ahead market clearing about marginal prices. The payment done by an aggregator participating in the Iberian market, over a whole year, has been calculated and compared with the payments done using different strategies, yielding better results. A thorough analysis is performed comparing the probabilistic strategy proposed in this paper with a more conventional one. Results show that although the deviations in the proposed strategy are higher than in the deterministic one, the overall payments are lower because the probabilistic method tends to produce positive imbalances. This work demonstrates that simple and independent probabilistic optimization problems report meaningful benefits for aggregators participating in power markets.

In the future, we will extend the proposed method to obtain an optimal bidding and scheduling to real cases of urban and semirural grids of small size represented by the aggregator and including electric vehicles, and heating and cooling loads. Moreover, we will perform the assessment of likely scheduling for future scenarios of resources available in these grids.

Acknowledgments: This work was supported by the European research project IDE4L (Ref. FP7-SMARTCITIES-2013-608860) and the Spanish project RESmart (Ref. ENE2013-48690-C2-1-R).

Author Contributions: Xiaolin Ayón, María Ángeles Moreno, and Julio Usaola conceived and designed the experiments; Xiaolin Ayón performed the experiments; María Ángeles Moreno and Xiaolin Ayón wrote the paper and overall all authors set and refined the manuscript.

Conflicts of Interest: The authors declare no conflict of interest.

Notations

The main notations used throughout the paper is reproduced below for quick reference. Other symbols are defined as required.

Indices and Numbers

dp, ip	Indices of day-ahead/intraday market prices scenarios running from 1 to N_{dp}/N_{ip}.
dr, ir	Indices of day-ahead/intraday scenarios of fixed demand minus renewable energy sources (RES) production running from 1 to N_{dr}/N_{ir}.
t	Index of time periods running from 1 to N_h.
w, iw	Indices of day-ahead/intraday market scenarios of the global random variables running from 1 to N_w/N_{iw}.
s	Index of generic scenarios (dp, ip, w, and iw).

Continuous Variables

$E_{B,t}^s$	Energy stored in the batteries in time period t and scenario s [kWh].
$P_{B,t}^{+s}, P_{B,t}^{-s}$	Power from/to the batteries in time period t and scenario s [kW].
$P_{shift,t}^s$	Shiftable demand in time period t and scenario s [kW].
$P_{n,t}^s$	Net power offered to the market in time period t and scenario s [kW].
Δ_t^s	Energy deviation incurred by the aggregator with respect to the schedule in time period t and scenario s [kWh].
$\Delta_t^{+s}, \Delta_t^{-s}$	Deficit of energy (positive deviation)/Excess of energy (negative deviation) incurred by the aggregator with respect to the schedule in time period t and scenario s [kWh].

Random Variables

$P_{fix,t}^s$	Fixed demand in time period t and scenario s [kW].
$P_{pv,t}^s$	Photovoltaic generation in time period t and scenario s [kW].
$P_{rnd,t}^s$	Fixed demand minus RES production in time period t and scenario s [kW].
$P_{w,t}^s$	Wind generation in time period t and scenario s [kW].
π_t^s	Market price in time period t and scenario s [c€/kWh].

Constants and Data

p^s	Probability of occurrence of scenario s.
d_t	Duration of the time period t (in hours).
E_B^{max}, E_B^{min}	Maximum/minimum energy in the storage system [kWh].
E_{shift}	Daily limit of total shiftable energy [kWh].
$P_{B,t}^{+d}, P_{B,t}^{-d}$	Scheduled power from/to the batteries for time period t after the flexible demand scheduling process [kW].
P_B^{+max}, P_B^{-max}	Maximum power from/to the batteries [kW].
P_{gen}, P_{cont}	Total installed capacity and contracted power of the aggregator's customers (prosumers) [kW].
$P_{n,t}^d, P_{n,t}^{id}$	Scheduled net power in day-ahead/intraday market for time period t [kW].
$P_{shift,t}^d$	Scheduled shiftable demand for time period t after the flexible demand scheduling process [kW].
P_{shift}^{max}	Maximum shiftable demand [kW].
r_t^+, r_t^-	Ratios between positive/negative imbalance price and day-ahead market price in time period t. A superscript *est* or *true* is added to indicate the estimated or true values.
η^+, η^-	Efficiencies of giving back/storing energy of the batteries.
ΔE	Energy losses in the storage system [kWh].
π_t^d, π_t^{id}	Actual day-ahead/intraday market price in time period t [c€/kWh].
$\Delta_t^{+true}, \Delta_t^{-true}$	Actual deficit of energy (positive deviation)/excess of energy (negative deviation) incurred by the aggregator with respect to the schedule in time period t [kWh].

References

1. CIRED Working Group on Smart Grids. Smart Grids on the Distribution Level—Hype or Vision? CIRED's Point of View, Final Report. 2013. Available online: http://www.cired.at/pdf/CIRED_WG_SmartGrids_FinalReport.pdf (accessed on 19 December 2016).
2. Pudjianto, D.; Ramsay, C.; Strbac, G. Virtual power plant and system integration of distributed energy resources. *IET Renew. Power Gener.* **2007**, *1*, 10–16. [CrossRef]
3. Momber, I.; Gómez, T.; Söder, L. PEV fleet scheduling with electricity market and grid signals charging schedules with capacity pricing based on DSO's long run marginal cost. In Proceedings of the IEEE 10th International Conference on European Energy Markets, Stockholm, Sweden, 27–31 May 2013.
4. Carrión, M.; Arroyo, J.M.; Conejo, A.J. A bilevel stochastic programming approach for retailer futures market trading. *IEEE Trans. Power Syst.* **2009**, *24*, 1446–1456. [CrossRef]
5. Carrión, M.; Philpott, A.B.; Conejo, A.J.; Arroyo, J.M. A stochastic programming approach to electric energy procurement for large consumers. *IEEE Trans. Power Syst.* **2007**, *22*, 744–754. [CrossRef]
6. Rahimiyan, M.; Baringo, L.; Conejo, A.J. Energy management of a cluster of interconnected price-responsive demands. *IEEE Trans. Power Syst.* **2014**, *29*, 645–655. [CrossRef]
7. Bornapour, M.; Hooshmand, R.A.; Khodabakhshian, A.; Parastegari, M. Optimal coordinated scheduling of combined heat and power fuel cell, wind, and photovoltaic units in micro grids considering uncertainties. *Energy* **2016**, *117*, 176–189. [CrossRef]
8. Dehghani, H.; Vahidi, B.; Hosseinian, S.H. Wind farms participation in electricity markets considering uncertainties. *Renew. Energy* **2017**, *101*, 907–918. [CrossRef]
9. Zhou, Y.; Wang, C.; Wu, J.; Wang, J.; Cheng, M.; Li, G. Optimal scheduling of aggregated thermostatically controlled loads with renewable generation in the intraday electricity market. *Appl. Energy* **2017**, *188*, 456–465. [CrossRef]
10. Soares, J.; Ghazvini, M.A.F.; Borges, N.; Vale, Z. A stochastic model for energy resources management considering demand response in smart grids. *Electr. Power Syst. Res.* **2017**, *143*, 599–610. [CrossRef]
11. Parisio, A.; Glielmo, L. Stochastic model predictive control for economic/environmental operation management of microgrids. In Proceedings of the IEEE European Control Conference, Zürich, Switzerland, 17–19 July 2013; pp. 2014–2019.
12. Ferruzzi, G.; Cervone, G.; Delle Monache, L.; Graditi, G.; Jacobone, F. Optimal bidding in a day-ahead energy market for Micro Grid under uncertainty in renewable energy production. *Energy* **2016**, *106*, 194–202. [CrossRef]

13. Puglia, L.; Bernardini, D.; Bemporad, A. A multi-stage stochastic optimization approach to optimal bidding on energy markets. In Proceedings of the IEEE 50th Conference on Decision and Control and European Control Conference, Orlando, FL, USA, 12–15 December 2011; pp. 1509–1514.

14. Shi, L.; Luo, Y.; Tu, G.Y. Bidding strategy of microgrid with consideration of uncertainty for participating in power market. *Int. J. Electr. Power Energy Syst.* **2014**, *59*, 1–13. [CrossRef]

15. Pandžić, H.; Morales, J.M.; Conejo, A.J.; Kuzle, I. Offering model for a virtual power plant based on stochastic programming. *Appl. Energy* **2013**, *105*, 282–292. [CrossRef]

16. Asensio, M.; Contreras, J. Risk-constrained optimal bidding strategy for pairing of wind and demand response resources. *IEEE Trans. Smart Grid* **2017**, *8*, 200–208. [CrossRef]

17. Morales, J.M.; Conejo, A.J.; Pérez-Ruiz, J. Short-term trading for a wind power producer. *IEEE Trans. Power Syst.* **2010**, *25*, 554–564. [CrossRef]

18. Nojavan, S.; Mohammadi-Ivatloo, B.; Zare, K. Robust optimization based price-taker retailer bidding strategy under pool market price uncertainty. *Int. J. Electr. Power Energy Syst.* **2015**, *73*, 955–963. [CrossRef]

19. Ottesen, S.Ø.; Tomasgard, A.; Fleten, S.E. Prosumer bidding and scheduling in electricity markets. *Energy* **2016**, *94*, 828–843. [CrossRef]

20. Moreno, M.Á.; Bueno, M.; Usaola, J. Evaluating risk-constrained bidding strategies in adjustment spot markets for wind power producers. *Int. J. Electr. Power Energy Syst.* **2012**, *43*, 703–711. [CrossRef]

21. Laia, R.; Pousinho, H.M.I.; Melíco, R.; Mendes, V.M.F. Bidding strategy of wind-thermal energy producers. *Renew. Energy* **2016**, *99*, 673–681. [CrossRef]

22. Xu, Z.; Hu, Z.; Song, Y.; Wang, J. Risk-averse optimal bidding strategy for demand-side resource aggregators in day-ahead electricity markets under uncertainty. *IEEE Trans. Smart Grid* **2017**, *8*, 96–105. [CrossRef]

23. Sanchez-Martin, P.; Alberdi-Alen, A. Day ahead and intraday stochastic decision model for EV charging points. In Proceedings of the IEEE Powertech Grenoble Conference, Grenoble, France, 16–20 June 2013.

24. Herranz, R.; Muñoz San Roque, A.; Villar, J.; Campos, F.A. Optimal demand-side bidding strategies in electricity spot markets. *IEEE Trans. Power Syst.* **2012**, *27*, 1204–1213. [CrossRef]

25. Nguyen, D.T.; Le, L.B. Optimal bidding strategy for microgrids considering renewable energy and building thermal dynamics. *IEEE Trans. Smart Grid* **2014**, *5*, 1608–1620. [CrossRef]

26. Faria, E.; Fleten, S.E. Day-ahead market bidding for a Nordic hydropower producer: Taking the Elbas market into account. *Comput. Manag. Sci.* **2011**, *8*, 75–101. [CrossRef]

27. Klaboe, G.; Fosso, O.B. Optimal bidding in sequential physical markets–A literature review and framework discussion. In Proceedings of the IEEE Powertech Grenoble Conference, Grenoble, France, 16–20 June 2013.

28. Baringo, L.; Conejo, A.J. Offering strategy via robust optimization. *IEEE Trans. Power Syst.* **2011**, *26*, 1418–1425. [CrossRef]

29. Ansari, M.; Al-Awami, A.T.; Sortomme, E.; Abido, M.A. Coordinated bidding of ancillary services for vehicle-to-grid using fuzzy optimization. *IEEE Trans. Smart Grid* **2015**, *6*, 261–270. [CrossRef]

30. Sousa, T.; Morais, H.; Castro, R.; Vale, Z. Evaluation of different initial solution algorithms to be used in the heuristics optimization to solve the energy resource scheduling in smart grids. *Appl. Soft Comput.* **2016**, *48*, 491–506. [CrossRef]

31. Wei, M.; Zhong, J. Optimal bidding strategy for demand response aggregator in day-ahead markets via stochastic programming and robust optimization. In Proceedings of the IEEE 12th International Conference on European Energy Markets, Lisbon, Portugal, 19–22 May 2015.

32. Chen, Z.; Wu, L.; Fu, Y. Real-time price-based demand response management for residential appliances via stochastic optimization and robust optimization. *IEEE Trans. Smart Grid* **2012**, *3*, 1822–1831. [CrossRef]

33. Fleten, S.E.; Wallace, S.W.; Ziemba, W.T. Hedging electricity portfolios via stochastic programming. In *Decision Making under Uncertainty*, 1st ed.; Greengard, C., Ruszczynski, A., Eds.; Springer: New York, NY, USA, 2002; Volume 128, pp. 71–93.

34. Fernandes, C.; Frías, P.; Reneses, J. Participation of intermittent renewable generators in balancing mechanisms: A closer look into the Spanish market design. *Renew. Energy* **2016**, *89*, 305–316. [CrossRef]

35. Conejo, A.J.; Carrión, M.; Morales, J.M. *Decision Making under Uncertainty in Electricity Markets*, 1st ed.; Springer: New York, NY, USA, 2010; pp. 211–230.

36. Ayon, X.; Usaola, J. An optimal scheduling for aggregators in smart grid. In Proceedings of the IEEE 12th International Conference on European Energy Markets, Lisbon, Portugal, 19–22 May 2015.

37. Daily and Intraday Electricity Market Operating Rules. pp. 67–71. Available online: http://www.omel.es/files/reglas_agosto_2012_ingles.pdf (accessed on 19 December 2016).

38. Chaves-Ávila, J.P.; Fernandes, C. The Spanish intraday market design: A successful solution to balance renewable generation? *Renew. Energy* **2015**, *74*, 422–432. [CrossRef]

39. Electricity Iberian Market Operator: OMIE, Market Results. Available online: http://www.omie.es/files/flash/ResultadosMercado.swf (accessed on 19 December 2016).

40. Spanish System Operator's Information System. Available online: http://www.esios.ree.es/es (accessed on 19 December 2016).

41. Christoph, W. Adequate intraday market design to enable the integration of wind energy into the European power systems. *Energy Policy* **2010**, *38*, 3155–3163.

42. Abedi, S.; Riahy, G.H.; Farhadkhani, M.; Hosseinian, S.H. *Improved Stochastic Modeling: An Essential Tool for Power System Scheduling in the Presence of Uncertain Renewables*; INTECH Open Access: Rijeka, Croatia, 2013; pp. 113–120.

43. Usaola, J.; Ramírez, V.; Gafurov, T.; Prodanovic, M. PV modelling for generation adequacy studies. In Proceedings of the 3rd Solar Integration Workshop, London, UK, 21–22 October 2013.

44. Plazas, M.A.; Conejo, A.J.; Prieto, F.J. Multimarket optimal bidding for a power producer. *IEEE Trans. Power Syst.* **2005**, *20*, 2041–2050. [CrossRef]

45. Spanish Electricity Market. Available online: http://www.omie.es/inicio/mercados-y-productos/mercado-electricidad/nuestros-mercados-de-electricidad/diario-e-intradia (accessed on 19 December 2016).

46. Rudion, K.; Orths, A.; Styczynski, Z.A.; Strunz, K. Design of benchmark of medium voltage distribution network for investigation of DG integration. In Proceedings of the IEEE Power Engineering Society General Meeting, Montreal, QC, Canada, 18–22 June 2006.

47. Spanish Consumption Profiles. Available online: http://www.ree.es/es/actividades/operacion-del-sistema-electrico/medidas-electricas (accessed on 19 December 2016).

48. Razali, N.M.; Hashim, A.H. Backward reduction application for minimizing wind power scenarios in stochastic programming. In Proceedings of the IEEE 4th International Power Engineering and Optimization Conference, Shah Alam, Malasya, 23–24 June 2010; pp. 430–434.

49. Tuohy, A.; Zack, J.; Haupt, S.E.; Sharp, J.; Ahlstrom, M.; Dise, S.; Grimit, E.; Mohrlen, C.; Lange, M.; Casado, M.G.; et al. Solar forecasting: Methods, challenges, and performance. *IEEE Power Energy Mag.* **2015**, *13*, 50–59. [CrossRef]

energies

MDPI

Article

Energy Trading and Pricing in Microgrids with Uncertain Energy Supply: A Three-Stage Hierarchical Game Approach

Kai Ma [1], Shubing Hu [1], Jie Yang [1,*,†], Chunxia Dou [1] and Josep M. Guerrero [2]

[1] School of Electrical Engineering, Yanshan University, Qinhuangdao 066004, China; kma@ysu.edu.cn (K.M.); shubingdf@163.com (S.H.); cxdou@ysu.edu.cn (C.D.)
[2] Department of Energy Technology, Aalborg University, Aalborg 9100, Denmark; joz@et.aau.dk
*   Correspondence: jyangysu@ysu.edu.cn; Tel.: +86-335-8047169
†   Current address: Hebei Avenue 438, Qinhuangdao 066004, China

Academic Editor: Pedro Faria
Received: 5 March 2017; Accepted: 6 May 2017; Published: 11 May 2017

Abstract: This paper studies an energy trading and pricing problem for microgrids with uncertain energy supply. The energy provider with the renewable energy (RE) generation (wind power) determines the energy purchase from the electricity markets and the pricing strategy for consumers to maximize its profit, and then the consumers determine their energy demands to maximize their payoffs. The hierarchical game is established between the energy provider and the consumers. The energy provider is the leader and the consumers are the followers in the hierarchical game. We consider two types of consumers according to their response to the price, i.e., the price-taking consumers and the price-anticipating consumers. We derive the equilibrium point of the hierarchical game through the backward induction method. Comparing the two types of consumers, we study the influence of the types of consumers on the equilibrium point. In particular, the uncertainty of the energy supply from the energy provider is considered. Simulation results show that the energy provider can obtain more profit using the proposed decision-making scheme.

Keywords: microgrid; uncertainty; hierarchical game; non-cooperative game (NCG); energy trading; pricing strategy

1. Introduction

In the microgrid, energy trading is an important segment [1,2]. The energy provider determines the energy purchase to meet the consumer demands. Meanwhile, in order to increase its profit, the energy provider faces a problem of pricing decision. With the development of renewable energy (RE), it is reasonable for the energy provider to use the renewable energy as supply [3,4]. Due to the introduction of the renewable energy, the energy provider has to predict the generating capacity of the renewable energy system, and then decides how much energy it needs to purchase from the electricity markets. The energy provider's prediction can have a deviation from the actual generation, which leads to the uncertainty (UC) of the energy supply [5–7].

There are some works in literature related to the interactions among the consumers. A non-cooperative game (NCG) was formulated among the consumers in [8–14]. In [9], the price-taking consumers and the price-anticipating consumers were considered. The price-taking consumers assume that their energy consumption cannot affect the electricity price, whereas the price-anticipating consumers believe that their energy consumption can change the electricity price. Recently, the Stackelberg game (SG) is formulated between the energy provider and the consumers in [15–20]. In addition, the authors in [20,21] took into account a two-stage Stackelberg game between the power station and the consumers. In order to reduce the cost of the energy purchased from the electricity markets, the renewable energy was taken into account in [22]. Most of these works mainly focus on the price-taking consumers, and they seldom take into account the renewable energy generation, thereby the uncertainty of the energy supply is not involved. The differences of the proposed work with the above literature are shown in Table 1.

Table 1. Differences of the proposed work with the literature.

Indexes	RE	UC	NCG	SG
[3,4]	✓	×	×	×
[5–7]	×	✓	×	×
[8–14]	×	×	✓	×
[15–21]	×	×	×	✓
This work	✓	✓	✓	✓

In this paper, we consider the uncertainty of the energy supply caused by the wind power generation. Furthermore, we both consider the price-taking consumers and the price-anticipating consumers. We model the interactions between the energy provider and the consumers as a three-stage hierarchical game. The energy provider, which is the hierarchical game's leader, determines the price and the energy purchase to maximize its profit. Finally, we obtain the equilibrium of the hierarchical game through the backward induction method [23,24].

The rest of the paper is organized as follows. Section 2 introduces the problem formulation. Section 3 shows the backward induction method of the three-stage hierarchical game for the energy provider and the price-taking consumers. In Section 4, the backward induction method of the three-stage game for the energy provider and the price-anticipating consumers is described. Section 5 gives the simulation and comparison results. Finally, the conclusions are summarized in Section 6.

2. Problem Formulation

We consider the energy trading and pricing problem in the microgrid consisting of one energy provider and a set $\mathbb{N} = \{1, ..., N\}$ of consumers. The energy provider and the consumers are integrated into a microgrid with renewable energy generation. The energy provider purchases energy from the electricity markets when the renewable energy supply is not enough. In that case, the energy supply includes the energy generated from the renewable energy source and the energy purchased from the electricity markets. In the microgrid, the system structure of the energy trading is given in Figure 1. Because of the uncertainty of the energy generated from renewable energy sources, the energy purchased from the electricity markets is uncertain. According to the interaction between the energy provider and the consumers, we establish a hierarchical game as below.

- Leader: the energy provider determines the energy purchase and the pricing strategy to maximize its profit.
- Followers: the consumers determine the energy demands to maximize their payoffs.

According to the types of the consumers, we consider two scenarios in this paper. In scenario A, there is no competition among the price-taking consumers, i.e., the consumers' energy consumption cannot affect the price announced by the energy provider. In scenario B, the interactions among the price-anticipating consumers are formulated into a non cooperative game, i.e., the consumers' energy consumption can change the price announced by the energy provider [9].

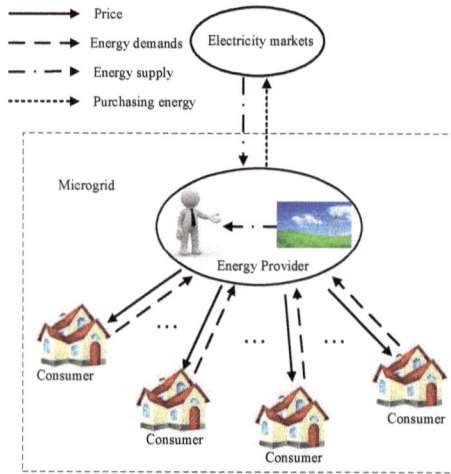

Figure 1. The system structure of energy trading.

3. Wind Power Generation Model

There exists a large body of literature on wind power forecasting, and the day-ahead wind forecast based on numerical weather prediction (NWP) models can enable relatively accurate wind forecasts [25,26]. Because the operating time moves closer to the near term, the computation complexity often renders NWP models intractable at a high spatial resolution [26]. An adaptive wavelet neural network was proposed for mapping the NWP's wind speed and wind direction forecasts to wind power forecasts in [27]. The authors in [28] proposed a novel statistical wind power forecast framework, which leverages the spatio-temporal correlation in wind speed and direction data among geographically dispersed wind farms. In [29], the author developed a feed-forward neural network approach for wind power generation forecasting to improve the wind forecasting accuracy. However, the wind power forecast is relatively complex, and the forecast errors cannot be avoided. Generally, the wind speed can be approximated as the Gamma distribution [30], inverse Gaussian [31], log-normal [32], and Weibull [33–36]. Alternatively, copula theory has recently been applied to wind speed and wind power as a way of modeling nonlinear dependence structures [37]. According to the wind speed, we establish the wind power generation model adopting the mixed copula function in this paper. Firstly, we introduce the copula theory.

The copula theory was proposed by Sklar in 1959 [38]. Supposing that $F(x_1, x_2, \cdots, x_N)$ is a joint distribution function whose marginal distributions are $F_1(x_1), F_2(x_2), \cdots, F_N(x_N)$, then there is a copula function C satisfies [39]:

$$F(x_1, x_2, \cdots, x_N) = C(F_1(x_1), F_2(x_2), \cdots, F_N(x_N)). \tag{1}$$

Defining that $F_1^{-1}(u_1), F_2^{-1}(u_2), \cdots, F_N^{-1}(u_N)$ are the pseudo inverse functions of $F_1(x_1), F_2(x_2), \cdots, F_N(x_N)$, respectively. Therefore, the copula function C can be obtained as the following:

$$C(u_1, u_2, \cdots, u_N) = F(F_1^{-1}(u_1), F_2^{-1}(u_2), \cdots, F_N^{-1}(u_N)), \tag{2}$$

where the marginal distributions of $C(u_1, u_2, \cdots, u_N)$ follow uniform distribution in $[0, 1]$ [40].

When $N = 2$, the copula function C is a binary function. $H(x_1, x_2)$ is a joint distribution function whose marginal distributions are $F(x_1)$ and $W(x_2)$, and $F^{-1}(u)$ and $W^{-1}(v)$ are the pseudo inverse function of $F(x_1)$ and $W(x_2)$, respectively. Therefore, the copula function C is expressed as the following:

$$C(u,v) = H(F^{-1}(u), W^{-1}(v)).$$ (3)

The mixed copula function was further proposed in [41,42]:

$$C_M(u,v) = \lambda_1 C_1(u,v,\gamma_1) + \lambda_2 C_2(u,v,\gamma_2) + \lambda_3 C_3(u,v,\gamma_3),$$ (4)

where $C_M(u,v)$ is the mixed copula function that is composed of $C_1(u,v,\gamma_1)$, $C_2(u,v,\gamma_2)$, and $C_3(u,v,\gamma_3)$. λ_1, λ_2, and λ_3 are weight coefficients of $C_1(u,v,\gamma_1)$, $C_2(u,v,\gamma_2)$, and $C_3(u,v,\gamma_3)$, respectively, and satisfy $\lambda_1 + \lambda_2 + \lambda_3 = 1$. γ_1, γ_2, and γ_3 are correlation coefficients and can measure the correlation degree of variables.

The results of [41] showed that the relevant structures of the mixed copula function are more flexible than a single copula function. The wind power generation model is established by the mixed copula function (see details in [38]).

4. Scenario A: The Three-Stage Game for Price-Taking Consumers

In this section, we prove the existence and uniqueness of the hierarchical equilibrium by using the backward induction method. First, we analyze the consumers' demands given the energy provider's pricing strategy and energy purchase. Then, we study the energy provider's pricing strategy given the consumers' energy demands and the energy purchase. Finally, we analyze the energy purchased from the electricity markets in the case of uncertain renewable generation, and then obtain the maximum profit of the energy provider.

4.1. Consumer's Energy Demands in Stage III

In Stage I, the energy provider needs to determine the energy purchased from the electricity markets. In Stage II, the energy provider announces the price to the consumers. In Stage III, the consumers determine their energy demands given the unit price p announced by the energy provider in Stage II. The payoff of consumer i is defined as the difference between the satisfaction level and the payment for energy purchase, i.e.,

$$u_i(p,c_i) = -h(c_i - c_i')^2 - pc_i,$$ (5)

where h is the consumers' cost coefficient, c_i is the actual energy demand of the consumer i, and c_i' is the energy demand of the consumer i to maintain normal operation of appliances. The consumers determine their demands to maximize their payoffs. p is a fixed price announced by the energy provider.

The first derivative of $u_i(p,c_i)$ with respect to c_i is:

$$\frac{\partial u_i(p,c_i)}{\partial c_i} = -2h(c_i - c_i') - p.$$ (6)

Letting $\partial u_i(p,c_i)/\partial c_i = 0$, we obtain:

$$-2h(c_i - c_i') - p = 0, \forall i \in N.$$ (7)

From Equation (7), we can obtain the energy demands of the consumer i.

$$c_i(p) = c_i' - \frac{p}{2h}.$$ (8)

Adding Equation (8) from 1 to N, we obtain the total energy consumption of all consumers:

$$\sum_{i \in N} c_i^*(p) = \sum_{i \in N} c_i' - \frac{pN}{2h}. \tag{9}$$

We assume that $Q = \sum_{i \in N} c_i'$ for convenience. When all consumers' total demands are $\sum_{i \in N} c_i^*(p)$, the payoffs of the consumers are at a maximum. Next, we consider how the energy provider makes the purchase strategy and pricing strategy in Stages I and II based on the total energy demands, respectively. In particular, we show that the energy provider can determine a price in Stage II such that the total energy demands (as a function of price) cannot exceed the total energy supply.

4.2. Optimal Pricing Strategy in Stage II

In Stage II, the energy provider determines the pricing strategy according to the consumers' energy demands, given the energy purchase in Stage I. The profit of the energy provider is denoted by:

$$W(p_s, \beta) = \min(p \sum_{i \in N} c_i^*(p), p(\beta + \beta_0)p_s) - \mu \beta_0 p_s - p_w \beta p_s, \tag{10}$$

which is the difference between the revenue and the total cost. We assume that p_s is the summation of the energy generated from the renewable energy source and the energy purchased from the electricity markets. $\beta_0 p_s$ indicates the energy purchase, β is the uncertainty factor of the renewable energy source, βp_s is the wind power generation, and μ is the energy provider's cost coefficient. In this paper, the energy provider's cost comes from the energy purchase and the wind power generation. p_w is the cost coefficient and $p_w \beta p_s$ is the wind power generating cost. Equation (10) denotes the fact that the revenue of the energy provider is determined by the consumers' demands subject to its available supply. In Stage II, the objective is to find the optimal price p that maximizes the energy provider's profit, i.e.,

$$W_{II}(p_s, \beta) = \max_{p \geq 0} W(p_s, \beta), \tag{11}$$

where $W_{II}(p_s, \beta)$ denotes the maximum profit of the energy provider in Stage II. Since the energy supply p_s from the energy provider is given in this stage, the total cost $\mu \beta_0 p_s$ is already fixed. Therefore, the maximum revenue can be achieved by optimizing the price:

$$\max_{p \geq 0} \min(p \sum_{i \in N} c_i^*(p), p(\beta + \beta_0)p_s). \tag{12}$$

Let us define the consumers' energy demands $D(p) = p \sum_{i \in N} c_i^*$ and the energy supply $S(p) = p \beta p_s$. Then, we have:

$$D(p) = -\frac{Np^2}{2h} + Qp, \tag{13}$$

$$S(p) = p(\beta + \beta_0)p_s. \tag{14}$$

From the above equations, we observe that $D(p)$ is a quadratic function, and $S(p)$ is a linear function. Thus, we can obtain the maximum point of $D(p)$ at $p_d = Qh/N$. The relationships between $S(p)$ and $D(p)$ are described in Figure 2.

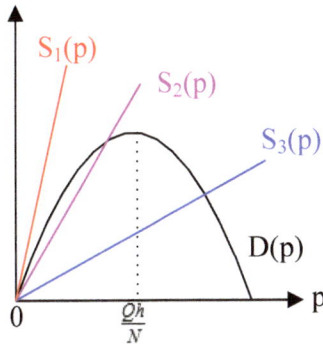

Figure 2. The relationships between $S(p)$ and $D(p)$.

- $S_1(p)$ (excessive supply): $S_1(p)$ doesn't intersect with $D(p)$, $p^* = p_d$;
- $S_2(p)$ (excessive supply): $S_2(p)$ has one intersection with $D(p)$, where $D(p)$ has a non-negative slope, $p^* = p_d$;
- $S_3(p)$ (conservative supply): $S_3(p)$ has one intersection with $D(p)$, where $D(p)$ has a negative slope, $p^* = p_h$, where p_h is the intersection point of $D(p)$ and $S(p)$ and p^* is the optimal price announced by the energy provider.

Letting $D(p) = S(p)$, we obtain a quadratic function with respect to p and make it equal to zero:

$$-\frac{Np^2}{2h} + (Q - (\beta + \beta_0)p_s)p = 0. \tag{15}$$

Solving the above Equation (15), we obtain the intersection point of $D(p)$ and $S(p)$:

$$p_h = \frac{2h(Q - (\beta + \beta_0)p_s)}{N}. \tag{16}$$

In the excessive supply regime, the maximum profit of the energy provider is at $p = p_d$:

$$W_{II}^{ES} = \frac{hQ^2}{2N} - \mu\beta_0 p_s - p_w\beta p_s. \tag{17}$$

In the conservative supply regime, the maximum profit of the energy provider is at $p = p_h$:

$$W_{II}^{CS} = \frac{2h(Q - (\beta + \beta_0)p_s)}{N}(\beta + \beta_0)p_s - \mu\beta_0 p_s - p_w\beta p_s. \tag{18}$$

The optimal pricing decision and the corresponding optimal profit at Stage II are given in Table 2

Table 2. Optimal pricing decision and profit in Stage II in scenario A.

Total Energy Obtained in Stages I and II	Optimal Price $p^*(p_s, \beta)$	Optimal Profit $W_{II}(p_s, \beta)$
Excessive Supply Regime: $p_s \geq \frac{Q}{2}$	$p^{ES} = p_d$	$W_{II}^{ES}(p_s, \beta)$ in Equation (17)
Conservative Supply Regime: $p_s < \frac{Q}{2}$	$p^{CS} = p_h$	$W_{II}^{CS}(p_s, \beta)$ in Equation (18)

4.3. Energy Supply Strategy in Stage I

In Stage I, the energy provider determines the energy purchase to maximize its profit by taking into account the uncertainty factor of the energy supply β [15]. The profit of the energy provider in the Stage I is given as follows:

$$W_I = \max_{p_s \geq 0} W_{II}(p_s, \beta), \tag{19}$$

where $W_{II}(p_s, \beta)$ is the energy provider's profit functions with respect to p_s and the uncertain factor β obtained in Stage II.

We assume that the wind power generation $P = \beta p_s$, and the minimum power and maximum power of the wind power generation are P_{min} and P_{max}, respectively. The probability density function of the wind power $f_{WP}(P)$ can be obtained in [38]. From Figure 2, we can obtain that the maximum consumers' demands $\sum_{i \in N} c_i^*(p)$ is $Q/2$ when the price p is Qh/N. Thus, we consider the following two intervals:

(1) Interval I: $p_s \in [0, \frac{Q}{2}]$. In this interval, the energy provider's profit function is:

$$\begin{aligned}
W_{II}^1(p_s) &= E_{P \in [P_{min}, P_{max}]}[W_{II}^{CS}(P)] \\
&= \int_{P_{min}}^{P_{max}} W_{II}^{CS}(P) f_{WP}(P) dP.
\end{aligned} \tag{20}$$

(2) Interval II: $p_s \in [\frac{Q}{2}, \infty]$. The energy provider's profit function is:

$$\begin{aligned}
W_{II}^2(p_s) &= E_{P \in [P_{min}, \frac{Q}{2}]}[W_{II}^{CS}(P)] + E_{P \in [\frac{Q}{2}, P_{max}]}[W_{II}^{ES}(P)] \\
&= \int_{P_{min}}^{\frac{Q}{2}} W_{II}^{CS}(P) f_{WP}(P) dP + \int_{\frac{Q}{2}}^{P_{max}} W_{II}^{ES}(P) f_{WP}(P) dP.
\end{aligned} \tag{21}$$

By comparing Interval I with Interval II, we can obtain the maximum profit of the energy provider and the optimal energy purchase in scenario A.

5. Scenario B: The Three-Stage Game for Price-Anticipating Consumers

Since the price is set by the energy provider based on the total energy consumption, the consumers are interactive with each other. Thus, we formulate a non-cooperative game among the consumers. The non-cooperative game has a unique Nash equilibrium if $p(c)$ is a linear rotational symmetric function, and $p(c)$ is formulated as follows [8]:

$$p(c) = \omega \sum_{i \in N} c_i + p_0, \tag{22}$$

where ω is a positive parameter to implement the elastic pricing, c_i is the actual energy demands of the consumer i, and p_0 is a basic price.

5.1. Consumer's Energy Demands in Stage III

In Stage I and Stage II, the energy provider determines the energy purchased from the electricity markets and the pricing strategy for the consumers, respectively. In Stage III, similar to Equation (5), we formulate the payoff of price-anticipating consumer i given the unit price $p(c)$ announced by the energy provider as follows:

$$u_i(p(c), c_i) = -(h(c_i - c_i')^2 + b_i) - p(c)c_i, \tag{23}$$

where h and c_i' were defined in Equation (5), and b_i is a base value of the satisfaction level of consumer i and is different for each consumer, which reflects the flexibility of the consumers. The first derivative of $u_i(c)$ with respect to c_i is:

$$\frac{\partial u_i(c)}{\partial c_i} = -2h(c_i - c_i') - \omega c_i - \omega \sum_{i \in N} c_i - p_0. \tag{24}$$

Letting $\partial u_i(c)/\partial c_i = 0$, we obtain:

$$-2h(c_i - c_i') - \omega c_i - \omega \sum_{i \in N} c_i - p_0 = 0, \forall i \in N. \tag{25}$$

Adding Equation (25) from 1 to N, we have:

$$-2h \sum_{i \in N} c_i + 2h \sum_{i \in N} c_i' - \omega \sum_{i \in N} c_i - \omega N \sum_{i \in N} c_i - \sum_{i \in N} p_0 = 0, \tag{26}$$

from which we obtain the total energy consumption of all consumers:

$$\sum_{i \in N} c_i^*(\omega) = \frac{2h \sum_{i \in N} c_i' - N p_0}{2h + \omega(N+1)}. \tag{27}$$

To simplify the calculation process, we make:

$$2h \sum_{i \in N} c_i' - \sum_{i \in N} p_0 = G, \tag{28}$$

and then we have:

$$\sum_{i \in N} c_i^*(\omega) = \frac{G}{2h + \omega(N+1)}. \tag{29}$$

5.2. Optimal Pricing Strategy in Stage II

In Stage II, the energy provider determines the pricing strategy according to the consumers' energy demands, given the energy purchase in Stage I. The profit of the energy provider is:

$$W(p_s, \beta) = \min(p(c) \sum_{i \in N} c_i^*(\omega), p(c)(\beta + \beta_0)p_s) - \mu \beta_0 p_s - p_w \beta p_s, \tag{30}$$

and the maximum profit of the energy provider is:

$$W_{II}(p_s, \beta) = \max_{\omega \geq 0} W(p_s, \beta), \tag{31}$$

where $W_{II}(p_s, \beta)$ denotes the maximum profit of the energy provider in Stage II. We can maximize the revenue of the energy provider by optimizing the price:

$$\max_{\omega \geq 0} \min(p(c) \sum_{i \in N} c_i^*(\omega), p(c)(\beta + \beta_0)p_s). \tag{32}$$

Let us define the consumers' total energy demands $D(\omega) = p(c) \sum_{i \in N} c_i^*$ and the energy supply $S(\omega) = p(c)(\beta + \beta_0)p_s$. Then,

$$D(\omega) = \frac{\omega G^2}{[2h + \omega(N+1)]^2} + \frac{p_0 G}{2h + \omega(N+1)}, \tag{33}$$

and the intersection point of $D(\omega)$, and the y-axis is $p_0 G / 2h$.

The first derivative of $D(\omega)$ with respect to ω is:

$$\frac{\partial D(\omega)}{\partial \omega} = \frac{G^2}{(2h + \omega(N+1))^2} - \frac{2\omega(N+1)G^2}{(2h + \omega(N+1))^3} - \frac{p_0(N+1)G}{(2h + \omega(N+1))^2}$$
$$= \frac{2hG^2 - \omega(N+1)G^2 - 2hp_0(N+1)G - p_0\omega(N+1)^2G}{(2h + \omega(N+1))^3}. \tag{34}$$

When

$$\omega < \frac{2h(G - p_0(N+1))}{G(N+1) + p_0(N+1)^2}, \tag{35}$$

$\frac{\partial D(\omega)}{\partial \omega} > 0$, so $D(\omega)$ is an increasing function. When

$$\omega > \frac{2h(G - p_0(N+1))}{G(N+1) + p_0(N+1)^2}, \tag{36}$$

$\frac{\partial D(\omega)}{\partial \omega} < 0$, so $D(\omega)$ is an decreasing function.

Letting $\partial D(\omega)/\partial \omega = 0$, we obtain:

$$\omega_0 = \frac{2h(G - p_0(N+1))}{G(N+1) + p_0(N+1)^2} \tag{37}$$

and

$$S(\omega) = \frac{\omega G(\beta + \beta_0)p_s}{2h + \omega(N+1)} + p_0(\beta + \beta_0)p_s, \tag{38}$$

and the intersection point of $S(\omega)$, and the y-axis is $p_0(\beta + \beta_0)p_s$.

The first derivative of $S(\omega)$ with respect to ω is:

$$\frac{\partial S(\omega)}{\partial \omega} = \frac{G(\beta + \beta_0)p_s}{2h + \omega(N+1)} - \frac{\omega G(\beta + \beta_0)p_s(N+1)}{(2h + \omega(N+1))^2}$$
$$= \frac{2hG(\beta + \beta_0)p_s}{(2h + \omega(N+1))^2}. \tag{39}$$

Since $\sum_{i \in N} c_i^* > 0$ and $\partial S(\omega)/\partial \omega > 0$, $S(\omega)$ is an increasing function. The relationships between $S(\omega)$ and $D(\omega)$ are described in Figure 3.

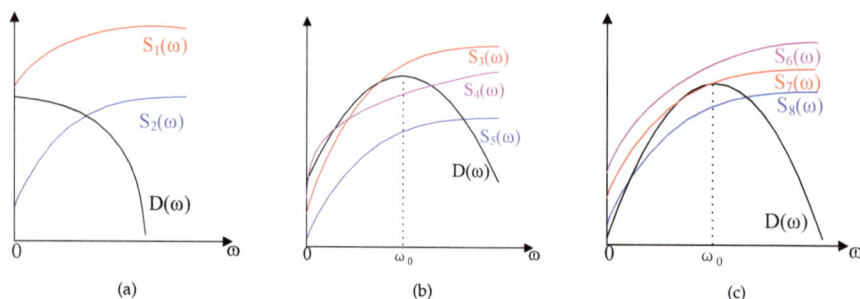

Figure 3. The relationships between $S(\omega)$ and $D(\omega)$ under the different conditions.

(a) When $G - p_0(N+1) < 0$, we can obtain the following conclusions from Figure 3a:

- $S_1(\omega)$ (excessive supply): $p_0(\beta + \beta_0)p_s \geq p_0G/2h$, $\omega^* = 0$,
- $S_2(\omega)$ (conservative supply): $p_0(\beta + \beta_0)p_s < p_0G/2h$, $\omega^* = \omega_p$,

where ω_p is the intersection point of $D(\omega)$ and $S(\omega)$, and ω^* is the optimal parameter of the elastic price.

Because $p(c)$ is a linear rotational symmetric function, the case with $\omega^* = 0$ is neglected.

(b) When $G - p_0(N+1) \geq 0$ and $p_0(\beta + \beta_0)p_s \geq p_0 G/2h$, we have the conclusions by analyzing Figure 3b:

- $S_3(\omega)$ (excessive supply): $S_3(\omega)$ has one intersection with $D(\omega)$, where $D(\omega)$ has a non-negative slope, $\omega^* = \omega_0$,
- $S_4(\omega)$ (conservative supply): $S_4(\omega)$ has three intersections with $D(\omega)$, $\omega^* = \omega_p$,
- $S_5(\omega)$ (conservative supply): $S_5(\omega)$ has one intersection with $D(\omega)$, where $D(\omega)$ has a negative slope, $\omega^* = \omega_p$.

(c) When $G - p_0(N+1) \geq 0$ and $p_0(\beta + \beta_0)p_s < p_0 G/2h$, we can get the conclusions from Figure 3c:

- $S_6(\omega)$ (excessive supply): $S_6(\omega)$ doesn't intersect with $D(\omega)$, $\omega^* = \omega_0$,
- $S_7(\omega)$ (excessive supply): $S_7(\omega)$ has one or two intersections with $D(\omega)$, where both intersections are located in the increasing interval of $D(\omega)$, $\omega^* = \omega_0$,
- $S_8(\omega)$ (conservative supply): $S_8(\omega)$ has two intersections with $D(\omega)$, where both intersections are located in the both sides of ω_0, respectively, $\omega^* = \omega_p$.

Letting $D(\omega) = S(\omega)$, we obtain a quadratic function with respect to ω and make it equal to zero:

$$(G(N+1) + p_0(N+1)^2)(\beta + \beta_0)p_s\omega^2 + [(4hp_0(N+1) + 2hG)(\beta + \beta_0)p_s$$
$$- (Gp_0(N+1) + G^2)]\omega + 4h^2 p_0(\beta + \beta_0)p_s - 2hp_0 G = 0. \tag{40}$$

For convenience, we define:

$$A = G(N+1) + p_0(N+1)^2,$$
$$B = 4hp_0(N+1) + 2hG,$$
$$C = Gp_0(N+1) + G^2,$$
$$D = 4h^2 p_0,$$
$$E = 2hp_0 G,$$
$$\Delta = \sqrt{(B(\beta + \beta_0)p_s - C)^2 - 4A(\beta + \beta_0)p_s(D(\beta + \beta_0)p_s - E)}.$$

Solving the above Equation (40), we obtain the intersection point of $D(\omega)$ and $S(\omega)$:

$$\omega_p = \frac{-(B(\beta + \beta_0)p_s - C) + \Delta}{2A(\beta + \beta_0)p_s}. \tag{41}$$

In the excessive supply regime, the maximum profit of the energy provider is at $\omega = \omega_0$:

$$W_{II}^{ES} = \frac{[G + p_0(N+1)]^2}{8h(N+1)} - \mu\beta_0 p_s - p_w\beta p_s. \tag{42}$$

In the conservative supply regime, the maximum profit of the energy provider is at $\omega = \omega_p$:

$$W_{II}^{CS} = \frac{(C - B(\beta + \beta_0)p_s + \Delta)G(\beta + \beta_0)p_s}{4Ah(\beta + \beta_0)p_s + (C - B(\beta + \beta_0)p_s + \Delta)(N+1)} + p_0(\beta + \beta_0)p_s - \mu\beta_0 p_s - p_w\beta p_s. \tag{43}$$

The optimal pricing decision and the corresponding optimal profit in Stage II are given in Table 3.

Table 3. Optimal pricing decision and profit in Stage II in scenario B.

Total Energy Obtained in Stages I and II	Optimal Parameter $p^*(p_s, \beta)$	Optimal Profit $W_{II}(p_s, \beta)$
Excessive Supply Regime: $p_s \geq \frac{A}{4h(N+1)}$	$\omega^{ES} = \omega_0$	$W_{II}^{ES}(p_s, \beta)$ in Equation (42)
Conservative Supply Regime: $p_s < \frac{A}{4h(N+1)}$	$\omega^{CS} = \omega_p$	$W_{II}^{CS}(p_s, \beta)$ in Equation (43)

5.3. Energy Supply Strategy in Stage I

In Stage I, the energy provider also determines the energy purchase to maximize its profit by taking into account the uncertainty of the energy supply. The profit of the energy provider in the Stage I is given by:

$$W_I = \max_{p_s \geq 0} W_{II}(p_s, \beta), \tag{44}$$

where $W_{II}(p_s, \beta)$ is the energy provider's profit function with respect to p_s and the uncertain factor β obtained in Stage II.

We assume that the wind power generation $P = \beta p_s$, and the minimum power and maximum power of the wind power generation are P_{min} and P_{max}, respectively. The probability density function of the wind power $f_{WP}(P)$ can be obtained in [38]. From Figure 3, we can obtain that the maximum consumers' demands $\sum_{i \in N} c_i^*(\omega)$ is $A/4h(N+1)$ when $\omega = \omega_0$. For convenience, we assume that $L = A/4h(N+1)$ and consider the following two intervals:

(1) Interval I: $p_s \in [0, \frac{A}{4h(N+1)}]$. In this interval, the energy provider's profit function is:

$$\begin{aligned} W_{II}^{1'}(p_s) &= E_{P \in [P_{min}, P_{max}]}[W_{II}^{CS}(P)] \\ &= \int_{P_{min}}^{P_{max}} W_{II}^{CS}(P) f_{WP}(P) dP. \end{aligned} \tag{45}$$

(2) Interval II: $p_s \in [\frac{A}{4h(N+1)}, \infty]$. The energy provider's profit function is:

$$\begin{aligned} W_{II}^{2'}(p_s) &= E_{P \in [P_{min}, \frac{A}{4h(N+1)}]}[W_{II}^{CS}(P)] + E_{P \in [\frac{A}{4h(N+1)}, P_{max}]}[W_{II}^{ES}(P)] \\ &= \int_{P_{min}}^{\frac{A}{4h(N+1)}} W_{II}^{CS}(P) f_{WP}(P) dP + \int_{\frac{A}{4h(N+1)}}^{P_{max}} W_{II}^{ES}(P) f_{WP}(P) dP. \end{aligned} \tag{46}$$

Similar to scenario A, we can obtain the maximum profit of the energy provider and the optimal amount of energy purchased from the electricity markets.

6. Simulation Results

This section presents simulation studies of the proposed scheme using MATLAB 7.11.0 (MathWorks, Natick, MA, USA). In the simulations, we assume that the wind power generation follows a uniform distribution in $[P_{min}, P_{max}]$, and we select that $h = 0.04$, $p_0 = 0.1$, $N = 100$, and $\mu = 10$. For the parameter c_i', we select a set of stochastic values. Then, we can obtain the profit of the energy provider under different β_0 for two scenarios as shown in Figures 4 and 5, respectively.

From Figures 4 and 5, we observe that the optimal energy supply decreases with the increase of the β_0 and the maximum profit is changed from Interval II to Interval I. In general, the wind power generation cost is less than the cost of purchasing energy. From the profit function of the energy provider, when β_0 increases, only by decreasing p_s can the profit of the energy provider be maximized.

Thus, it is verified that the proposed method is effective, and the simulation values of Figures 4 and 5 are shown in Table 4.

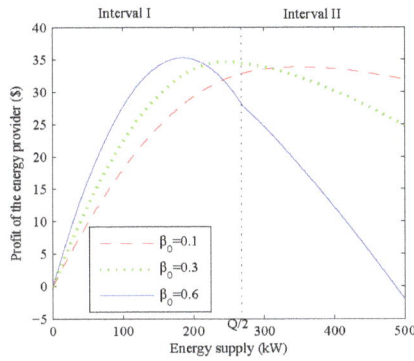

Figure 4. Scenario A: the profit of the energy provider under different β_0.

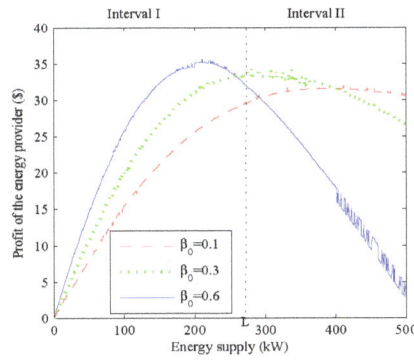

Figure 5. Scenario B: the profit of the energy provider under different β_0.

Table 4. Simulation values of the two scenarios.

β_0	Scenario A		Scenario B	
	P_s	Profit	P_s	Profit
0.1	349	33.79	399.6	31.52
0.3	246	34.61	288.6	33.27
0.6	186	35.2	209	35.13

Taking $\beta_0 = 0.1$ as an example, the comparisons between the two scenarios are shown in Figure 6. From Figure 6 and Table 4, we observe that the energy provider can obtain more profit in scenario A.

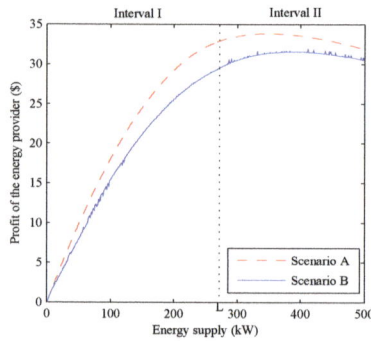

Figure 6. Comparisons between Scenario A with Scenario B.

To explain the effect of the uncertainty, taking $\beta_0 = 0.6$ under scenario A as an example, we show the profit of the energy provider under the certain and uncertain energy supply in Figure 7. It is observed that the energy provider can achieve the higher profit under the certain energy supply. In reality, the uncertainty of the energy supply is necessary because the energy generated from the renewable energy sources is uncertain.

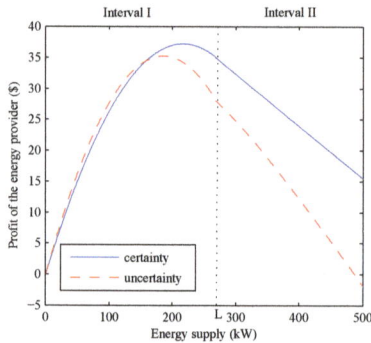

Figure 7. The effect of the uncertainty of the energy supply.

7. Conclusions

In this paper, we establish a model for energy trading and pricing in the microgrid. We formulate a hierarchical game between the energy provider with the renewable energy generation and the consumers, e.g., the price-taking consumers and the price-anticipating consumers. In the hierarchical game, the energy provider acts as the leader and the consumers act as the followers. The equilibrium point of the hierarchical game is obtained through the backward induction method. Furthermore, we also consider the uncertainty of the energy supply in the problem formulation. The simulation results show that the optimal energy supply can be obtained based on the reasonable pricing strategy and purchase strategy. Comparing the price-taking consumers with the price-anticipating consumers, we can obtain that the energy provider obtains more profit from the price-taking consumers. From the simulation results, we also can obtain that the energy provider's profit reduces because of the uncertainty of the energy supply.

However, we do not consider that the consumers can sell the energy to the energy provider when the consumers have photovoltaic (PV) panels and a storage system. In that case, the energy demands of the consumers will be uncertain, and the payoff of the consumer includes two additional parts: one part is the PV generation cost, and the other part is the uncertainty of the energy demands. In order to compute the payoff of the consumer, we need to know the distribution that the PV generation follows. Then, we can get the average payoff of the consumer by expectation, and the optimal energy demands are obtained by the derivation method. Simultaneously, the profit of the energy provider needs to introduce two additional parts that denote buying the energy from the consumers and selling the energy to the electricity markets, which will be considered in the future.

Acknowledgments: This research was supported in part by the National Natural Science Foundation of China under Grants 61573303, 61503324, and 61573300, in part by the Natural Science Foundation of Hebei Province under Grants F2016203438, E2017203284, and E2016203092, in part by a project funded by the China Postdoctoral Science Foundation under Grants 2015M570233 and 2016M601282, in part by a project funded by the Hebei Education Department under Grant BJ2016052, and in part by the Technology Foundation for Selected Overseas Chinese Scholar under Grant C2015003052

Author Contributions: Kai Ma and Shubing Hu wrote the paper; Jie Yang analyzed the data; and Chunxia Dou and Josep M. Guerrero contributed the analysis tools.

Conflicts of Interest: The authors declare no conflict of interest.

References

1. Chen, H.; Li, Y.; Han, Z.; Vucetic, B. A stackelberg game-based energy trading scheme for power beacon-assisted wireless-powered communication. In Proceedings of the IEEE International Conference on Acoustics, Speech and Signal Processing (ICASSP), Brisbane, Australia, 19–24 April 2015; pp. 3177–3181.
2. Misra, S.; Bera, S.; Ojha, T.; Mouftah, H.T.; Anpalagan, A. ENTRUST: Energy trading under uncertainty in smart grid systems. *Comput. Netw.* **2016**, *110*, 232–242.
3. Belgana, A.; Rimal, B.P.; Maier, M. Open energy market strategies in microgrids: A Stackelberg game approach based on a hybrid multiobjective evolutionary algorithm. *IEEE Trans. Smart Grid* **2015**, *6*, 1243–1252.
4. Jia, L.; Tong, L. Dynamic pricing and distributed energy management for demand response. *IEEE Trans. Smart Grid* **2016**, *7*, 1128–1136.
5. Duan, L.; Huang, J.; Shou, B. Investment and pricing with spectrum uncertainty: A cognitive operator's perspective. *IEEE Trans. Mob. Comput.* **2011**, *10*, 1590–1604.
6. Hu, M.C.; Lu, S.Y.; Chen, Y.H. Stochastic–multiobjective market equilibrium analysis of a demand response program in energy market under uncertainty. *Appl. Energy* **2016**, *182*, 500–506.
7. Nie, S.; Huang, C.Z.; Huang, G.H.; Li, Y.P.; Chen, J.P.; Fan, Y.R.; Cheng, G.H. Planning renewable energy in electric power system for sustainable development under uncertainty—A case study of Beijing. *Appl. Energy* **2016**, *162*, 772–786.
8. Ma, K.; Hu, G.; Spanos, C.J. Distributed energy consumption control via real-time pricing feedback in smart grid. *IEEE Trans. Control Syst. Technol.* **2014**, *22*, 1907–1914.
9. Ma, K.; Hu, G.; Spanos, C.J. A cooperative demand response scheme using punishment mechanism and application to industrial refrigerated warehouses. *IEEE Trans. Ind. Inform.* **2015**, *11*, 1520–1531.
10. Mohsenian-Rad, A.H.; Wong, V.W.S.; Jatskevich, J.; Schober, R.; Leon-Garcia, A. Autonomous demand-side management based on game-theoretic energy consumption scheduling for the future smart grid. *IEEE Trans. Smart Grid* **2010**, *1*, 320–331.
11. Baharlouei, Z.; Hashemi, M.; Narimani, H.; Mohsenian-Rad, H. Achieving optimality and fairness in autonomous demand response: Benchmarks and billing mechanisms. *IEEE Trans. Smart Grid* **2013**, *4*, 968–975.
12. Yu, M.; Hong, S.H. Supply–demand balancing for power management in smart grid: A Stackelberg game approach. *Appl. Energy* **2016**, *164*, 702–710.
13. Gao, B.; Ma, T.; Tang, Y. Power transmission scheduling for generators in a deregulated environment based on a game-theoretic approach. *Energies* **2015**, *8*, 13879–13893.
14. Liu, N.; Wang, C.; Lin, X.; Lei, J. Multi-party energy management for clusters of roof leased PV prosumers: A game theoretical approach. *Energies* **2016**, *9*, 536.
15. Bu, S.; Yu, F.R. A game-theoretical scheme in the smart grid with demand-side management: Towards a smart cyber-physical power infrastructure. *IEEE Trans. Emerg. Top. Comput.* **2013**, *1*, 22–32.
16. Maharjan, S.; Zhu, Q.; Zhang, Y.; Gjessing, S. Dependable demand response management in the smart grid: A stackelberg game approach. *IEEE Trans. Smart Grid* **2013**, *4*, 120–132.

17. Soliman, H.M.; Leon-Garcia, A. Game-theoretic demand-side management with storage devices for the future smart grid. *IEEE Trans. Smart Grid* **2014**, *5*, 1475–1485.

18. Maharjan, S.; Zhu, Q.; Zhang, Y.; Gjessing, S.; Basar, T. Demand response management in the smart grid in a large population regime. *IEEE Trans. Smart Grid* **2016**, *7*, 189–199.

19. Lee, J.; Guo, J.; Choi, J.K.; Zukerman, M. Distributed energy trading in microgrids: a game-theoretic model and its equilibrium analysis. *IEEE Trans. Ind. Electron.* **2015**, *62*, 3524–3533.

20. Yoon, S.G.; Choi, Y.J.; Park, J.K.; Bahk, S. Demand response design based on a Stackelberg game in smart grid. In Proceedings of the International Conference on ICT Convergence, Jeju, Korea, 14–16 October 2013; pp. 177–178.

21. Fadlullah, Z.M.; Quan, D.M.; Kato, N.; Stojmenovic, I. GTES: An optimized game-theoretic demand-side management scheme for smart grid. *IEEE Syst. J.* **2014**, *8*, 588–597.

22. Yang, B.; Li, J.; Han, Q.; He, T.; Chen, C.; Guan, X. Distributed control for charging multiple electric vehicles with overload limitation. *IEEE Trans. Parallel Distrib. Syst.* **2016**, *27*, 3441–3454.

23. Abegaz, B.W.; Mahajan, S.M. Optimal dispatch control of energy storage systems using forward-backward induction. In Proceedings of the 2015 International Conference on Clean Electrical Power (ICCEP), Taormina, Italy, 16–18 June 2015; pp. 731–736.

24. Cho, J.; Kleit, A.N. Energy storage systems in energy and ancillary markets: A backwards induction approach. *Appl. Energy* **2015**, *147*, 176–183.

25. Mahoney, W.P.; Parks, K.; Wiener, G.; Liu, Y.; Myers, W.L.; Sun, J.; Monache, L.D.; Hopson, T.; Johnson, D.; Haupt, S.E. A wind power forecasting system to optimize grid integration. *IEEE Trans. Sustain. Energy* **2012**, *3*, 670–682.

26. Constantinescu, E.M.; Zavala, V.M.; Rocklin, M.; Lee, S.; Anitescu, M. A computational framework for uncertainty quantification and stochastic optimization in unit commitment with wind power generation. *IEEE Trans. Power Syst.* **2011**, *26*, 431–441.

27. Kanna, B.; Singh, S.N. Long term wind power forecast using adaptive wavelet neural network. In Proceedings of the 2016 IEEE Uttar Pradesh Section International Conference on Electrical, Computer and Electronics Engineering (UPCON), Varanasi, India, 9–11 December 2016; pp. 671–676.

28. Xie, L.; Gu, Y.; Zhu, X.; Genton, M.G. Short-term spatio-temporal wind power forecast in robust look-ahead power system dispatch. *IEEE Trans. Smart Grid* **2014**, *5*, 511–520.

29. Finamore, A.R.; Galdi, V.; Calderaro, V.; Piccolo, A.; Conio, G.; Grasso, S. Artificial neural network application in wind forecasting: An one-hour-ahead wind speed prediction. In Proceedings of the 5th IET International Conference on Renewable Power Generation (RPG), London, UK, 21–23 September 2016; pp. 1–6.

30. Sherlock, R.H. Analyzing winds for frequency and duration on atmospheric pollution. *Am. Meteorol. Soc.* **1951**, *4*, 42–49.

31. Bardsley, W.E. Note on the use of the inverse Gaussian distribution for wind energy applications. *J. Appl. Meteorol.* **1980**, *19*, 1126–1130.

32. Luna, R.E.; Church, H.W. Estimation of long-term concentrations using a 'universal' wind speed distribution. *J. Appl. Meteorol.* **1974**, *13*, 910–916.

33. Hennessey, J.P.J. A comparison of the Weibull and Rayleigh distributions for estimating wind power potential. *Wind Eng.* **1978**, *2*, 156–164.

34. Justus, C.G.; Hargraves, W.R.; Yalcin, A. Nationwide assessment of potential output from wind-powered generators. *J. Appl. Meteorol.* **1976**, *15*, 673–678.

35. Stewart, D.A.; Essenwanger, O.M. Frequency distribution of wind speed near the surface. *J. Appl. Meteorol.* **1978**, *17*, 1633–1642.

36. Takle, E.S.; Brown, J.M. Note on the use of Weibull statistics to characterize wind-speed data. *J. Appl. Meteorol.* **1978**, *17*, 556–559.

37. Liu, S.; Li, G.; Xie, H.; Wang, X. Correlation characteristic analysis for wind speed in different geographical hierarchies. *Energies* **2017**, *10*, 237.

38. Pan, X.; Wang, L.; Xu, Y.; Zhang, L.; Liu, W.; Wu, R. A wind farm power modeling method based on mixed Copula. *Dianli Xitong Zidonghua/Autom. Electr. Power Syst.* **2014**, *38*, 17–22.

39. Du, M.; Yi, J.; Mazidi, P.; Cheng, L.; Guo, J. A parameter selection method for wind turbine health management through SCADA data. *Energies* **2017**, *10*, 253.

40. Ma, C. Robust exponential stability of reaction-diffusion generalized Cohen-Grossberg neural networks with distributed delays. *J. Xinjiang Norm. Univ.* **2007**, *26*, 18–24.

41. Li, X.M.; Shi, D.J. Research on dependence structure between shanghai and shenzhen stock markets. *Appl. Stat. Manag.* **2006**, *25*, 729–736.

42. Hu, L. Dependence patterns across financial markets: A mixed copula approach. *Appl. Financ. Econ.* **2006**, *16*, 717–729.

energies

MDPI

Article

Implementation of a Real-Time Microgrid Simulation Platform Based on Centralized and Distributed Management

Omid Abrishambaf [1], Pedro Faria [1,*], Luis Gomes [1], João Spínola [1], Zita Vale [1] and Juan M. Corchado [2]

[1] GECAD—Research Group on Intelligent Engineering and Computing for Advanced Innovation and Development, IPP—Polytechnic Institute of Porto, Rua DR. Antonio Bernardino de Almeida, 431, Porto 4200-072, Portugal; ombaf@isep.ipp.pt (O.A.); lufog@isep.ipp.pt (L.G.); jafps@isep.ipp.pt (J.S.); zav@isep.ipp.pt (Z.V.)

[2] BISITE—Bioinformatics, Intelligent Systems and Educational Technology Research Center, University of Salamanca, Salamanca 37008, Spain; corchado@usal.es

* Correspondence: pnfar@isep.ipp.pt; Tel.: +351-228-340-511; Fax: +351-228-321-159

Academic Editor: Chunhua Liu
Received: 26 February 2017; Accepted: 7 June 2017; Published: 14 June 2017

Abstract: Demand response and distributed generation are key components of power systems. Several challenges are raised at both technical and business model levels for integration of those resources in smart grids and microgrids. The implementation of a distribution network as a test bed can be difficult and not cost-effective; using computational modeling is not sufficient for producing realistic results. Real-time simulation allows us to validate the business model's impact at the technical level. This paper comprises a platform supporting the real-time simulation of a microgrid connected to a larger distribution network. The implemented platform allows us to use both centralized and distributed energy resource management. Using an optimization model for the energy resource operation, a virtual power player manages all the available resources. Then, the simulation platform allows us to technically validate the actual implementation of the requested demand reduction in the scope of demand response programs. The case study has 33 buses, 220 consumers, and 68 distributed generators. It demonstrates the impact of demand response events, also performing resource management in the presence of an energy shortage.

Keywords: demand response; distributed generation; microgrid; real-time simulation

---

1. Introduction

The increment on the penetration of the distributed generation (DG) resources encounters the current power grid with management and reliability challenges [1]. For overcoming these issues, the entire power network can be distributed into several small power grids, which are the sub set of the main power network. This solution is attainable via the concepts defined in smart grids, such as microgrids [2]. The microgrid refers to a group of DG units, renewable energy resources (RERs), and the local loads that can rely upon the main distribution network [3]. Basically, the RERs consist of photovoltaic (PV) systems and wind turbines [4].

The real-time measurements of different nodes of a microgrid are an essential issue for managing and controlling the grid through both the centralized and distributed methods. This can be released by phasor measurement units (PMU). The PMU are synchronized time based instruments, which collects highly precise phasor data of the power system [5]. The PMU plays a key role in the real-time monitoring of the smartgrids and microgrids that utilizes the global positioning system (GPS) to provide the concurrent measurements [6]. Typical PMU devices are able to provide 30 samples per second [7]. This enables the grid operator to be informed from the synchronized time based voltage and current phasor measurements in different nodes of the grid, in order to control and manage the power stability and delivery [8].

Additionally, if the DG resources are integrated with demand response (DR) programs, the microgrid conceptions can be fully addressed. DR programs are defined as altering the electricity consumption profiles based on the incentives payment provided by the network operator due to technical reasons or economic purposes. Incentive-based and price-based are two major classifications of DR programs [9]. In this context, virtual power players (VPPs) play a key role for aggregating the DG and DR small size resources, in order to be used in electricity markets as a large scale resource [10].

In order to control and manage the resources available in the microgrid, two main methods can be proposed: centralized and distributed control. In the centralized control method, a powerful central controller unit is responsible to manage and control the microgrid, where communication between this unit and each single component of the network is required [11]. However, in the distributed control method, the decisions take place locally and are based on the real-time information exchanged by the network components [12]. Both methods have several advantages and disadvantages. For example, the centralized control requires high initial cost and needs a widespread scheming; however, it provides better efficiency. In the meantime, the centralized network can be implemented step by step from the bottom levels to the top levels [13].

This paper presents the development and implementation of a real-time microgrid simulation platform managed by centralized and distributed controlling decision support. In this platform it is attempted to provide a realistic microgrid implementation using real and laboratorial hardware equipment. The microgrid players included in this platform consist of two renewable DG units (PV and wind turbine), and a low and a medium consumer load (laboratorial equipment), which are connected to each other as well as the main power grid through four power lines. The local demand of the microgrid can be supplied from the energy provided by the DG and the grid as well. For the centralized control method, a real-time simulator model has been employed in order to manage the system, and for the distributed control manner, a local controller is associated for each player in order to perform the decision making locally and achieve the microgrid goals.

There are several related research works, which implemented and surveyed the microgrid models based on centralized or distributed decision support. In [13], the authors examined two implemented microgrid topologies, one centralized and one distributed model, which combine solar panels and batteries for 20 residential houses. In [14], the authors provided an optimal solution for dispatching of the local resources in the medium voltage (MV) microgrids that temporary or permanently operate in islanded mode. In the optimization problem, they considered that all the power produced by renewable generators (PV and wind) is used, in order to minimize the microgrid operation costs as well as the pollutant emissions of the programmable generators. In [15], a new distributed controlling method was proposed for secondary frequency and voltage control and stability in a microgrid while it is operating in islanded mode. In this method, the authors utilized localized data as well as nearest-neighbor communication to implement the secondary control operations while there is no necessity of information about the loads and microgrid methodology. In [16], a unified controlling method is addressed for the cooperation of distributed energy resources (DERs) and the DR to support the voltage and frequency of an islanded microgrid in which it minimized the overall operation costs of the grid through an optimization problem. In the proposed algorithm, the frequency deviation was considered as a new state variable in the model. In this way, the model enables us to calculate the required set points for the DERs and the amount of power that should be curtailed by the controllable loads available in the grid. In [17], a simulation based analysis of dynamical behavior of a residential DC microgrid laboratory setup in distributed and centralized voltage control configurations is presented. In [18], the authors described the control algorithm of a utility connected microgrid, based on independent control of active and reactive power and operating in centralized and distributed operation mode. In addition to these works, a significant number of published works have been focused on the multi-agent based and distributed control models for the energy management of the microgrids [19–21].

There are a lot of laboratories and test beds implemented for development and validation of the capabilities of smartgrids and microgrids by utilizing the real-time simulation facilities [22]. Austrian Institute of Technology (AIT), Vienna, Austria, includes three configurable three-phase low-voltage grids and the real-time simulation with hardware-in-the-loop (HIL) setups in order to experiment with the real-time simulation platform for advanced power-HIL and controller-HIL analysis, and the validation of energy management systems and distribution supervisory control and data acquisition (SCADA). OFFIS—Institute for Information Technology, Oldenburg, Germany has an automation laboratory, which includes OPAL-RT simulator for executing a highly detailed and dynamic power grid. The OFFIS utilizes this laboratory for centralized and decentralized controlling methods and parallel simulation. Laboratoire de Genie Electrique de Grenoble, Grenoble, France, includes a real-time power-HIL simulation laboratory equipped with two real-time multiprocessors digital simulators. This enables them to focus on power system protection relays, testing different types of equipment, namely wind turbine emulator and hydro turbine, and testing the industrial converters for PV systems. Commissariat A L'energie Atomique et AUX Energies Alternatives, France, has a microgrid platform including several renewable and conventional generators, energy storage systems, controllable loads, and electrical vehicles. The main core of this platform is a HIL simulator, which enabled the facility to validate and examine the microgrid operation and protection, voltage and frequency control, energy storage systems management etc. Distribution Network and Protection Laboratory, Glasgow, UK, consists of a three-phase power grid including several multiple controllable voltage supplies, flexible and controllable loads. There are several real-time simulators in this laboratory, which are utilized for surveying protection concepts, automation equipment, and new solutions for distributed power system control.

The main objective of the present paper is to develop and implement a real-time microgrid simulation platform using several real and laboratorial hardware equipment. Such a platform supports real-time simulation skills and HIL means in order to address the validation of demand response and distributed generation optimization. A microgrid accommodates such resources and is managed by a VPP that aims at minimizing the operation costs, using both distributed and centralized control methods. An upstream network is modeled in MATLAB/Simulink, using mathematical and non-physical models. The use of real-time simulation and HIL scenarios brings the ability of controlling and managing the real resources from the simulation environment with non-real management scenarios, such as optimization models.

The problem statement is related to how a microgrid business model can be examined and validated in terms of management and control, before massive implementation. Implementing a completely realistic microgrid model only for testing and validating, would not be a cost-effective solution. Furthermore, it would not be available for everyone, since only a limited number of companies or research institutes could be equipped with that type of test bed. The microgrid platform designed in this paper is flexible in terms of controlling methods and is up to the operator to choose.

In this way, namely when comparing with [23], the contribution of the present paper relies on the presented approach that integrates all the above referred aspects of the work, namely with improved aspects as the optimization of resource use.

In both centralized and distributed control methods, the different nodes of the microgrid (accommodating consumers and generators) will be measured through the several energy meters mounted on the various locations of the grid. The sampling period of these energy meters are one sample per second, which have enough accuracy for optimization problems and DR program applications, and the high precision measurement devices, such as PMUs, may not be required for these kinds of applications. This microgrid is also able to be configured in islanded or grid-connected mode. Since the energy transaction between the microgrid and the main grid is considered, the autonomous mode is out of the focus of the paper. Another topic out of the present work focus is the market congestion in the connection of the microgrid with the upstream network. It is not included on the economic model since the main focus is given to minimize microgrid operation costs. The considered VPP is selling electricity in the market and the network has enough capacity for the energy transactions.

This paper is structured as follows: after this introductory section, the development and implementation of the proposed microgrid simulation platform is described on Section 2. Then, a case study is defined and executed with the presented model in Section 3 and its results are described in Section 4. Finally, Section 5 clarifies the main conclusions of the work.

2. Real-Time Microgrid Simulation Platform

This section describes the real-time implementation of a microgrid simulation platform based on two controlling methods: centralized and distributed. This system has been implemented in GECAD laboratory [24]. In this model, several laboratorial hardware resources have been employed in order to simulate a realistic microgrid. The present model is designed and implemented in a way that the controlling methods can be selected by the user/operator. This enables the operator to choose the centralized or distributed control method, depending on its application.

Since the proposed system employs the real-time simulator as well as several real hardware equipment, it enables the systems or platforms that include network simulation models to use the real data in their simulation models. Therefore, the present microgrid model will be used as a part of a network simulation model used by DR program simulation platform developed in [25], called SPIDER—simulation platform for the integration of demand response. This platform has been designed to widely support the decision-making for different types of network players, which are involved in the DR programs. As a general description of SPIDER, it surveys and specifies the data-mining methods, which are appropriate for the consumers who intend to participate in DR programs. Data-mining algorithms are applied in the module "model optimization" (with orange highlights in Figure 1). In fact, this module includes several types of algorithms for DR implementation, such as the energy resource optimization, data-mining for aggregation of resources, forecasting online tools, etc. For example, if data-mining is applied, whenever a new scenario is computed in the simulation, the system automatically includes the scheduling of resource results as input to the aggregation of the resources. After the data-mining is computed, the simulation proceeds to step "4" (as can be seen in Figure 1). A data-mining algorithm used for energy resource aggregation and remuneration can be found in [26].

SPIDER is an essential instrument for validating and analyzing the business and economic aspects of the DR programs, and surveying their influence in the electricity network. For this purpose, SPIDER uses MATLAB/Simulink [27] tools in order to simulate the basis platform for the grid simulation. Figure 1 illustrates the overall view of the SPIDER simulation platform with the proposed microgrid simulation configuration using the centralized control method. In this system, several softwares have been employed in order to exchange data between different sections.

The platform starts the process from network simulation in Simulink; afterward, JAVA application programming interface (API) is used in order to transmit the information of the network simulation to the optimization block. Then, TOMLAB [28] tool is used for the optimization in the SPIDER, and its optimization results transfer to the network simulation block using JAVA API as well. Full details about the SPIDER and its infrastructures can be found in [25].

The microgrid model proposed in this paper has been demonstrated in the top of the Figure 1, as depicted by green color. The model includes four nodes; two nodes dedicated for the consumers, and the other two devoted to renewable DG units. This microgrid has the capability of supplying the local loads by its own DG units, and transacts energy with the main grid in order to feed the loads in the moments that there is not enough generation from the energy resources. In addition, it can inject the excess of the produced power to the main gird.

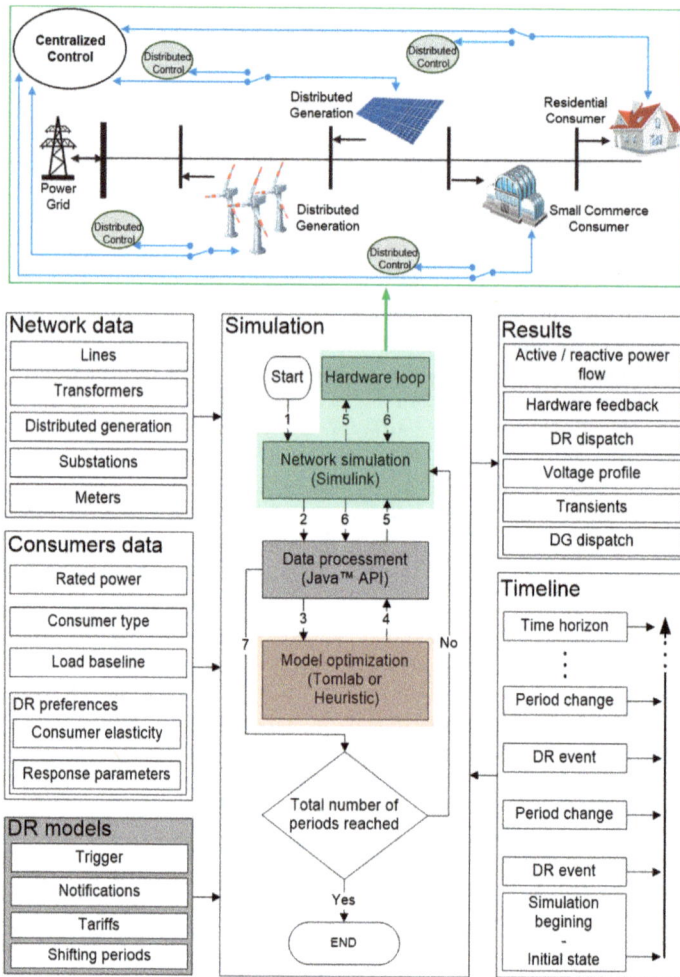

Figure 1. Proposed microgrid configuration for the simulation platform for the integration of demand response (SPIDER) simulation platform implementation.

As can be seen in the top of Figure 1, there are four switches; one for each player, which enables the microgrid operator to select the controlling method. If the centralized controlling method is selected, the central controller unit is responsible for managing the network players and controlling the consumption and generation of the resources. For this purpose, the central controller transmits the controlling commands to each player by using independent communication channels line. However, if distributed control method is selected, the central control unit is eliminated and the local controllers manage the network by transmitting and sharing information between each other. It should be noted that the status of all switches should be equal (all centralized or all distributed). The following sub sections describe how the microgrid is controlled by the centralized and the distributed methods.

2.1. Centralized Control Model

In this section, the central controller unit, network players, and the controlling methods, will be explained. The microgrid model proposed in this section is an improved and reformed version of the model proposed in [23]. In the previous work, there were low and medium consumer units playing the role of residential and small commerce facility consumers, and a wind turbine emulator playing the role of a home-scale wind turbine. However, in this paper, a 7.5 kW PV system and four power lines have been added to the system in order to implement a comprehensive laboratorial microgrid model.

Figure 2 presents the centralized microgrid control model proposed in this paper. As can be seen, the central controller unit is located at the top of the model and the other network players are connected to this unit. This unit is OP5600, the real-time simulator machine [29], a powerful instrument to produce real-time simulations even with a high complexity degree while enabling HIL. OP5600 is based on the MATLAB/Simulink and indeed it runs the Simulink models in real-time. Additionally, there are several Digital/Analog I/O slots embedded on the OP5600, which enable the user to control real hardware devices from Simulink models and also receive feedback. This is how HIL integrates the real data with the Simulink models.

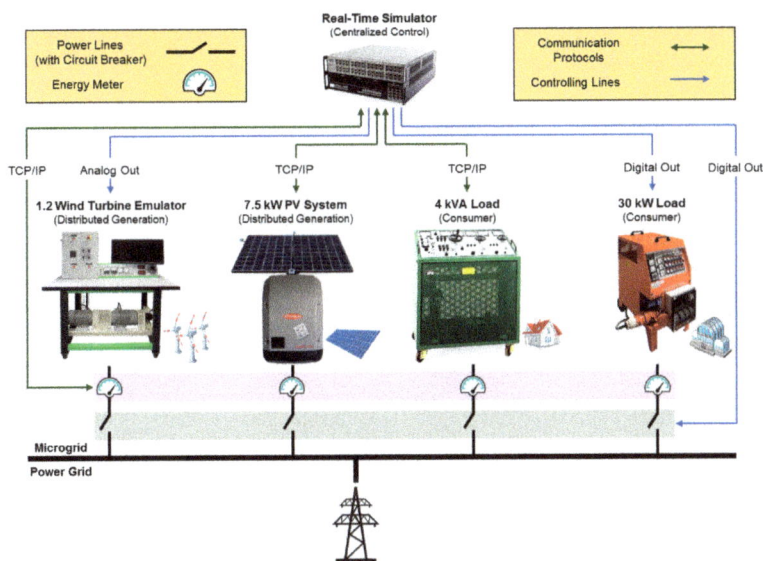

Figure 2. Centralized microgrid model architecture.

The other network players consist of a 4 kVA and a 30 kW load playing the role of low and medium consumer units, and a 1.2 kW wind turbine emulator and a 7.5 kW PV system as DG units in the microgrid. All of these network players were not operating automatically in their factory configuration. However, several automation projects have been implemented on them, in order to control and manage them remotely and automatically [23,30]. For concentrating on the innovative perspectives of the model, only the most related sections of the system are described here.

As shown in Figure 2, the first DG unit is referred to the 1.2 kW wind turbine emulator. This emulator consists of an inductive three-phase generator coupled with a three phase asynchronous motor with variable speed. The speed controller unit allows the variation of the wind speed and consequently the speed variation of the wind turbine rotor. This emulator is controlled through the analog outputs of the OP5600 (Simulink).

The second DG unit is a 7.5 kW PV system, which is already installed on the GECAD laboratory and currently is producing energy. For acquiring and monitoring the real-time generation data in OP5600 and Simulink model, Modbus/TCP (transmission control protocol) protocol has been used.

The third node is related to the 4 kVA load, the low consumer player of the microgrid, which plays the role of a domestic consumer in the microgrid. This load includes three independent sections of resistive, inductive, and capacitive. The automation process was focused on the resistive part. In the factory setting, it had a steering connected to a gauge in the resistive section, which enables the user to increase or decrease the consumption of the load. Currently, a programmable logic controller (PLC) connected to a 12 V DC motor controls the resistive gauge. This enables the 4 kVA load to receive the desired amount of consumption from OP5600 through Modbus/TCP communication protocol, and to adjust its consumption based on the received value.

The last node is connected to the 30 kW load, the medium consumer player of the microgrid that represents the consumption of a small commerce. By default, it had an integral control panel equipped with several selector switches, which enables the user to control the consumption. However, in order to control this unit automatically, four relays have been mounted on the load and are connected to the digital output of the OP5600. Therefore, the central controller unit is able to control the consumption of this resource through the Simulink model.

The power lines is the section that is not included in the previous microgrid model, and is proposed in this paper. As can be seen in Figure 2, there are four power lines that connects each node of the microgrid to the main power network. In each line, there is a circuit breaker and an energy meter. The circuit breakers are connected via digital output channels of the OP5600, and it enables the user to interrupt the line and disconnect the resource from the main grid through the Simulink model. Furthermore, the energy meters measure the power flow in the lines and transmit the real-time active power data to the Simulink model using Modbus/TCP protocol.

The existing platform can be improved in order to accommodate transient and stability studies which would require the use of PMUs instead of energy meters. In fact, the existing meters in the platform provide acceptable accuracy and sampling per second; however, it doesn't allow the synchronizing of measurements by GPS.

2.2. Distributed Control Fashion

As it was shown in Figure 1, there are four switches for the microgrid players where the user can choose how the microgrid be controlled. Figure 3 illustrates the microgrid distributed control method. In this condition, the central controller unit (OP5600) will be excluded from the microgrid point of view, and the local controllers manage the network.

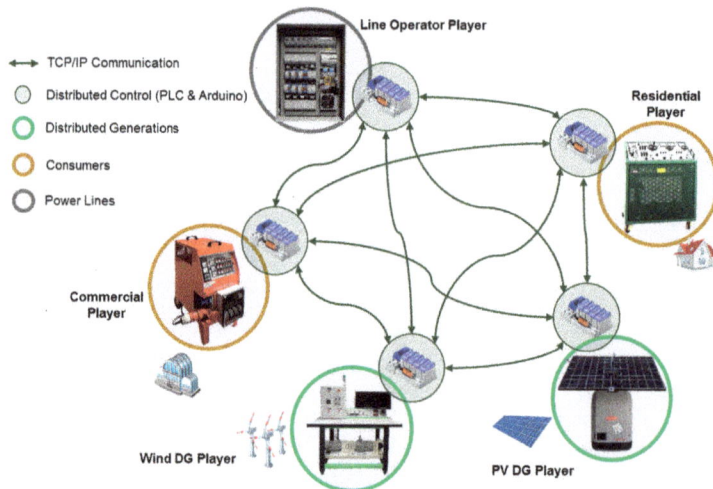

Figure 3. The distributed control based microgrid model.

In distributed control, there are five main players: a residential player, using a 4 kVA controllable load; a commercial player, using a 30 kW controllable load; a line operator player, that controls the power lines; a PV DG player; and wind DG player. A PLC is dedicated for each player. This enables the microgrid to accomplish decision making locally and communicate with other microgrid players, through TCP/IP (internet protocol) communication protocol, to achieve the microgrid's goals. The players are responsible for constantly exchanging messages in order to report their latest status in the network.

The microgrid players have dedicated PLCs with several Digital/Analog I/O slots used for their control and management. Residential player uses digital output slot to control the load motor. The commercial player is equipped with digital output slot in order to control the relays, and wind DG player employs an analog output slot for controlling the speed variation of the wind turbine rotor, and finally the line operator player uses the digital output slot to control the status of the circuit breakers of the lines.

The main task of the residential and commercial players are to control and adjust their consumption based on the overall system's goals. Furthermore, the PV DG player contains the data regarding the PV production and is accountable for informing the other agents with the latest value of the PV generation. Meanwhile, the wind DG player undertakes requesting the wind speed data from an external resource, such as a local weather station, and generates power depending on the received wind speed value. Finally, there are two main objectives for the line operator player since it contains all of the energy meters and the circuit breakers employed in the power lines. As mentioned, the first goal of the line operator player is to supervise the circuit breakers in the power lines. The second purpose is to request the real-time amount of the active power measured by the energy meter of each power line and transmit them to the other players. In this way, the other players, namely residential player, commercial player, and wind DG player, will be aware of their real-time amount of consumption or generation.

In the distributed control method, adaptability of the system is improved compared with the centralized control, since the response time to any changes is reduced. Furthermore, the distributed control method brings reconfigurability and flexibility features to the overall microgrid. Suppose that, in a simple way, the PV DG player transmits a signal to the other players saying that its instant amount of generation changed to 4500 W. The wind DG player also broadcasts a message saying that there is wind generating energy. Therefore, the line operator player responds to the wind DG player that their current output generation is supposedly 500 W. In the meantime, the residential and commercial player reply that they are consuming energy with a certain value, and the line operator player broadcasts their total amount of consumption, which is, supposedly 9500 W. Therefore, the microgrid supplies the rest of the required power from the power grid; hence, there is not enough energy production by DG units.

3. Case Study

In this section, a case study is presented and implemented by the microgrid model provided in this paper in order to test and validate the system capabilities. In this case study, it is considered that the user intends to use the centralized microgrid control model.

Figure 4 represents a 33 bus distribution network, including 220 consumers and 68 DG units. The distribution network was implemented MATLAB/Simulink, being compatible with OP5600. The microgrid model is a node connected to bus #10 of this network. Furthermore, the Simulink model developed in OP5600 for real-time controlling of the microgrid players is shown in Figure 4.

In this case study, we consider that a VPP owns the microgrid and its resources containing the consumers (with or without DR programs), and the energy generators. Therefore, the VPP aggregates the DG and DR resources since in the proposed microgrid they are considered as small size resources. Additionally, the VPP is capable to transact energy to the main grid, which means it can absorb energy while it has high demand and low generation, or inject power to the grid while it has more generation than consumption. This enables the VPP to have active participation in the electricity markets.

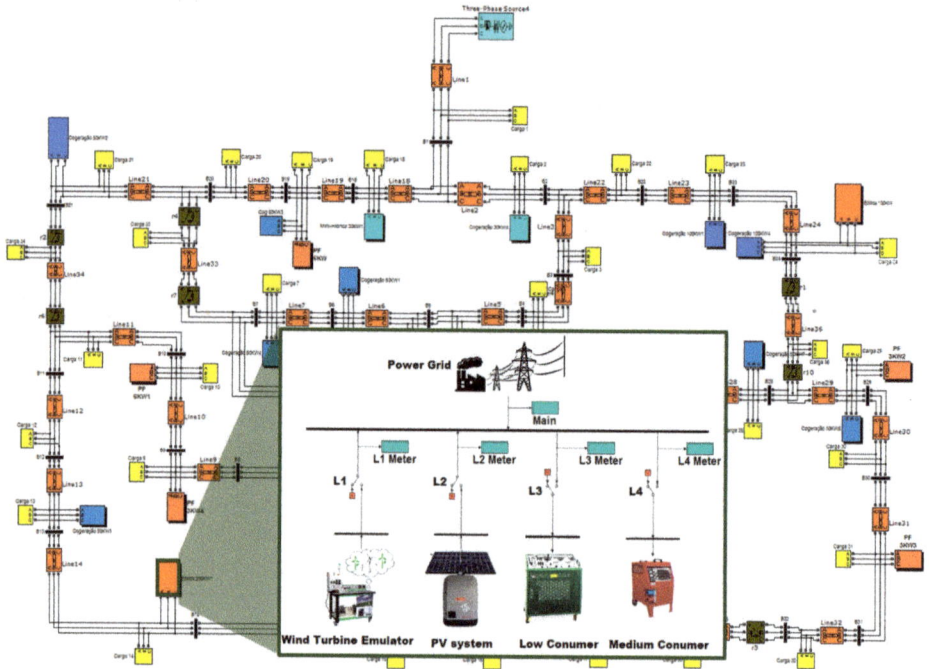

Figure 4. The microgrid model used in the 33 bus network.

VPP also can define several DR programs for the microgrid consumers in order to reduce or shift the consumption to one or more specific periods based on incentives and/or the prices offered to them. Technical or economic reasons can also be the motivation for the VPP to define DR programs. While the DR programs defined reduction or shifting, the VPP can use an optimization for the generation and demand resources in order to economically make a decision and execute the load shifting scenarios. The number of DR programs that VPP executed is a fundamental matter, which should be taken into account.

The shifting periods in this model are the amount of power that can be shifted from a period to other periods. Additionally, the number of periods that the shifted consumption will be entered, and also the amount of load reduction, which will not be shifted, should be considered.

The optimization problem used in this paper for the VPP has been adapted from [31], and only the most applicable information has been mentioned in this part. The objective function of this optimization is to minimize the operation costs of the VPP, considering the generation and shifting costs in each period t for all periods in the defined time horizon T. Equation (1) demonstrates the objective function of the optimization problem. The constraints of the model include:

- Balance equation containing the DR balance in each period of t, the energy production, and the consumption demand, which contains the shifted load from period t to period i, and the incoming consumption in period t shifted from period i. This is represented in Equation (2);
- The maximum DR capacity considering the consumption reduction executed in period t, which can be shifted to period i after or before t, presented in Equation (3).
- The maximum generation capacity limit in each period t, performed by Equation (4).

Minimize

$$OC = \sum_{t=1}^{T} \left[P_{DG(t)} \times C_{DG(t)} + \sum_{t-I \leq i}^{i \leq t+I} P_{DR(t,i)} \times C_{DR(t,i)} \right] \tag{1}$$

$$Load_{(b,t)} - \sum_{t-I \leq i}^{i \leq t+I} P_{DR(b,t,i)} + \sum_{t-I \leq i}^{i \leq t+I} P_{DR(b,i,t)} - P_{DG(b,t)} =$$

$$\sum_{j=1}^{B} V_{(b,t)} \cdot V_{(j,t)} \cdot \left[G_{(b,j)} \cdot \cos\left(\theta_{(b,t)} - \theta_{(j,t)}\right) + G_{(b,j)} \cdot \sin\left(\theta_{(b,t)} - \theta_{(j,t)}\right) \right] \tag{2}$$

$$\forall 1 \leq t \leq T, \forall 1 \leq b \leq B$$

$$P_{DR(b,t,i)} \leq P_{DR(b,t,i)}^{max}; \forall 1 \leq t \leq T, \forall -I \leq i \leq I, \forall 1 \leq b \leq B \tag{3}$$

$$P_{DG(b,t)} \leq P_{DG(b,t)}^{max}; \forall 1 \leq t \leq T, \forall 1 \leq b \leq B \tag{4}$$

$$V_{(b,t)}^{min} \leq V_{(b,t)} \leq V_{(b,t)}^{max}; \forall 1 \leq t \leq T, \forall 1 \leq b \leq B \tag{5}$$

$$\theta_{(b,t)}^{min} \leq \theta_{(b,t)} \leq \theta_{(b,t)}^{max}; \forall 1 \leq t \leq T, \forall 1 \leq b \leq B \tag{6}$$

TOMLAB, which is based on MATLAB, are used in order to solve the proposed optimization problem. Therefore, the optimized results can be easily provided to the microgrid central controller unit (OP5600) as inputs, and consequently, it controls the real hardware equipment in real-time based on these inputs. The output of the economic energy resource scheduling optimization model is a requested amount of power for each consumer to reduce its demand in a certain period. However, the actual implementation of this demand reduction request in a real load will depend on the electrical grid conditions. This is in fact one of the advantages of using real-time simulation (in this paper OP5600) and laboratorial equipment for consumption modeling. In this way, we validate the actual demand reduction in order to be included in the simulation results, namely for remuneration purposes.

4. Results

In this section, the results of the proposed methodology will be executed using the microgrid model and its results illustrated. We consider that the case study consists of 10 periods with a one minute time interval. The consumption and generation profiles of the microgrid aggregated by the VPP during this 10 min is shown in Figure 5. As can be seen, the blue area is the total power aggregated by the VPP during the 10 periods, and the red line indicates the total consumption. The aggregated power supply includes the PV generation, wind production, and the incoming power from the main network to the microgrid.

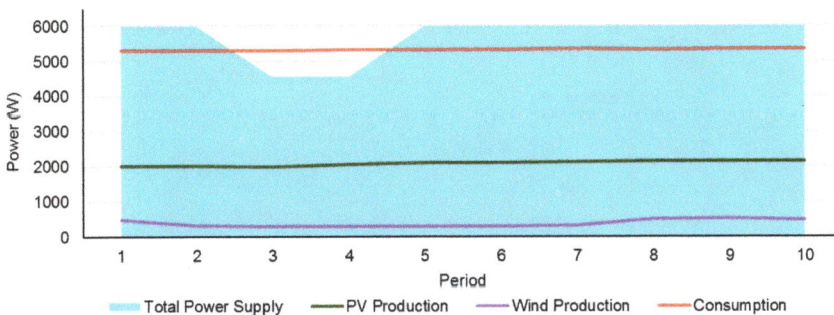

Figure 5. The microgrid model used in the 33 bus network.

The data used in the case study is for the day 13 January 2017 (Friday), between 11:30 AM to 11:40 AM. The PV curve is the real-time generation profile adopted from GECAD database. The wind generation is the simulated profile by the wind turbine emulator based on the real-time wind speed data, acquired from [32], and the consumption curve is also the real-time consumption of the GECAD building, emulated by the 4 kVA and 30 kW load.

As Figure 5 demonstrates, the microgrid meets a drop on generation in the periods 3 and 4. The reason for this lack of generation is considered to be a fault or any other cause in the main grid. Therefore, this is an opportunity for the VPP to start the optimization problem in order to optimally schedule the consumption shifting of the resources. The results of the optimization problem is depicted in Figure 6. The shifted periods have been scheduled in order to minimize the operation costs of the VPP.

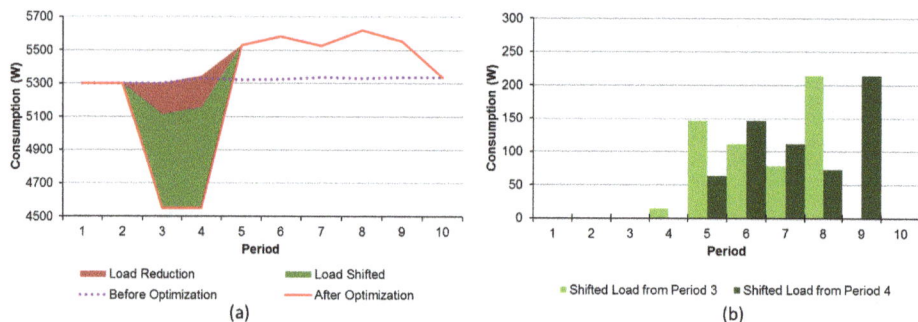

Figure 6. The optimization results for the consumption resources.

Figure 6a illustrates the load reduction and shifting that have occurred during the periods 3 and 4, where the VPP faced a lack of generation and shifted to the periods after period 4. The red area is the reduced consumption by the consumers, and the green area is the shifted consumption to the other periods. Also, as can be seen in Figure 6b, the incoming consumption in the periods of 8 and 9 are much higher compared with the other periods. This can be because of the DR programs and the economic advantages.

While TOMLAB outputted the results of optimization, they will be provided to the OP5600 real-time simulator as inputs. Consequently, the real-time simulator starts to control and manage the HIL equipment in order to implement the optimization results in real-time. Figures 7–9 show the final results of the real-time simulation during 10 min. All of the results illustrated in these figures are adapted from OP5600 and MATLAB/Simulink.

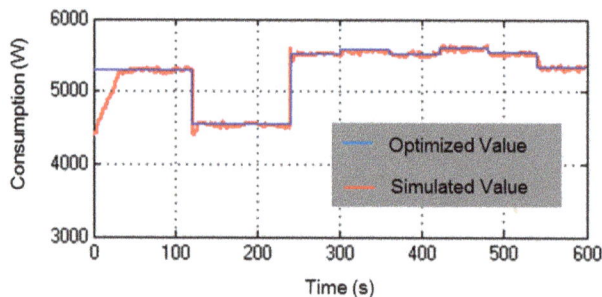

Figure 7. Real-time simulation of the consumption profile using the consumers of the microgrid.

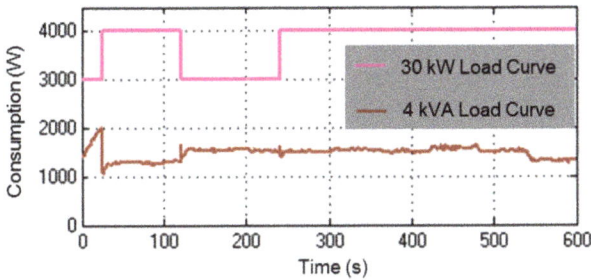

Figure 8. The separated consumption curve of each player of the microgrid.

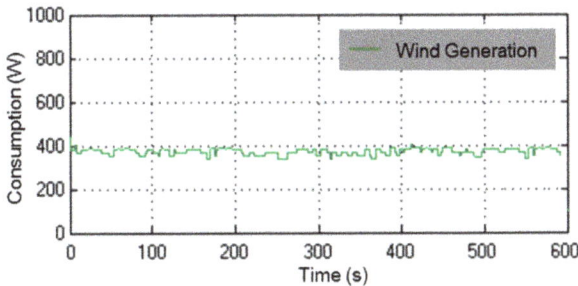

Figure 9. Real-time simulation of wind production curve.

As Figure 7 illustrates, the total amount of consumption of the microgrid has been reduced and shifted to other periods based on the optimization results, which occurred between the second of 120 to 240 (periods 3 and 4). Also, the denotative consumption profiles of the microgrid have been illustrated in Figure 8. It is obvious that the residential player shifted its consumption; however, the commercial player reduced consumption based on the data received from the OP5600 real-time simulator. Furthermore, Figure 9 represents the wind production simulated by the wind DG player. This generation curve has been simulated based on the real-time wind speed data provided to the emulator from the OP5600 real-time simulator.

5. Conclusions

Microgrids are a particular case of distribution networks, namely in the context of smartgrids. Demand response and distributed generation are very relevant resources in the scope of microgrids and smartgrids. As discussed in the present paper, the realistic simulation of the impact of these resources is very important in order to validate the technical and business model's impact in smartgrids management.

In this paper, important improvements have been added to SPIDER, a simulation platform that accommodates real-time simulation skills adequate for demand response and distributed generation. The innovative content provides details on the integration of both centralized and distributed control approaches, and also includes the emulation of generation and load components which allowed us to more realistically simulate the microgrid and validate the computational models.

The case study presented here has briefly demonstrated the platform skills in order to validate a business model for optimal resource scheduling in the microgrid, and its connection to the upstream distribution network. A VPP managed the resources aiming at minimizing the operation costs. It has been shown that the results obtained by the scheduling algorithm benefit with the integration in the real-time simulation platform in order to check the actual simulated consumption and generation values which include the variability of these resources. Moreover, the presented results are the ones actually measured in the load, and generation emulation devices which are shown to have relevant information that was not given by the electrical network simulation model. The main one is that when the load schedule is changed, the actual consumption devices take some time in order to reach the desired consumption.

Acknowledgments: The present work was done and funded in the scope of the following projects: Project AVIGAE (ANI | P2020 003401), funded by Portugal 2020 "Fundo Europeu de Desenvolvimento Regional" (FEDER) through POCI (Programa Operacional Competitividade e Internacionalização); UID/EEA/00760/2013 funded by FEDER Funds through COMPETE program and by National Funds through FCT; and ITEA project nr. 12004—SEAS—Smart Energy Aware Systems.

Author Contributions: Omid Abrishambaf wrote and organized the paper, and discussed the work with the rest of authors; Pedro Faria adapted the SPIDER platform and the network model used for the developed approach; Luis Gomes implemented the simulation components interoperability and JAVA API; João Spínola developed the energy resources optimization model; Zita Vale raised and developed the overall concept; Juan M. Corchado contributed for the distributed control nature.

Conflicts of Interest: The authors declare no conflict of interest.

References

1. Siano, P. Demand response and smart grids—A survey. *Renew. Sustain. Energy Rev.* **2014**, *30*, 461–478. [CrossRef]
2. Gouveia, C.; Moreira, J.; Moreira, C.; Peças Lopes, J. Coordinating storage and demand response for microgrid emergency operation. *IEEE Trans. Smart Grid* **2013**, *4*, 1898–1908. [CrossRef]
3. Khatibzadeh, A.; Besmi, M.; Mahabadi, A.; Reza Haghifam, M. Multi-agent-based controller for voltage enhancement in AC/DC hybrid microgrid using energy storages. *Energies* **2017**, *10*, 169. [CrossRef]
4. Abrishambaf, O.; Gomes, L.; Faria, P.; Afonso, J.; Vale, Z. Real-time simulation of renewable energy transactions in microgrid context using real hardware resources. In Proceedings of the 2016 IEEE/PES Transmission and Distribution Conference and Exposition (T&D), Dallas, TX, USA, 3–5 May 2016; pp. 1–5.
5. Wen, M.H.F.; Li, V.O.K. Optimal phasor data compression unit installation for wide-area measurement systems—An integer linear programming approach. *IEEE Trans. Smart Grid* **2016**, *7*, 2644–2653. [CrossRef]
6. Albuquerque, R.J.; Paucar, V.L. Evaluation of the PMUs measurement channels availability for observability analysis. *IEEE Trans. Power Syst.* **2013**, *28*, 2536–2544. [CrossRef]
7. Chen, J.; Shrestha, P.; Huang, S.; Sarma, N.D.R.; Adams, J.; Obadina, D.; Ballance, J. Use of synchronized phasor measurements for dynamic stability monitoring and model validation in ERCOT. In Proceedings of the 2012 IEEE Power and Energy Society General Meeting, San Diego, CA, USA, 22–26 July 2012; pp. 1–7.
8. Ge, Y.; Flueck, A.J.; Kim, D.K.; Ahn, J.B.; Lee, J.D.; Kwon, D.Y. Power system real-time event detection and associated data archival reduction based on synchrophasors. *IEEE Trans. Smart Grid* **2015**, *6*, 2088–2097. [CrossRef]
9. Faria, P.; Vale, Z. Demand response in electrical energy supply: An optimal real time pricing approach. *Energy* **2011**, *36*, 5374–5384. [CrossRef]
10. Gomes, L.; Faria, P.; Morais, H.; Vale, Z.; Ramos, C. Distributed, agent-based intelligent system for demand response program simulation in smart grids. *IEEE Intell. Syst.* **2014**, *29*, 56–65.
11. Qiang, W.; Zhang, W.; Xu, Y.; Khan, I. Distributed control for energy management in a microgrid. In Proceedings of the 2016 IEEE/PES Transmission and Distribution Conference and Exposition (T&D), Dallas, TX, USA, 3–5 May 2016; pp. 1–5.
12. Abrishambaf, R.; Hashemipour, M.; Bal, M. Structural modeling of industrial wireless sensor and actuator networks for reconfigurable mechatronic systems. *Int. J. Adv. Manuf. Technol.* **2012**, *64*, 793–811. [CrossRef]
13. Werth, A.; Kitamura, N.; Matsumoto, I.; Tanaka, K. Evaluation of centralized and distributed microgrid topologies and comparison to open energy systems (OES). In Proceedings of the 2015 IEEE 15th International Conference on Environment and Electrical Engineering (EEEIC), Rome, Italy, 10–13 June 2015; pp. 492–497.
14. Conti, S.; Nicolosi, R.; Rizzo, S.A.; Zeineldin, H.H. Optimal dispatching of distributed generators and storage systems for MV islanded microgrids. *IEEE Trans. Power Deliv.* **2012**, *27*, 1243–1251. [CrossRef]

15. Simpson-Porco, J.W.; Shafiee, Q.; Dörfler, F.; Vasquez, J.C.; Guerrero, J.M.; Bullo, F. Secondary frequency and voltage control of islanded microgrids via distributed averaging. *IEEE Trans. Ind. Electron.* **2015**, *62*, 7025–7038. [CrossRef]

16. Bayat, M.; Sheshyekani, K.; Hamzeh, M.; Rezazadeh, A. Coordination of distributed energy resources and demand response for voltage and frequency support of MV microgrids. *IEEE Trans. Power Syst.* **2016**, *31*, 1506–1516. [CrossRef]

17. Gulin, M.; Vasak, M.; Pavlovic, T. Dynamical behaviour analysis of a DC microgrid in distributed and centralized voltage control configurations. In Proceedings of the 2014 IEEE 23rd International Symposium on Industrial Electronics (ISIE), Istanbul, Turkey, 1–4 June 2014; pp. 2365–2370.

18. Colet-Subirachs, A.; Ruiz-Alvarez, A.; Gomis-Bellmunt, O.; Alvarez-Cuevas-Figuerola, F.; Sudria-Andreu, A. Centralized and distributed active and reactive power control of a utility connected microgrid using IEC61850. *IEEE Syst. J.* **2012**, *6*, 58–67. [CrossRef]

19. Karavas, C.; Kyriakarakos, G.; Arvanitis, K.; Papadakis, G. A multi-agent decentralized energy management system based on distributed intelligence for the design and control of autonomous polygeneration microgrids. *Energy Convers. Manag.* **2015**, *103*, 166–179. [CrossRef]

20. Zhao, J.; Dörfler, F. Distributed control and optimization in DC microgrids. *Automatica* **2015**, *61*, 18–26. [CrossRef]

21. Basir Khan, M.; Jidin, R.; Pasupuleti, J. Multi-agent based distributed control architecture for microgrid energy management and optimization. *Energy Convers. Manag.* **2016**, *112*, 288–307. [CrossRef]

22. The ERIGrid European Project. Transnational Access Procedure and Rules. Available online: https://erigrid.eu/resources/ (accessed on 16 April 2016).

23. Abrishambaf, O.; Gomes, L.; Faria, P.; Vale, Z. Simulation and control of consumption and generation of hardware resources in microgrid real-time digital simulator. In Proceedings of the 2015 IEEE PES Innovative Smart Grid Technologies Latin America (ISGT LATAM), Montevideo, Uruguay, 5–7 October 2015; pp. 799–804.

24. GECAD Website. Available online: www.gecad.isep.ipp.pt (accessed on 24 February 2017).

25. Faria, P.; Vale, Z.; Baptista, J. Demand response programs design and use considering intensive penetration of distributed generation. *Energies* **2015**, *8*, 6230–6246. [CrossRef]

26. Faria, P.; Spinola, J.; Vale, Z. Aggregation and remuneration of electricity consumers and producers for the definition of demand-response programs. *IEEE Trans. Ind. Inf.* **2016**, *12*, 952–961. [CrossRef]

27. Mathworks Website. Available online: www.mathworks.com (accessed on 24 February 2017).

28. TOMLAB Optimization Environment. Available online: www.tomopt.com/tomlab (accessed on 24 February 2017).

29. OPAL-RT Website. Available online: www.opal-rt.com (accessed on 24 February 2017).

30. Abrishambaf, O.; Ghazvini, M.; Gomes, L.; Faria, P.; Vale, Z.; Corchado, J. Application of a home energy management system for incentive-based demand response program implementation. In Proceedings of the 2016 27th International Workshop on Database and Expert Systems Applications (DEXA), Porto, Portugal, 5–8 September 2016; pp. 153–157.

31. Faria, P.; Vale, Z. Optimization of generation and aggregated consumption shifting for demand response programs definition. In Proceedings of the IEEE PES Innovative Smart Grid Technologies, Istanbul, Turkey, 12–15 October 2014; pp. 1–6.

32. Meteo ISEP Website. Available online: meteo.isep.ipp.pt (accessed on 24 February 2017).

energies

MDPI

Article

Diffusion Strategy-Based Distributed Operation of Microgrids Using Multiagent System

Van-Hai Bui [1], Akhtar Hussain [1] and Hak-Man Kim [1,2,*]

[1]   Department of Electrical Engineering, Incheon National University, 12-1 Songdo-dong, Yeonsu-gu, Incheon 406840, Korea; buivanhaibk@inu.ac.kr (V.-H.B.); hussainakhtar@inu.ac.kr (A.H.)
[2]   Research Institute for Northeast Asian Super Grid, Incheon National University, 12-1 Songdo-dong, Yeonsu-gu, Incheon 406840, Korea
*   Correspondence: hmkim@inu.ac.kr; Tel.: +82-32-835-8769; Fax: +82-32-835-0773

Received: 19 May 2017; Accepted: 28 June 2017; Published: 2 July 2017

Abstract: In distributed operation, each unit is operated by its local controller instead of using a centralized controller, which allows the action to be based on local information rather than global information. Most of the distributed solutions have implemented the consensus method, however, convergence time of the consensus method is quite long, while diffusion strategy includes a stochastic gradient term and can reach convergence much faster compared with consensus method. Therefore, in this paper, a diffusion strategy-based distributed operation of microgrids (MGs) is proposed using multiagent system for both normal and emergency operation modes. In normal operation, the MG system is operated by a central controller instead of the distributed controller to minimize the operation cost. If any event (fault) occurs in the system, MG system can be divided into two parts to isolate the faulty region. In this case, the MG system is changed to emergency operation mode. The normal part is rescheduled by the central controller while the isolated part schedules its resources in a distributed manner. The isolated part carries out distributed communication using diffusion between neighboring agents for optimal operation of this part. The proposed method enables peer-to-peer communication among the agents without the necessity of a centralized controller, and simultaneously performs resource optimization. Simulation results show that the system can be operated in an economic way in both normal operation and emergency operation modes.

Keywords: consensus algorithm; diffusion strategy; distributed system; energy management system; microgrid operation; optimal operation

---

1. Introduction

Microgrid (MG) system is a small-scale electrical distribution system integrating multiple loads and multiple distributed sources of generation, such as controllable distributed generators (CDGs), renewable distributed generators (RDGs), and energy storage systems (ESSs) [1]. An MG system can operate efficiently and safely in both grid-connected and islanded modes [2]. In grid-connected mode, the MG system can either buy electric power from the utility grid, or sell electric power to the utility grid. The power balance can be maintained by the utility grid in each time interval. However, in islanded mode, MG system is operated without the utility grid. The balancing between supply and demand is maintained by MG's resources and by performing load shedding in some peak intervals. The major considerations of an MG system are minimizing operation cost, preserving customer privacy, and enhancing the system reliability. Energy management system (EMS) is used to optimally schedule the power resources, such as CDGs, ESSs, and the amount of trading with the power grid to fulfill the load demands [3].

There are two fundamentally different approaches for the design of such an energy management system, which are centralized and decentralized approaches. The centralized EMS is to assign the responsible for coordinating CDGs, ESSs, loads and the utility grid connection to a central entity [4,5]. The centralized method requires all components to communicate with an MG energy management system (MG-EMS), and an optimization problem is solved at the central location. Then optimal solutions are sent to the individual components. A centralized control for optimizing MG system operation has been proposed considering two market policies for demand-side bidding options by [5]. A centralized EMS has been developed for optimal operation of an isolated MG system using the model predictive control technique [6]. The authors in [7] have introduced an optimization method for a cooperative multi-microgrids (MMGs) with sequentially coordinated operations.

Another approach is based on multi-agent systems (MAS) in which the decisions are made in a decentralized/distributed way [8,9]. The decentralized approaches do not need a central controller and each unit is controlled by its local controller, which allows the control actions to be simply based on local information. A fundamental problem in distributed control systems is the need for all the nodes to reach a consensus. The consensus problem has been widely applied in several areas, such as social science and computer science [10]. Consensus algorithms and their applications have been extensively studied in the MG system and control area [11]. A dynamic consensus algorithm based distributed optimization method has been proposed to improve the system efficiency and offer higher expandability and flexibility [12]. A fully distributed control strategy based on the consensus algorithm has been proposed for the optimal resource management in an islanded MG system [13]. A distributed energy management approach based on the consensus + innovations method has been presented in [14] to coordinate local generations, demands, and storage devices within the MG system. An analysis on convergence of the incremental cost consensus algorithm has been analyzed for a smart grid under different communication network topologies [15].

Each of the EMS architecture has its own merits and demerits. MG systems can reduce operation cost (global optimization), utilize efficient components of MG system, and reduce the amount of external trading by applying centralized methods [16]. However, once failure of the central controller occurs, the MG systems may fail, which decreases the reliability of the system. Alternatively, decentralized methods do not need a central controller and each unit is controlled by its local control system, which allows the control action to be simply based on local information rather than global information. However, the method can increase operation cost, unawareness of the system level resources, and excessive power trading with the utility grid in grid-connected mode [16].

In the literature, either only centralized or only distributed EMS architectures are considered. In the case of distributed EMSs, the distributed information sharing between neighboring agents is established through consensus [12–15]. Therefore, the authors in [17] have proposed a new method for optimal MG control scheme using a fully distributed diffusion strategy. The diffusion strategy includes a stochastic gradient term to expedite the process and reach convergence much faster compared with consensus. Additionally, by including the gradient of the cost function in the formulation, diffusion strategy can reach the economic dispatch point through distributed optimization.

To take the advantages of both EMS architectures, this paper proposes a new operation strategy for improving the system reliability using diffusion strategy for both normal and emergency operation mode. In normal operation, the MG system is operated by an MG-EMS. If any fault occurs in the MG system, which lead the system could be divided into two parts: normal and isolated parts. The normal part is still operated by the MG-EMS. In conventional operation, the isolated part is isolated from the main system. In many cases this part is out of service and waits to reclose, which reduces the system reliability. By applying the proposed strategy, this part is considered as a distributed system and is operated normally with new schedules by using diffusion strategy. The distributed communication is applied in this part by using diffusion between neighboring agents for optimal operation of isolated part. The proposed method enables peer-to-peer communication among the agents without the necessity of a centralized controller, and simultaneously performs resource optimization. The isolated part is updated every interval with new faulty/recovered equipment.

By using the proposed algorithm, the rescheduling is converged faster than consensus algorithm by implementing an additional gradient term.

The rest of this paper is organized as follows. In Section 2, the MG configuration and the proposed algorithm for the MG operation are presented using diffusion strategy. Communications in the MG system and diffusion strategy are introduced in detail in the subsections of Section 2. The mathematical model for both operation modes of MG system is introduced in Section 3. The performance of the proposed method is evaluated and the simulation results are analyzed and discussed in Section 4. Section 5 concludes this paper.

2. System Model

2.1. Microgrid System Configuration

In this paper, an MG system is considered as a portion of electric network including diesel generators (DGs), battery energy storage system (BESS), renewable energy generators (RDGs), and loads, as shown in Figure 1. In normal operation, the MG system is operated by a centralized controller (MG-EMS). In this mode, each component informs its information to MG-EMS and receives its optimal schedule from the MG-EMS. The output power of DG units and charging/discharging amount of BESS are decided by the MG-EMS. The amount of power exchange between MG system and the utility grid is also determined by the MG-EMS to minimize the total operation cost and fulfill shortage power in the system. On the other hand, when an event occurs in the MG system, such as short circuit, over current of power electronic interfaces [18,19], the corresponding circuit breaker (CB) is opened to isolate the fault, thus some parts are disconnected from main system. MG system is divided into two parts: normal and isolated parts. In the worst case (losing both electrical connection and communication), the MG system has two separate parts and MG-EMS cannot control all components in the system. Therefore, the operation mode of MG is changed to emergency mode. In this mode, the normal part is still rescheduled by the MG-EMS while the isolated part could be out of service. Because the occurrence time of fault is not known, the operation problem is becoming a real-time problem and the convergence time is important for survival of the isolated part. Thus, all components in the isolated part could be rescheduled to another operation point as soon as possible without MG-EMS. By applying the proposed strategy, components (agents) in isolated system can communicate with each other to share their information and determine a new operation point in a short time. The system information is updated every time with the new faulty/recovered equipment. Whenever the system information is changed, the isolated part can reschedule without the MG-EMS.

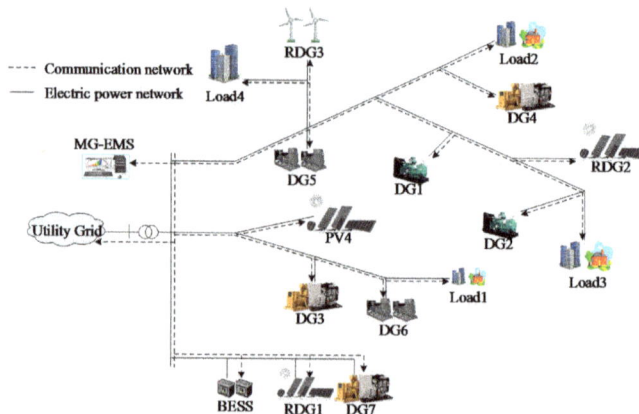

Figure 1. An illustration of a typical microgrid system.

2.2. Algorithm for Microgrid System Operation

The step-by-step procedure for performing one round of optimization is shown in Figure 2. In normal operation, agents inform their information to a centralized EMS (MG-EMS), such as buying and selling prices (hourly day-ahead market price signals), generation capabilities of DGs, RDGs, and BESS along with load profiles of MG system, which are taken as input data. After receiving all information, MG-EMS performs optimization and informs to participating agents with optimal results. The output power of each DG unit is decided by the comparison among the market price signals and its generation cost to minimize the operation cost of the MG system. The amount of exchanging power with the utility grid is decided to maintain the power balance in the system and maximize the profit. In peak intervals, the shortage power is fulfilled by importing electric power from the utility grid while the surplus power from cheap resources is sold to the utility grid in off-peak intervals. BESS is used to shift the surplus power from off-peak intervals to peak interval. The BESS is charged with cheap resources and discharged at expensive intervals for reducing the operation cost. If any event occurs in the MG system, faulty part is isolated from the system. The isolated part is considered as a distributed system, which is operated without MG-EMS. Thus, the MG system is changed to emergency operation mode.

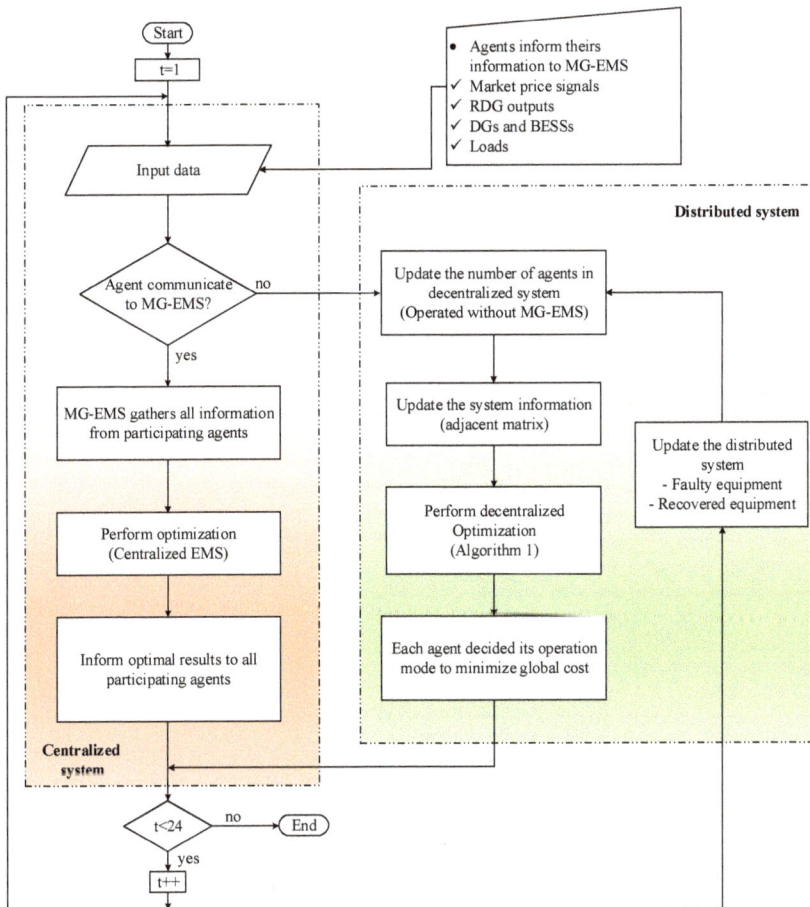

Figure 2. Flowchart for operation of the microgrid system.

In emergency operation mode, there are three possible disconnection scenarios:

1. Losing the communication with the MG-EMS but maintaining the electrical connection with the normal part of MG.
2. Maintaining the communication with the MG-EMS but losing the electrical connection with the MG.
3. Losing both communication with the MG-EMS and the electrical connection with the normal part of MG.

In this paper, isolated/islanded refers to failure in both electrical and communication system. Power failure refers to failure in only electrical system and communication failure refers to failure in only communication system.

The normal part is rescheduled from the occurrence time of event by MG-EMS while the operation of the other part is rescheduled based on failure scenario, as explained above. In the case of Scenario 2, the MG-EMS can control all components in the MG system, so the two separated parts of the MG system are operated by MG-EMS without sharing power between two parts. In the case of Scenario 1 and Scenario 3, the MG-EMS cannot communicate with all components of faulty part and agents cannot receive the operation information from the MG-EMS. Therefore, each component (agent) shares its information with its neighbor agents and performs distributed optimization by using the proposed algorithm. The distributed optimization method for isolated part is explained in detail in Algorithm 1. The system information is updated every time with the new faulty/recovered equipment. Whenever the system information is changed or a new time interval is started, the MG system is rescheduled by applying the proposed algorithm.

2.3. Diffusion Strategy for Distributed Optimization

The proposed diffusion framework for an agent j is shown in Figure 3. In this figure, the agent j is only required to interact with the neighboring agents N_j (highlighted in Figure 3). In the case of fault, a corresponding circuit breaker (CB) should be opened to isolate the fault. It can lead to a situation where an agent and its neighboring agents are not electrically connected and are part of two separated sections. To prevent the diffusion of information between these neighboring agents, the agent checks the status of the corresponding CB and decides the neighboring agents for sharing its information based on the location of fault point. In this way, all the interactions are done in the network. The method is able to cope with different network topologies, and does not require the global information or relying on the central manager. The system topology is represented by an adjacent matrix A by using Metropolis rule [17,20], as given by Equation (1).

$$
a_{ij} = \begin{cases} \frac{1}{\max(n_i, n_j)} & i \in N_j\{j\} \\ 1 - \sum_{i \in N_j\{j\}} a_{ij} & i = j \\ 0 & \text{otherwise} \end{cases} \tag{1}
$$

where n_i, n_j are the number of neighboring agents of agent i and j, and $a_{ij} = 0$ when agent i and j is not connect.

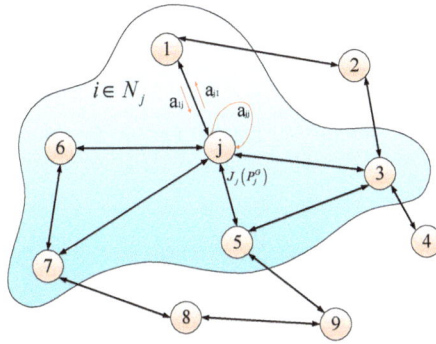

Figure 3. Neighborhood agents of agent *j* (the highlighted area).

To minimize the operation cost in the distributed system, global cost objective function of the system is defined by Equation (2), where J_i is the cost function of agent *i*. The real-valued vector of arguments $w \in \mathbb{R}n$, representing the output power of each dispatchable agent (diesel generator). Therefore, the objective function (3) minimizes the global cost J_{glob}, and it is obtained by summing all individual cost functions. Each dispatchable agent has its own cost function. Therefore, each agent could have different objective functions in the case for the MG system. All J_i need to be differentiable, convex, and at least one J_i needs to be strong convex for reaching only one global minimum solution [21,22]. In this paper, the operation cost of each diesel generator (DG) unit is represented by a quadratic cost function and are all strong convex, as shown in Equation (4), where *a*, *b*, and *c* are the quadratic coefficients. Thus, J_{glob} is also strong convex and can reach optimization point. From the generation cost functions, the derivative of the cost function is given in Equation (5), which is also known as the marginal cost function. This equation is used for optimization diffusion algorithm to reach economic dispatch. The detailed mechanisms for performing distributed optimization has been explained more detail in the Algorithm 1.

$$J_{glob}(w) = \sum_{i=1}^{n} J_i(w) \tag{2}$$

$$\min_{w} J_{glob}(w) \tag{3}$$

$$C_g^{DG}\left(P_{g,t}^{DG}\right) = a + b \cdot P_{g,t}^{DG} + c \cdot \left(P_{g,t}^{DG}\right)^2 \tag{4}$$

$$\frac{\partial C_g^{DG}\left(P_g^{DG}\right)}{\partial P_g^{DG}} = b + 2 \cdot c \cdot P_g^{DG} \tag{5}$$

In the proposed algorithm, each agent follows two steps: (1) information sharing diffusion for sharing the shortage amount; and (2) optimization diffusion for minimizing the operation cost. In Step 1, Combine-Then-Adapt (CTA) diffusion strategy is implemented to share information in the distributed system, as given in first two equations of Algorithm 1. At each iteration, the agent *i* updates its current state $(x_{k-1,i})$ to a new state $(x_{k,i})$ using the local stochastic gradient at this iteration. The local stochastic gradient available can be calculated from the difference of intermediate state ϕ at this iteration [17]. After finishing Step 1, the shortage amount in the system is known in the distributed system. In Step 2, namely decentralized optimization, similar to Step 1, CTA diffusion strategy is used for distributed optimization, as shown in last three equations of Algorithm 1. However, in this step, the gradient of the cost function is used instead of the stochastic gradient $(\nabla J_i\left(P_{k,i}^G\right) = \nabla C_i\left(P_{k,i}^G\right))$.

Algorithm 1 Diffusion strategy for distributed optimization

1: Initial values

2: Updated adjacent matrix A (Equation (1))

3: Step 1: Determine shortage power in system

4: **while** error < available value **do**

5:     | **for all** $i <$ N **do**

6:     | $\phi_{k-1,i} = \sum_{j \in N_i} a_{i,j} x_{k-1,j}$

7:     | $x_{k,i} = \phi_{k-1,i} - \mu\left(\phi_{k-1,i} - \phi_{k-2,i}\right)$

8:     | **end**

9: **end while**

10: Determine demand in whole system: P_N

11: Step 2: Decentralized optimization

12: **while** error < available value **do**

13:    | **for all** $i <$ N **do**

14:    | $\phi_{k-1,i} = \sum_{j \in N_i} a_{i,j} P^{G}_{k-1,j}$

15:    | $P^{G}_{k,i} = \phi_{k-1,i} - \mu \nabla J_i\left(\phi_{k-1,i}\right)$

16:    | **end**

17:    | Update based on condition:

18:    | $\sum_{i=1}^{N} P^{G}_{k,i} = P_N$

19:    | Update error

20: **end while**

2.4. Interaction among Agents in the Microgrid System

In the normal operation, all agents communicate with the MG-EMS agent by using agent communications language (ACL) messages. The interaction among agents of the proposed strategy is illustrated in Figure 4. Firstly, a message is sent by MG-EMS agent to market agent to inquire about the market price signals. The market agent sends the day-ahead buying and selling prices for each hour of the day to MG-EMS agent. The MG-EMS agent will inform its local resources about the market price signals along with call for proposal (cfp) messages. The local elements of MG system propose their proposals for operation scheduling. Based on the proposals received from its local elements, the MG-EMS agent decides to accept/reject the proposals from its local agents. After receiving the acceptance/rejection of their proposals, each local agents implement its operation scheduling and informs the MG-EMS. Finally, the MG-EMS decides the amount of buying/selling power with the utility grid based on the amount of shortage/surplus power in the MG system. Communication between all the agents is realized through ACL by using a modified contract net protocol (MCNP) [23].

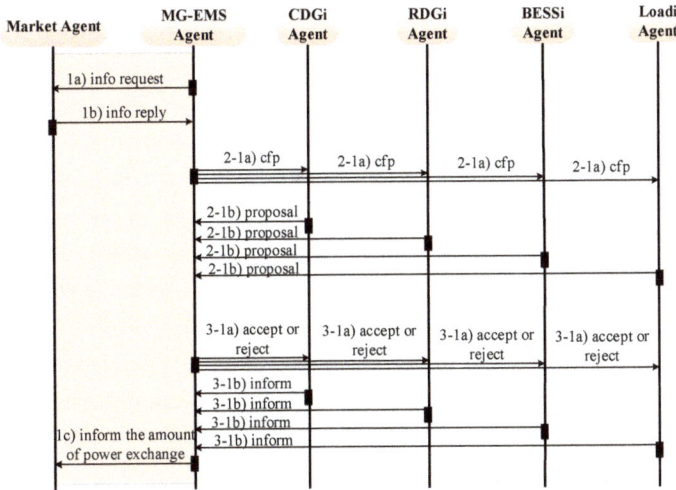

Figure 4. Agents communication in normal operation.

In the emergency operation, isolated agents cannot communicate to the MG-EMS agent. The operation of normal agents is rescheduled by MG-EMS agent, similar to normal mode. On the other hands, each isolated agent communicates to its neighbor agents for sharing its information and performing distributed optimization based on its receiving information. The state diagram of each agent in the isolated part is shown in Figure 5. Firstly, each agent receives information from all its neighbor agents and updates the information of the amount of shortage power in the isolated part. The updated information will be sent to the neighbor agents. After the shortage power information of all agents has reached convergence, the amount of shortage power in the system is determined. Then, the economic dispatch is started by using the optimization diffusion. Each agent shares its information to its neighbor agents and updates its generation output to fulfill the amount of shortage power in the isolated part. When the information sharing converges once again, the generation amount of each DG unit is determined for isolated part.

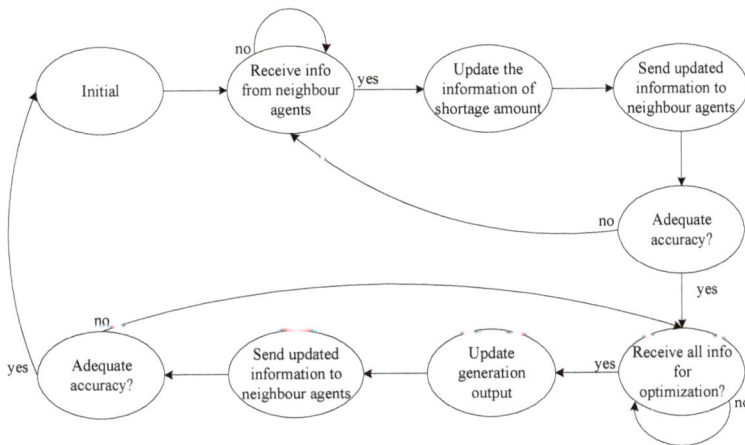

Figure 5. State diagram of each agent in isolated part.

2.5. Scheduling Horizons

The two possible operation modes of the proposed microgrid are normal mode and emergency mode. Each of the operation modes has a different scheduling horizon as depicted in Figure 6. The scheduling horizon for normal mode is 24 h (T) and operation is based on the day-ahead model. If any event occurs at time h, the MG-EMS will switch its operation mode to emergency mode. The scheduling horizon of the emergency mode is from $t = h$ to the end of the day (T). If any event occurs, the MG system is divided into two parts: normal part and isolated part. In normal part, the operation of all components is rescheduled by the MG-EMS while the isolated part is considered as a distributed system. To reschedule the operation of all components in isolated part, each component (agent) will communicate to its neighbor agents and perform distributed optimization to make new operation point. At each time interval, the number of normal/isolated equipment (agents) is updated in the MG system. The scheduling for all components is determined by using the MG-EMS (normal part) and diffusion strategy (isolated part). Finally, if the event is fully cleared, the MG-EMS will switch back to normal operation mode depending on its input values. Generally, a scheduling horizon of one day is considered for scheduling of microgrids [24,25]. Therefore, in this study, simulations are conducted for one day. However, the formulated mathematical models can be used to extend the simulations for longer durations by iterating the scheduling window.

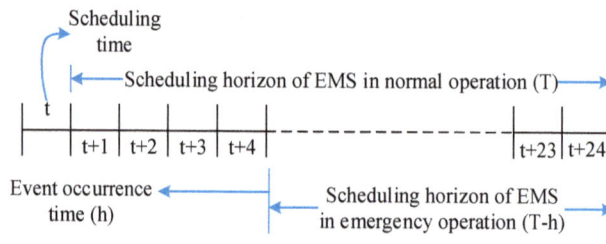

Figure 6. Scheduling horizons of different operation modes of the proposed microgrid.

3. Problem Formulation

A hierarchical control structure has three levels based on the required time frame: primary, secondary, and tertiary controls [26,27].

- Primary control is designed to preserve voltage, frequency stability and plug and play capability of distributed energy resources (DERs).
- Secondary control is designed to compensate the voltage and frequency deviation caused by primary control.
- Tertiary control is designed for optimal operation of MGs and/or deciding the amount of power sharing with the utility grid (in grid-connected mode).

In this paper, we focus on the tertiary control and set long-term set points based on the status of the distributed energy resource units, market price signals, and other system requirements. It is responsible for managing MG system in an economical way. Therefore, the constraints related to the primary control and secondary control are assumed to be fulfilled, and out of the scope for this paper.

In the following section, the day-ahead operation planning of an MG system is determined by solving an optimization problem for both normal and emergency operation mode. The proposed model is formulated for 24 h with any uniform interval of time t. However, in the proposed day-ahead scheduling model, t has been assumed to be one hour.

3.1. Normal Mode

3.1.1. Objective Function

The objective of the normal operation is to minimize the operation cost of the MG system, as shown in Equation (6). The first term of the objective function contains generation cost, start-up cost, and shut-down cost of DGs. The second term contains profit gained by trading electricity with the utility grid:

$$
\min \sum_{t=1}^{T} \sum_{g=1}^{G} \left(C_g^{DG} \left(P_{g,t}^{DG} \right) + y_{g,t} \cdot C_g^{SU} + z_{g,t} \cdot C_g^{SD} \right)
$$
$$
+ \sum_{t=1}^{T} PR_t^{Buy} \cdot P_t^{Buy} - \sum_{t=1}^{T} PR_t^{Sell} \cdot P_t^{Sell}
$$

(6)

In this paper, the operation cost of each DG unit is represented by a quadratic cost function, as shown in Equation (4).

3.1.2. Constraints for Operation of MG System

Power generated by RDGs, DG units, discharged amount, and buying amount from the utility grid should be balanced with load, charging amounts, and selling amount from the utility grid at each interval, as shown by (7):

$$
\sum_{r}^{R} P_{r,t}^{RDG} + \sum_{g}^{G} P_{g,t}^{DG} + P_t^{BD} - P_t^{BC} + P_t^{Buy} - P_t^{Sell} = \sum_{l}^{L} P_{l,t}^{Load}
$$

(7)

Equations (8)–(11) show the constraints for each diesel generator units. Equation (8) represents the operation bounds of DG unit g at time t. The on-off mode of DG is determined by Equation (9). The start-up and shutdown status is determined based on the on-off mode of each DG unit, as shown in Equations (10) and (11). Equations (12) and (13) depicts the ramp-up and ram-down constraints for gth DG unit:

$$
u_{g,t} \cdot P_{g,\min}^{DG} \leq P_{g,t}^{DG} \leq u_{g,t} \cdot P_{g,\max}^{DG}
$$

(8)

where

$$
u_{g,t} = \begin{cases} 1 & \text{DG is on} \\ 0 & \text{DG is off} \end{cases}
$$

(9)

$$
y_{g,t} = \max\{ (u_{g,t} - u_{g,t-1}), 0 \}
$$

(10)

$$
z_{g,t} = \max\{ (u_{g,t-1} - u_{g,t}), 0 \}
$$

(11)

$$
P_{g,t}^{DG} - P_{g,t-1}^{DG} \leq RU_g \cdot (1 - y_{g,t}) + P_{g,\min}^{DG} \cdot y_{g,t}
$$

(12)

$$
P_{g,t-1}^{DG} - P_{g,t}^{DG} \leq RD_g \cdot (1 - z_{g,t}) + P_{g,\min}^{DG} \cdot z_{g,t}
$$

(13)

The BESS model in MG system can be represented by using Equations (14)–(17). The bounds of charging/discharging amount is given by Equations (14) and (15). Each interval, the state of charge (SOC) of BESS is updated according to the charging/discharging amount and previous interval's SOC, as given by (16). The SOC of the BESS at any time interval t is constrained by Equation (17):

$$
0 \leq P_t^{BC} \leq P_{cap}^{B} \cdot \left(1 - SOC_{t-1}^{B} \right) \cdot \frac{1}{1 - P_{loss}^{BC}}
$$

(14)

$$
0 \leq P_t^{BD} \leq P_{cap}^{B} \cdot SOC_{t-1}^{B} \cdot \left(1 - P_{loss}^{BD} \right)
$$

(15)

$$
SOC_t^{B} = SOC_{t-1}^{B} - \frac{1}{P_{cap}^{B}} \cdot \left(\frac{1}{1 - P_{loss}^{BD}} \cdot P_t^{BD} - \left(1 - P_{loss}^{BC} \right) \cdot P_t^{BC} \right)
$$

(16)

$$SOC_{min}^B \leq SOC_t^B \leq SOC_{max}^B \tag{17}$$

3.2. Emergency Mode

The scheduling horizon of emergency operation is from the event occurrence time (at interval h) to the end of the day (at interval T). In this mode, the MG-EMS will reschedule for all components in the normal part. The components in isolated part perform distributed optimization by using diffusion strategy.

3.2.1. Objective Function

Whenever a fault occurs, the corresponding circuit breaker (CB) is opened to isolate the fault. By using the information of fault location, which is sent from the corresponding CB, each single agent is able to know that it is electrically islanded or not from the MG. Therefore, the number of agents is determined in each part. In normal part, all components are rescheduled by EMS. In isolated part, each agent determines its new operation point by using diffusion strategy. The cost objective function of the MG system in emergency operation is given by Equation (18). The first term of Equation (18) represents the operation cost, start-up cost, and shut-down cost of DG units in the normal part. The second term of Equation (18) shows profit of exchanging electric power between the utility grid. The total operation cost of isolated part is presented by Equation (19), which includes the operation cost, start-up cost, shut-down cost of DG units, and the penalty of load shedding in this system. The output power of DGs and load shedding amount are determined by using diffusion strategy.

$$\min \sum_{t=h}^{T} \sum_{g_1=1}^{G_1} \left(C_{g_1}^{DG} \left(P_{g_1,t}^{DG} \right) + y_{g_1,t} \cdot C_{g_1}^{SU} + z_{g_1,t} \cdot C_{g_1}^{SD} \right)$$
$$+ v_t \cdot \left(\sum_{t=h}^{T} PR_t^{Buy} \cdot P_t^{Buy} - \sum_{t=h}^{T} PR_t^{Sell} \cdot P_t^{Sell} \right) \tag{18}$$

where
$$v_t = \begin{cases} 1 & \text{in grid} - \text{connected mode} \\ 0 & \text{in islanded mode} \end{cases}$$

$$OC^{Isolated\ part} = \sum_{t=h}^{T} \sum_{g_2=1}^{G_2} \left(C_{g_2}^{DG} \left(P_{g_2,t}^{DG} \right) + y_{g_2,t} \cdot \left| C_{g_2}^{SU} + z_{g_2,t} \cdot C_{g_2}^{SD} \right) + \sum_{t=h}^{T} C_t^{Pen} \cdot P_t^{Sh} \tag{19}$$

3.2.2. Constraints for Operation of MG System

Equation (20) shows that the power generated by RDGs, DGs, BESS discharging amount, and power bought from the utility grid should be balanced with charging amount, load amount, and the amount of power sold to the utility grid. Similarly, the power balancing between supplies and loads is given by Equation (22) for isolated part considering the load shedding in peak intervals. In addition to Equations (20)–(22), Equation (18) is also constrained to Equations (8)–(17).

$$\sum_{r_1}^{R_1} P_{r_1,t}^{RDG} + \sum_{g_1}^{G_1} P_{g_1,t}^{DG} + k_t \cdot \left(P_t^{BD} - P_t^{BC} \right) + v_t \cdot \left(P_t^{Buy} - P_t^{Sell} \right) = \sum_{l_1}^{L_1} P_{l_1,t}^{Load} \tag{20}$$

where
$$k_t = \begin{cases} 1 & \text{BESS in normal part} \\ 0 & \text{BESS in isolated part} \end{cases} \tag{21}$$

$$\sum_{r_2}^{R_2} P_{r_2,t}^{RDG} + \sum_{g_2}^{G_2} P_{g_2,t}^{DG} + (1 - k_t) \cdot \left(P_t^{BD} - P_t^{BC} \right) = \sum_{l_2}^{L_2} P_{l_2,t}^{Load} - P_t^{Sh} \tag{22}$$

4. Numerical Simulations

In this study, the test MG system considered for simulations is similar to Figure 1. Firstly, the normal operation mode is considered and the MG system is operated based on day-ahead scheduling. A fault is considered at interval 10, the fault occurs during a very short time. It is isolated as soon as possible by opening the corresponding circuit breakers. In this way, some parts of the MG could be isolated from main system and could not communicate with the MG-EMS. Therefore, the MG system is changed to emergency mode operation. The proposed strategy operation is used to operate the MG system in an economical way for both normal and isolated parts after isolating the fault. The isolated part is operated by using diffusion strategy while normal part is operated by MG-EMS. At interval 15, one part is recovered from the event. The number of agents in normal/isolated part is updated and the normal part is rescheduled by MG-EMS while the isolated part is rescheduled by using diffusion strategy. Finally, at interval 20, the event is fully cleared and the entire MG system is rescheduled by MG-EMS to the end of day ($t = 24$). The proposed model has been implemented on a computer with an Intel(R) Core i5(TM) 2500 CPU @ 3.30 GHz and 8 GB of RAM memory using Java (Oracle Corporation, Redwood City, CA, USA), JADE (Oracle Corporation, Redwood City, CA, USA) with integration of IBM ILOG CPLEX (International Business Machines Corporation, Armonk, NY, USA).

4.1. Input Parameters

Figure 7 shows the hourly generation amount of RDGs, the hourly electric loads of the MG system, and the market price signals, which are taken as input data. The maximum value and the initial value of BESS are 200 kWh and 50 kWh, respectively. The charging/discharging loss of BESS is 5%. The parameters related to DG units of the MG system are also shown in Table 1.

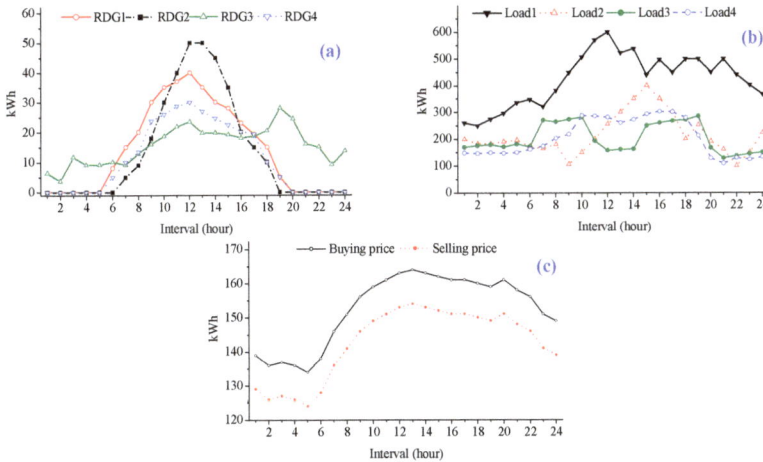

Figure 7. Input data: (**a**) renewable generations; (**b**) load amount; and (**c**) market price signals.

Table 1. Parameters related to DG units and BESS of the microgrid system.

Parameters	DG 1	DG 2	DG 3	DG 4	DG 5	DG 6	DG 7
Min.	0	0	0	0	0	0	0
Max.	150	150	100	200	200	150	100
a	561	310	300	561	310	300	570
b	7.92	7.88	7.9	7.92	7.85	7.9	7.98
c	0.00125	0.00194	0.00198	0.00125	0.002	0.0025	0.0014

4.2. Piecewise Method Linearization

The cost objective functions (6) and (18) are nonlinear, therefore piecewise linearization method has been used to transform them into linear counterparts. As shown in Figure 8, the cost function of generator is approximated by using a set of piecewise blocks. The analytic representation of this linear approximation is [28,29]:

$$
C_g^{DG}\left(P_{g,t}^{DG}\right) = \frac{C_g^{DG}\left(P_{g,1}^{DG}\right) - C_g^{DG}\left(P_{g,\min}^{DG}\right)}{P_{g,1}^{DG} - P_{g,\min}^{DG}} \cdot \lambda_1(t) + \frac{C_g^{DG}\left(P_{g,2}^{DG}\right) - C_g^{DG}\left(P_{g,1}^{DG}\right)}{P_{g,2}^{DG} - P_{g,1}^{DG}} \cdot \lambda_2(t) + \cdots
$$
$$
+ \frac{C_g^{DG}\left(P_{g,\max}^{DG}\right) - C_g^{DG}\left(P_{g,n}^{DG}\right)}{P_{g,\max}^{DG} - P_{g,n}^{DG}} \cdot \lambda_n(t)
\tag{23}
$$

where $\dfrac{C_g^{DG}\left(P_{g,i}^{DG}\right) - C_g^{DG}\left(P_{g,i-1}^{DG}\right)}{P_{g,i}^{DG} - P_{g,i-1}^{DG}}$ is considered as constant cost in segment i

$$
0 \le \lambda_1(t) \le P_1 - P_{\min}
\tag{24}
$$

$$
0 \le \lambda_2(t) \le P_2 - P_1
\tag{25}
$$

$$
0 \le \lambda_n(t) \le P_{\max} - P_n
\tag{26}
$$

$$
P_{g,t}^{DG} = \lambda_1(t) + \lambda_1(t) + \ldots + \lambda_n(t)
\tag{27}
$$

The generation cost is approximated by Equation (23), where the generation amount is divided into many segments between $P_{\min}^{DG}$, $P_{\max}^{DG}$. The per unit generation cost is consedered as a constant value in each segment, which is calculated as the slope of the approximated generation cost cuver in each segment. The approximated generation cost is shown in Figure 8 (red curve). Equations (24)–(26) represent the generation amount of gth DG at interval t in each segment. Finally, the total output power of the DG is calculated by Equation (27).

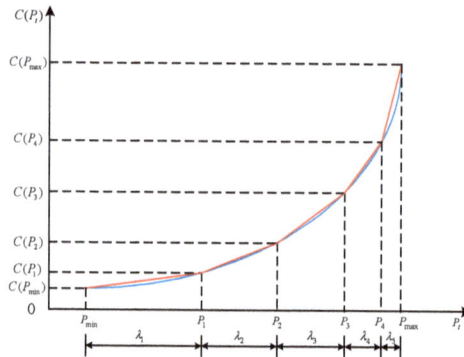

Figure 8. Piecewise linear generation cost of DG units.

4.3. Normal Operation Mode

In normal operation mode, the entire MG system is operated by MG-EMS based on day-ahead scheduling for minimizing the total operation cost. The simulated scenario for evaluating the performance of proposed strategy is shown in Figure 9. It can be observed from Figure 9a that the generation amount of DGs is determined by the comparison between generation cost of DG units and market price signals. The DG units having lower generation costs (DG1–DG4, and DG7) are always set to maximum. During off-peak price intervals (intervals 1–6), the DG units having higher generation costs (DG5 and DG6) are set to minimum and the shortage amount is fulfilled by buying electricity from the utility grid in order to minimize the MG's operation cost. In peak price intervals (intervals 8–10), DGs are set to maximum and surplus is sold to the utility grid to increase the profit. The amount of exchanged power with the utility grid is depicted in Figure 9b. Electricity is bought from the utility grid for fulfilling the shortage power and avoiding the use of expensive resources. Similarly, electricity is sold to the utility grid to increase the profit by selling electric power from the cheap resources. BESS are charged during the off-peak intervals (intervals 5 and 6) and are discharged during the peak price intervals (intervals 12–14) as shown in Figure 9c. The BESS is used either to fulfill the local demand of MG system or to trade power with the utility grid.

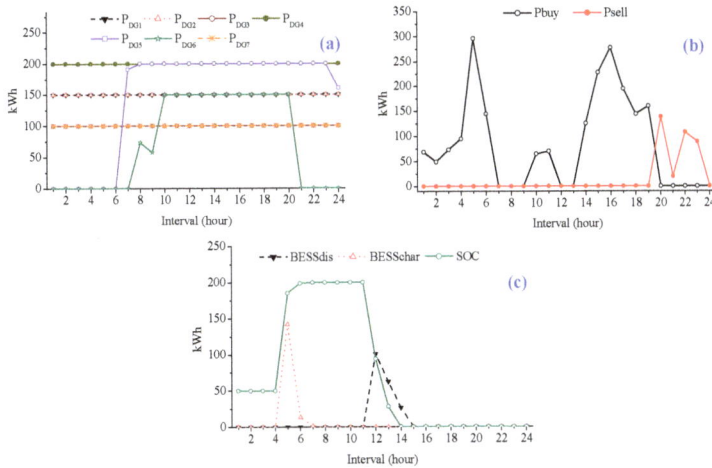

Figure 9. Operation results of normal case: (**a**) output power of DG units; (**b**) the amount of exchanging power; and (**c**) the charging/discharging amount and the SOC of BESS.

4.4. Emergency Operation Mode

In emergency mode, two cases are simulated. In the first case, an event is considered at interval 10, point N1, which divides the MG system into two parts: normal part and isolated part, (Figure 10). In the second case, at interval 15, one part is recovered from the event (Figure 14).

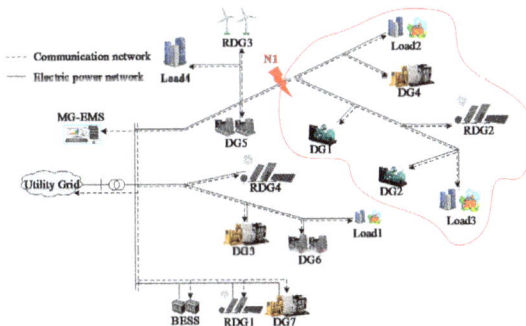

Figure 10. Case 1: Simulated scenario for evaluating proposed strategy during emergency mode operation.

4.4.1. Case 1: Event at Interval 10 at N1

According to the proposed algorithm, the normal part (outside the red area) in MG system is rescheduled by MG-EMS from interval 10 to 24. The output power of DG units (DG3, and DG5–DG7) is rescheduled from interval 10 to 24, as shown in Figure 11a. Figure 11b shows the amount of exchanging power with the utility grid. Due to the isolation of DG units (DG1, DG2, and DG4), the generated amount from these DG units cannot supply to the demand of normal part in the MG system. Therefore, the amount of buying power is increased to fulfill the shortage power in the MG system while the amount of selling power is decreased to zero. BESS is rescheduled with the new initial value of SOC, which is taken at occurrence time of the event, i.e., $SOC_{initial} = SOC(10)$. In this case, the amount of shortage in the system is high. Therefore, the BESS is only discharged to reduce the amount of buying power from the utility grid.

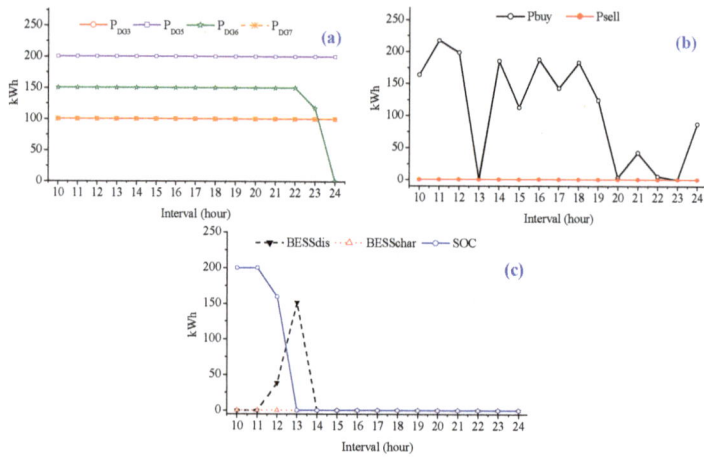

Figure 11. Operation results of normal part: **(a)** output power of DG units; **(b)** the amount of exchanging power; and **(c)** the charging/discharging amount and the SOC of BESS.

The faulty part (in the red area) is considered as a distributed system. The output power of DG units (DG1, DG2, and DG4) should fulfill the load amount in this area considering three different failure scenarios, as mentioned in Section 2.2. In Scenario 2, the faulty part (power failure part) is also rescheduled by MG-EMS without power sharing between the two areas. The output power of DG units is shown in Figure 12a for minimizing the operation cost of the entire system. To maintain the power balance in this area, in some peak intervals (15–17, and 19), load shedding should be implemented. The amount of load shedding is shown in Figure 12b for maintaining the power balance.

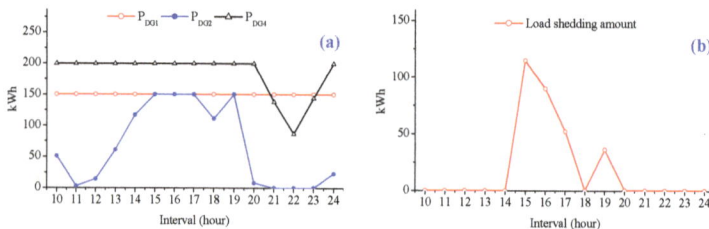

Figure 12. Operation results of power failure part for remaining intervals (rescheduling by MG-EMS): **(a)** output power of DG units; and **(b)** load shedding amount.

To reschedule for this area with Scenarios 1 and 3, each agent shares its information with its neighbor agents to get the shortage amount in the system. By applying Algorithm 1, the shortage power and the new operation point of these DG units are determined for interval 10 (at occurrence time of the event), as shown in Figure 13. The shortage power information of all agents converges to the value 66.833 kW, which is the average of the shortage amount requirement. The shortage power (PN) is equal to the difference between total loads (load 1 and load 2) and the output power of RDG2. Based on the information of the shortage amount in the distributed system, the output power of DG units is determined to maintain the power balance and to minimize the operation cost. At interval 10, the output power of DG1, DG2 and DG4 is 148 kW, 105 kW and 148 kW, respectively. The output power of each DG depends on the operation cost of that DG.

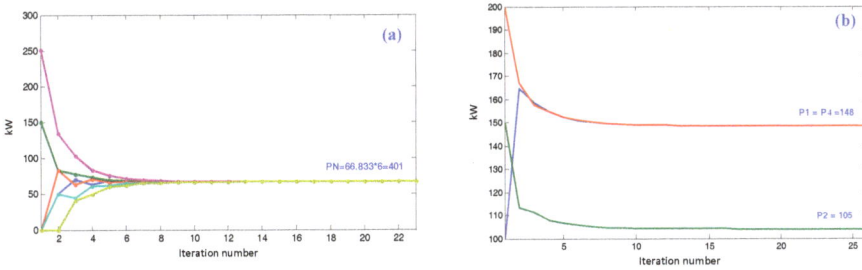

Figure 13. Operation results of isolated part at interval 10: (**a**) information sharing (shortage amount); and (**b**) distributed optimization.

Similarly, the output power of these DG units is rescheduled by applying the proposed algorithm for remaining intervals (from interval 10 to 24), as shown in Figure 14a. The output power of DG units is determined by sharing information among agents to minimize the total operation cost and maintain the balance of supply and demand. In off-peak intervals, generation amount can fulfill the load amount. However, in peak intervals (intervals 14–17, and 19), although all DG units are set to their maximum values, they cannot fulfill all loads. To maintain the power balance in the distributed system, load shedding should be implemented. The amount of load shedding is shown in Figure 14b.

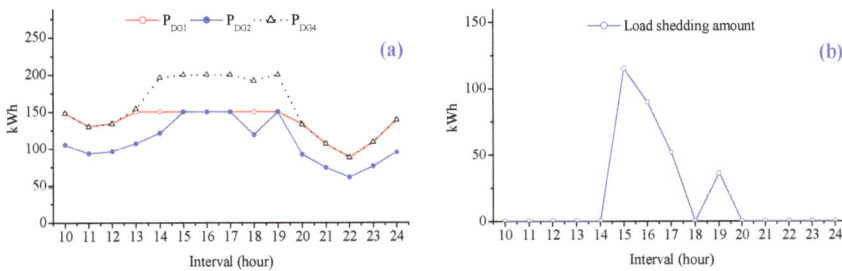

Figure 14. Operation results of isolated part for remaining intervals: (**a**) output power of DG units; and (**b**) load shedding amount.

4.4.2. Case 2: Recovered One Part from the Event at Interval 15

In this case, we assume that some equipment (DG4 and load 2) are recovered from the event at interval 15 as shown in Figure 15. According to the proposed algorithm, the normal part is updated with new recovered equipment and rescheduled by MG-EMS from interval 15 to 24. The amount of DG units is set based on the optimal values from MG-EMS, as shown in Figure 16a. Due to the reconnection of cheap resource (DG4), the output power of expensive resource (DG6) has been decreased to minimize the operation cost in off-peak intervals (intervals 22 and 23). The shortage power is fulfilled by buying electric power from the utility gird while the surplus power is sold for getting profit in some off-peak intervals. The amount of exchanging power with the utility grid is depicted in Figure 16b. At interval 15, BESS is fully discharged (SOC(15) = 0) while the price for charging is high. Therefore, the SOC of BESS is set to zero for reaming intervals.

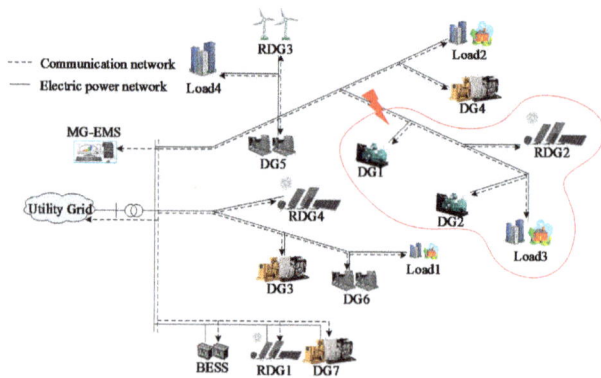

Figure 15. Case 2: Simulated scenario for evaluating proposed strategy during emergency mode operation.

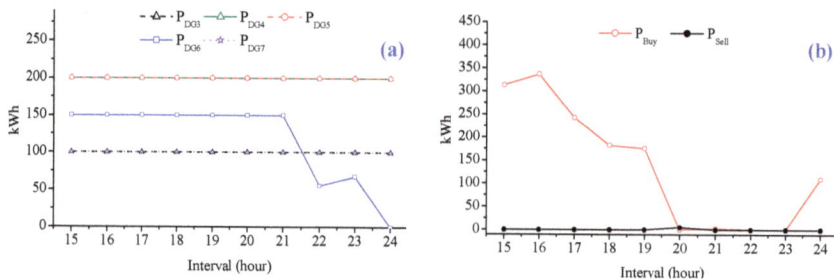

Figure 16. Operation results of normal part: (**a**) output power of DG units; and (**b**) the amount of exchanging power.

In the distributed system, the number of agents is updated considering the number of recovered equipment (agents). Similarly, in Scenario 2, the MG-EMS reschedules to minimize total operation cost for both normal and power failure areas. The output power of DG units for remaining intervals is illustrated by Figure 17d. The DG (DG1) having low operation cost is always set to the maximum value before using the DG (DG2) having higher operation cost. In Scenarios 1 and 3, each agent shares its information with its neighbor agents to get the information of shortage amount in the system. At interval 15, the amount of shortage power came out to be PN that is equal to the difference between load 3 and the output power of RDG2, as shown in Figure 17a. After determining the shortage amount, the proposed distributed optimization method is used to determine the output power of DG units (DG1 and DG2). The output power of DG1 and DG2 is decided to maintain the power balance in the isolated part and to minimize the operation cost, as given in Figure 17b. Similarly, the DG units are rescheduled for all remaining intervals (intervals 15–24) as shown in Figure 17c. In this case, the generated amount can fulfill the load amount in all remaining intervals. Therefore, the amount of load shedding is reduced to zero.

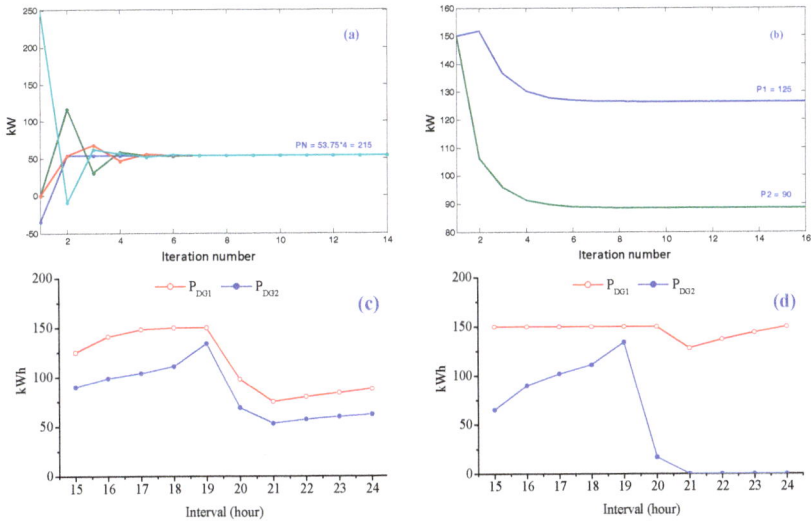

Figure 17. Operation results of isolated part at interval 15: (**a**) information sharing (shortage amount); (**b**) distributed optimization; (**c**) the output power of DG units for remaining intervals; and (**d**) the output power of DG units for remaining intervals (rescheduling by MG-EMS).

4.5. Fault Recovery at Interval 20

At interval 20, all isolated part is recovered from the event. The schedules of all components are also recovered, similar to normal operation. The MG system is operated by MG-EMS based on day-ahead scheduling. Therefore, the operation of DG units, the amount of exchanging power, the amount of BESS charging/discharging, and the SOC of BESS are similar to Figure 9a–c from interval 20 to the end of day.

4.6. Comparison between Consensus Algorithm and Diffusion Strategy

In this section, the comparison between consensus algorithm and diffusion strategy is presented to show the advantage of the proposed algorithm. In the isolated part, the DG units are rescheduled by using the proposed algorithm based on the diffusion strategy for each case of emergency operation mode, which is shown in Figures 12 and 16. Figure 18a,b shows the output power of DG units by using the consensus algorithm in the isolated part for interval 10 (Case 1) and interval 15 (Case 2). By comparing the results of the consensus algorithm and the proposed diffusion strategy, it can be conclude that the proposed strategy has converged faster than the consensus algorithm.

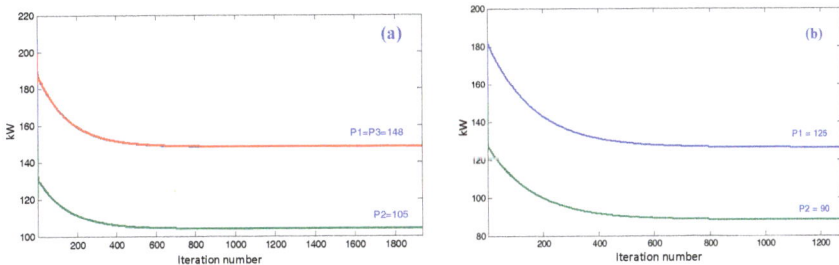

Figure 18. Operation results of isolated part using consensus algorithm: (**a**) Case 1; and (**b**) Case 2.

Table 2 shows the summary of the results of both the consensus algorithm and the diffusion strategy for first interval operation of distributed system. In the Case 1, the number of agents in distributed system is 6. The reductions of iteration and calculation time are 97.4% and 31.25%, respectively, compared with the consensus algorithm. In Case 2, due to the recovered equipment, the number of agents is reduced to 4. The reductions of iteration and calculation time are 97.6% and 28.6%, respectively, compared with the consensus algorithm. Therefore, it can be conclude that the diffusion strategy is better when the size of the distributed system is increasing.

Table 2. The comparison for applying consensus algorithm and diffusion strategy.

Parameters	Case 1		Case 2	
	Consensus Algorithm	Diffusion Strategy	Consensus Algorithm	Diffusion Strategy
The number of agents	6	6	4	4
The number of iterations	1900	49	1250	30
Calculation time (s)	0.16	0.11	0.14	0.1
Reduction of iteration (%)	0	97.4	0	97.6
Reduction of calculation time (%)	0	31.25	0	28.6

It can be observed in Table 2 that, in a small distributed system, the calculation time of both diffusion and consensus approaches are very fast. However, when the size of distributed system is increased, consensus algorithm can lead to drastic increase in number of iterations while the numbers of iterations are gradually increasing for the diffusion strategy. Therefore, in the case of a large system, the impact of the proposed method is more significant compared with the consensus algorithm [17].

5. Conclusions

A novel operation strategy for enhancing the reliability of microgrids is proposed using diffusion strategy. The MG system can operate in an economic way in both normal and emergency operation modes. In normal operation mode, the MG system is operated by MG-EMS for minimization of operation cost. In emergency operation mode, the MG system is divided into normal and isolated parts. Normal part is still operated by MG-EMS while the isolated part is considered as a distributed system. The proposed strategy maintains this part, which is also operated in an economic way by using diffusion strategy for minimizing global cost without a central controller. The numerical results have shown that a multiagent system based on the diffusion strategy has a desirable performance compared with consensus method and can be easily applied for microgrid optimization. During emergency operation, in isolated part, the number of iterations and the calculation time was reduced by 97% and 28.6%, respectively, as compared with the consensus algorithm. The proposed operation strategy is suitable to apply for a distributed system. By applying the proposed strategy, the system can be operated in an economical way without an energy management system.

Another application of the proposed method is in large-scale microgrids, where, after the occurrence of a fault, the isolated part could be out of service. It leads to a large amount of load to be interrupted, thus, the proposed strategy can be used to solve this problem by using diffusion strategy. In this way, each agent can communicate with neighboring agents to determine a new operation point.

Acknowledgments: This work was partially supported by the Power Generation & Electricity Delivery Core Technology Program of the Korea Institute of Energy Technology Evaluation and Planning (KETEP), granted financial resource from the Ministry of Trade, Industry & Energy, Republic of Korea. (No. 20141020402350) and partially supported by the Korea Institute of Energy Technology Evaluation and Planning (KETEP) and the Ministry of Trade, Industry & Energy (MOTIE) of the Republic of Korea (No. 20151210200080).

Author Contributions: The paper was a collaborative effort between the authors. The authors contributed collectively to the theoretical analysis, modeling, simulation, and manuscript preparation.

Conflicts of Interest: The authors declare no conflict of interest.

Abbreviations

t	Index of time, running from 1 to T
g, g_1, g_2	Index for total, normal part, and isolated part DGs, respectively ($g \in g_1, g_2$)
l, l_1, l_2	Index for total, normal part, and isolated part loads, respectively ($l \in l_1, l_2$)
r, r_1, r_2	Index for total, normal part, and isolated part DRGs, respectively ($r \in r_1, r_2$)
$u_{g,t}, y_{g,t}, z_{g,t}$	Commitment, startup, and shutdown status identifier of gth DG at t
$C_g^{DG}\left(P_{g,t}^{DG}\right)$	Generation cost of DG unit g at t
C_g^{SU}, C_g^{SD}	Start-up and shutdown cost of DG unit g at t
PR_t^{Buy}, PR_t^{Sell}	Price for buying and selling power to/from the utility grid at t
$P_{g,t}^{DG}$	Amount of power generated by DG g at t
P_t^{Buy}, P_t^{Sell}	Total amount of power bought from and sold to the utility grid at t
$P_{l,t}^{Load}$	Electric load l of microgrid at t
$P_{r,t}^{RDG}$	Amount of power generated by RDG unit r at t
$P_{g,\min}^{DG}, P_{g,\max}^{DG}$	Minimum and maximum generation amount of gth DG unit
RU_g, RD_g	Ramp-up and Ramp-down time of gth DG unit
P_t^{BC}, P_t^{BD}	Amount of electrical energy charged/discharged to/from BESS at t
$P_{loss}^{BC}, P_{loss}^{BD}$	Charging and discharging losses of BESS unit
P_{cap}^{B}, SOC_t^{B}	Capacity and SOC of BESS unit
$SOC_{\min}^{B}, SOC_{\max}^{B}$	Lower and upper limits for SOC of BESS unit
P_t^{Sh}, C^{Pen}	Total amount of shed load and penalty cost for load shedding at t
$\lambda_i(t)$	The generation amount of DG in segment i at t
P_i, P_{i-1}	The value to determine the segment i from P_i to P_{i-1}

References

1. Hatziargyriou, N.; Asano, H.; Iravani, R.; Marnay, C. Microgrids. *IEEE Power Energy Mag.* **2007**, *5*, 78–94. [CrossRef]
2. Kim, H.M.; Lim, Y.; Kinoshita, T. An intelligent multiagent system for autonomous microgrid operation. *Energies* **2012**, *5*, 3347–3362. [CrossRef]
3. Bui, V.H.; Hussain, A.; Kim, H.M. A multiagent-based hierarchical energy management strategy for multi-microgrids considering adjustable power and demand response. *IEEE Trans. Smart Grid* **2016**. [CrossRef]
4. Tsikalakis, A.G.; Hatziargyriou, N.D. Centralized control for optimizing microgrids operation. *IEEE Trans. Energy Convers.* **2008**, *23*, 241–248. [CrossRef]
5. Conti, S.; Nicolosi, R.; Rizzo, S.A.; Zeineldin, H.H. Optimal dispatching of distributed generators and storage systems for MV islanded microgrids. *IEEE Trans. Power Deliv.* **2012**, *27*, 1243–1251. [CrossRef]
6. Olivares, D.E.; Cañizares, C.A.; Kazerani, M. A centralized energy management system for isolated microgrids. *IEEE Trans. Smart Grid* **2014**, *5*, 1864–1875. [CrossRef]
7. Song, N.O.; Lee, J.H.; Kim, H.M.; Im, Y.H.; Lee, J.Y. Optimal energy management of multi-microgrids with sequentially coordinated operations. *Energies* **2015**, *8*, 8371–8390. [CrossRef]
8. Colson, C.M.; Nehrir, M.H. Algorithms for distributed decision-making for multi-agent microgrid power management. In Proceedings of the 2011 IEEE Power and Energy Society General Meeting, Detroit, MI, USA, 24–29 July 2011; pp. 1–8.
9. Cha, H.J.; Won, D.J.; Kim, S.H.; Chung, I.Y.; Han, B.M. Multi-agent system-based microgrid operation strategy for demand response. *Energies* **2015**, *8*, 14272–14286. [CrossRef]
10. Lynch, N.A. *Distributed Algorithms*; Morgan Kaufmann Publishers Inc.: San Francisco, CA, USA, 1996.
11. Ren, W.; Beard, R.W.; Atkins, E.M. A survey of consensus problems in multi-agent coordination. In Proceedings of the 2005 American Control Conference, Portland, OR, USA, 8–10 June 2005; pp. 1859–1864.
12. Meng, L.; Dragicevic, T.; Guerrero, J.M.; Vasquez, J.C. Dynamic consensus algorithm based distributed global efficiency optimization of a droop controlled DC microgrid. In Proceedings of the 2014 IEEE International Energy Conference (ENERGYCON), Cavtat, Croatia, 13–16 May 2014; pp. 1276–1283.

13. Xu, Y.; Li, Z. Distributed optimal resource management based on the consensus algorithm in a microgrid. *IEEE Trans. Ind. Electron.* **2015**, *62*, 2584–2592. [CrossRef]
14. Hug, G.; Kar, S.; Wu, C. Consensus + innovations approach for distributed multiagent coordination in a microgrid. *IEEE Trans. Smart Grid* **2015**, *6*, 1893–1903. [CrossRef]
15. Zhang, Z.; Chow, M.Y. Convergence analysis of the incremental cost consensus algorithm under different communication network topologies in a smart grid. *IEEE Trans. Power Syst.* **2012**, *27*, 1761–1768. [CrossRef]
16. Hussain, A.; Bui, V.H.; Kim, H.M. A resilient and privacy-preserving energy management strategy for networked microgrids. *IEEE Trans. Smart Grid* **2016**. [CrossRef]
17. De Azevedo, R.; Cintuglu, M.H.; Ma, T.; Mohammed, O.A. Multi-agent based optimal microgrid control using fully distributed diffusion strategy. *IEEE Trans. Smart Grid* **2017**, *8*, 1997–2008. [CrossRef]
18. Laaksonen, H.; Kauhaniemi, K. Fault type and location detection in islanded microgrid with different control methods based converters. In Proceedings of the 19th International Conference on Electricity Distribution (CIRED), Vienna, Australia, 21–24 May 2007; pp. 372–376.
19. Ali, H.; Reza, I. Microgrid Protection. *Proc. IEEE* **2017**, *105*, 1332–1353.
20. Sayed, A.H. Adaptive networks. *Proc. IEEE* **2014**, *102*, 460–497. [CrossRef]
21. Chen, J.; Sayed, A.H. Diffusion adaptation strategies for distributed optimization and learning over networks. *IEEE Trans. Signal Process.* **2012**, *60*, 4289–4305. [CrossRef]
22. Chen, J.; Sayed, A.H. Distributed Pareto optimization via diffusion strategies. *IEEE J. Sel. Top. Signal Process.* **2013**, *7*, 205–220. [CrossRef]
23. FIPA. The Foundation for Intelligent Physical Agents Standards. Available online: http://www.fipa.org (accessed on 30 June 2017).
24. Liu, Y.; Hou, X.; Wang, X.; Lin, C.; Guerrero, J.M. A coordinated control for photovoltaic generators and energy storages in low-voltage AC/DC hybrid microgrids under islanded mode. *Energies* **2016**, *9*, 651. [CrossRef]
25. Wang, Z.; Wang, J. Self-healing resilient distribution systems based on sectionalization into microgrids. *IEEE Trans. Power Syst.* **2015**, *30*, 3139–3149. [CrossRef]
26. Bidram, A.; Davoudi, A. Hierarchical structure of microgrids control system. *IEEE Trans. Smart Grid* **2012**, *3*, 1963–1976. [CrossRef]
27. Yazdanian, M.; Mehrizi-Sani, A. Distributed control techniques in microgrids. *IEEE Trans. Smart Grid* **2014**, *5*, 2901–2909. [CrossRef]
28. Carrión, M.; Arroyo, J.M. A computationally efficient mixed-integer linear formulation for the thermal unit commitment problem. *IEEE Trans. Power Syst.* **2006**, *21*, 1371–1378. [CrossRef]
29. Tenfen, D.; Finardi, E.C.; Delinchant, B.; Wurtz, F. Lithium-ion battery modelling for the energy management problem of microgrids. *IET Gener. Transm. Distrib.* **2016**, *10*, 576–584. [CrossRef]

![energies logo] *energies*

MDPI

Article

Optimal Energy Management for Microgrids with Combined Heat and Power (CHP) Generation, Energy Storages, and Renewable Energy Sources

Guanglin Zhang [1,*], Yu Cao [1], Yongsheng Cao [1], Demin Li [1] and Lin Wang [2,*]

[1] College of Information Science and Technology, Engineering Research Center of Digitized Textile and Fashion Technology, Ministry of Education, Donghua University, Shanghai 201620, China; yucao@mail.dhu.edu.cn (Y.C.); yongshengcao@mail.dhu.edu.cn (Y.C.); deminli@dhu.edu.cn (D.L.)
[2] Department of Automation, Shanghai Jiaotong University, Shanghai 200240, China
*   Correspondence: glzhang@dhu.edu.cn (G.Z.); wanglin@sjtu.edu.cn (L.W.); Tel.: +86-21-6779-2332 (G.Z.)

Academic Editor: Pedro Faria
Received: 26 July 2017; Accepted: 25 August 2017; Published: 29 August 2017

Abstract: This paper studies an energy management problem for a typical grid-connected microgrid system that consists of renewable energy sources, Combined Heat and Power (CHP) co-generation, and energy storages to satisfy electricity and heat demand simultaneously. We formulate this problem into a stochastic non-convex optimization programming to achieve the minimum microgrid's operating cost, which is difficult to solve due to its non-convexity and coupling feature of constraints. Existing approaches such as dynamic programming (DP) assume that all the system dynamics are known, which results in a high computational complexity and thus are not feasible in practice. The focus of this paper is on the design of a real-time energy management strategy for the optimal operation of microgrids with low computational complexity. Specifically, derived from a modified Lyapunov optimization technique, an online algorithm with random inputs (e.g., the charging/discharging of energy storage devices, power from the CHP system, the electricity from external power grid, and the renewables generation, etc.), which requires no statistic system information, is proposed. We provide an implementation of the proposed energy management algorithm and prove its optimality theoretically. Based on real-world data traces, extensive empirical evaluations are presented to verify the performance of our algorithm.

Keywords: microgrids; renewable energy; storage; scheduling; co-generation

---

1. Introduction

Microgrids stand a good chance of becoming a future power grid paradigm that uses centralized power grids as well as local generated energy [1]. They can be operated with or without a grid connection. Microgrids usually consist of distributed renewable energy, decentralized energy storage devices (e.g., PHEVs), a local CHP System (e.g., gas-fired generators), and flexible loads.

With environmental concerns growing, a future power grid is expected to integrate more renewable energy (e.g., solar or wind) to reduce the discharge of greenhouse gas. For instance, the European Commission intends to include 20% renewables into the EU energy profile by 2020 [2], and California aims to get 33% of retail sales from renewables by 2020 [3]. As we know, the generation of renewable energy is intermittent and non-dispatchable. If we simply integrate large amounts of renewable energy, the system will encounter some reliability problems. Besides, renewable energy supply is a stochastic process, which brings a new dimension of uncertainty to energy management. Therefore, how to integrate the generation of renewables efficiently and ensure the reliability of our system simultaneously is of great importance for microgrids.

Energy storage devices are utilized to smooth energy fluctuations and reduce the system cost in a more environmentally friendly way by intelligent charging/discharging, which plays an important role in microgrids [4–7]. Apart from its power management capability, energy storage devices can act as a backup in microgrids when the external grid breaks down, which will reduce the negative effects with a quick response [8,9]. The distributed storage plays a significant role in the design and evolution of a power grid, and particularly increases additional design choices for reducing the operating cost of microgrids [10–12]. The hybrid energy storage system is considered for primary frequency control using a dynamic droop method in an isolated microgrid power system [13]. Online energy management algorithms are developed to investigate the operating cost reduction for microgrids with an energy storage system [14].

Apart from renewables generation and energy storage devices, CHP systems are becoming very popular in the microgrids industry [15,16]. CHP systems can generate both electricity and thermal energy simultaneously, which can achieve a much higher energy efficiency than generating electricity and heat separately [17]. The characteristics of local generations and local consumptions of micrigrid make it more flexible in the utilization of renewable energy and CHP generation, which extends the adaptability of a traditional centralized grid. The power management strategy between different elements should be considered in order to design feasible control algorithms for microgrid systems [18]. Furthermore, with the augmentation of CHP generation technology, microgrids can often be much more economical than the traditional grid by using centralized grid supply and separate heat supply [19,20]. The integration analysis of hybrid energy storage system and novel CHP systems in residential scenarios are also investigated [21].

In this paper, we consider the grid connected microgrid. We aim to propose an intelligent scheduling action (e.g., charging/discharging of energy storage device, power drawn from centralized grid, power obtained from local generator, etc.) to achieve the operating minimum cost of microgrids while considering all the random inputs of the system. We first formulate the problem of achieving the minimum operating cost in microgrids as a stochastic non-convex programming. Considering that the dependence between power level of the battery pack and heat level of the water tank leads to this problem's non-convexity, we study the relationship between them and convert it into stochastic convex optimization programming. Then we adopt the Lyapunov optimization [22] approach to design an online algorithm of some random system inputs (e.g., the charging/discharging of the energy storage devices, power from the local generator, the electricity from the power grid, and the renewable energy generation from different sources, etc.), which requires no statistic information of our system.

The contributions of this paper are summarized as follows:

1. We formulate a stochastic non-convex programming for the online scheduling problem to minimize the microgrid's cost, which captures the randomness in stochastic renewables, power and heat demands, charge level of energy storage, co-generation and physical constraints as well.
2. To solve this stochastic non-convex optimization problem, we convert it into the subproblem with convex property. Then we design an online algorithm to reduce the operating cost of microgrids by using the Lyapunov optimization approach which relies on no future knowledge about the system inputs with stochastic distribution. In this way, we can get the optimal average cost.
3. Through our evaluations by using practical data traces, we can see that by the proposed algorithm, we can achieve an approving empirical optimality ratio.

2. System Model and Problem Statement

The following components are typically included in our designed system: centralized power grid (supply power to the electricity load and charge the battery in an on-site way), large capacity battery (power energy storage), local co-generator (generate both heat and power energy simultaneously), external gas station (supply heat demand), thermal storage device (heat energy storage), and renewables generation (e.g., wind or solar). The system model is shown in Figure 1. For convenience of analysis, we assume that the system operates in discrete time with time slot $t \in \{0, 1, 2, \cdots\}$. We then divide the time slots into frames of size T. Figure 2 gives the time structure of slots and frames. Let T_m denote the set of time slots in time frame m, i.e., $T_m \triangleq \{mT, \cdots, (m+1)T-1\}$. This structure of time slot and time frame is defined to illustrate the time scales of the system operation for easy theoretical analysis. Therefore, we have two time scales in the system.

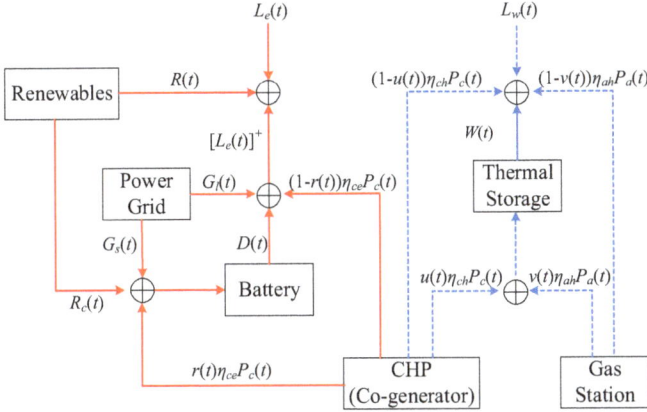

Figure 1. Illustration of the System Model.

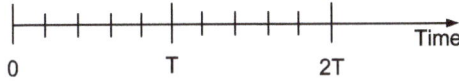

Figure 2. Illustration of time slot and time frame.

2.1. System Model

(1) Local co-generation (CHP): We assume that electricity and heat energy can be generated simultaneously by our local generator. Here we use an idealized model, we will investigate a more practical CHP model in our future work. η_{ce} and η_{ch} are the conversion efficiencies from fuel to the electricity and thermal energy, respectively. At each time slot t, the co-generator generates electricity and thermal energy, whose amounts are denoted as $\eta_{ce}P_c(t)$ and $\eta_{ch}P_c(t)$, respectively. The generated electricity can be used for power supply to directly satisfy the net power demand $(1-r(t))\eta_{ce}P_c(t)$ or be charged into the battery $r(t)\eta_{ce}P_c(t)$. Similarly, the generated thermal energy $\eta_{ch}P_c(t)$ can be used for direct heat supply for users' heat demand $(1-u(t))\eta_{ch}P_c(t)$ or be charged into the thermal tank $u(t)\eta_{ch}P_c(t)$, respectively. $y(m_t)$ represents the on/off decision of the generator: $y(m_t) = 1$ represents switching on and $y(m_t) = 0$ denotes switching off in frame m_t, which $m_t = \lceil t/T \rceil + 1$ is defined as the number of slots in a frame.

(2) Centralized power grid: We assume that the power grid and microgrid are connected. The power can be acquired from the centralized power grid in an on-demand manner to meet electricity demands. The system obtains power in the amount of $G_1(t)$ for satisfying demands directly and the power in the amount of $G_s(t)$ for charging the battery, respectively. $G_{1,max}$ are defined as the upper bound of direct power supply from external power grid and $G_{s,max}$ denotes the upper bound of charging power for the battery from the external power grid, respectively. Then we have $0 \leq G_1(t) + G_s(t) \leq G_{max}$ and $0 \leq G_1(t) \leq G_{1,max}$, $0 \leq G_s(t) \leq G_{s,max}$. Supposing that the power demand can be satisfied by power grid alone, we assume that $L_{e,max} \leq G_{max}$ holds at any time slot, where $L_{e,max}$ is the upper bound of $L_e(t)$.

(3) External gas station: The heat energy from the external gas station, $\eta_{ah}P_a(t)$, can be used for direct heat supply and energy charging of the thermal tank. Using $v(t)$ to denote the fraction for charging, we denote the amount for heat supply and heat charging as $(1 - v(t))\eta_{ah}P_a(t)$ and $v(t)\eta_{ah}P_a(t)$, respectively. Under the online control algorithm that we will propose later, the heat demand can be satisfied with the energy from the co-generation and gas station while the total cost can be minimized in an intelligent way which schedules the energy properly.

(4) Renewable energy: Let $R(t)$ denote the renewable energy harvested at time t. In our model, the renewable energy is used as electricity supply for users first because it is free. If we have excess renewable energy when the power demand has been satisfied, we use this part of energy, which is called $R_c(t)$ to charge into the battery. In addition, the amount of renewable energy harvested in a time slot is bounded, and thus we have $0 \le R(t) \le R_{max}, \forall t \in T$. The excess renewable power that is charged to the battery cannot exceed the total amount of harvested renewable energy. Therefore, we have $0 \le R_c(t) \le [-L_e(t)]^+$, where $[-L_e(t)]^+ = \max\{R(t) - L_e(t), 0\}$. Note that, although the system model we consider in this paper only involves the electricity renewable energy, heat renewable energy is applicable as well.

(5) Power and heat demands: In our microgrid system, the total demand includes the demand for power and heat. $L_e(t)$ represents power demand at time slot t, which must be satisfied once requested. The net power demand $[L_e(t)]^+$, which is the excess of power demand over renewable energy at time slot t, equals the subtraction of power demand and renewable energy, and can be expressed as $[L_e(t)]^+ = \max\{L_e(t) - R(t), 0\}$. Let $L_{e,max}$ denote the maximum net power demand in a time slot, then we have $0 \le [L_e(t)]^+ \le L_{e,max}$. The power can be acquired from power grid, local co-generator as well as the battery, denoted as $G_l(t)$, $(1 - r(t))\eta_{ce}P_c(t)$ and $D(t)$ respectively, to balance $[L_e(t)]^+$. It can be presented as follows:

$$[L_e(t)]^+ = G_l(t) + D(t) + (1 - r(t))\eta_{ce}P_c(t)y(m_t) \tag{1}$$

Similarly, the heat can be acquired from external natural gas station, co-generation as well as the thermal tank, denoted as $(1 - v(t))\eta_{ah}P_a(t)$, $(1 - u(t))\eta_{ch}P_c(t)$ and $W(t)$ (more details about $W(t)$ can be found in the thermal tank model) respectively, to balance the heat demand. Thus, at every time slot, we have:

$$L_w(t) \le (1 - v(t))\eta_{ah}P_a(t) + (1 - u(t))\eta_{ch}P_c(t)y(m_t) + W(t) \tag{2}$$

Let $L_{w,max}$ denote the maximum heat demand in a time slot. An additional constraint $L_{w, max} \le \eta_{ah}P_{a, max}$ has to be added to assure the balance of heat demand and supply at any time slot, where $P_{a, max}$ is the maximum heat output of the external gas station. Here η_{ah} are defined as the conversion efficiencies from gas to the thermal energy. Let parameter $r(t)$ denote the fraction of co-generated power that is used for charging. Then $(1 - r(t))$ denotes the fraction of co-generated power that is used for direct power supply. $u(t)$ is defined as the dispatch ratio from CHP to thermal tank, $v(t)$ denotes the dispatch ratio from thermal source to thermal tank. It should be noted that any stochastic information of the net power demands and heat demands is not required in our proposed algorithm. Here we use "$\le$" instead of "$=$" to insure mathematical rigorous. Actually, it could be "$=$" for both Equations (1) and (2) in our optimization problem. However, in the operation of the optimization problem, it should be consider the feasibility of the mathematical solution. In the algorithm design point of view, there is no difference for "$=$" and "$\le$" for Equation (2). We pointed out that if we change "$\le$" to "$=$" in Equation (2), the solution is the same. In fact, we can also change Equation (1) into "$\le$" and Equation (2) into "$=$" in the problem formulation. It has equivalent solutions for the two optimization problem.

2.2. Battery Model and Thermal Tank Model

(1) Battery model: The dynamics of battery's state of charge (SOC) level $B(t)$ is given as follows:

$$B(t+1) = B(t) - \eta_d D(t) + \eta_c[R_c(t) + G_s(t) + r(t)\eta_{ce}P_c(t)y(m_t)] \tag{3}$$

where η_d stands for discharging efficiency of battery and η_c denotes the charging efficiency of it. We can find that the battery must satisfy constraints of capacity and charge/discharge in any slot t.

$$0 \leq B(t) \leq B_{max}, \ 0 \leq D(t) \leq D_{max}, \ B(t) \cdot D(t) = 0 \tag{4}$$

$$0 \leq G_s(t) + r(t)\eta_{ce}P_c(t)y(m_t) + R_c(t) \leq TC_{char} \tag{5}$$

where B_{max} is the capacity of the battery and D_{max} is the maximum discharging power of the battery in each time slot and TC_{char} is the maximum charging power of the battery in each frame.

(2) Thermal tank model: We utilize a thermal tank to store the excess heat for later use. With the charging and discharging of tank at each time slot, the heat state evolves over time:

$$T(t+1) = T(t) - \eta_\beta W(t) + \eta_\alpha[u(t)\eta_{ch}P_c(t)y(m_t) + v(t)\eta_{ah}P_a(t)] \tag{6}$$

where $T(t)$ is the thermal tank's heat energy state at time slot t. Because heat energy stored in the thermal tank can not exceed the capacity of thermal tank, we have:

$$0 \leq T(t) \leq T_{max}, \ 0 \leq W(t) \leq W_{max}, \ T(t) \cdot W(t) = 0 \tag{7}$$

$$0 \leq v(t)\eta_{ah}P_a(t) + u(t)\eta_{ch}P_c(t)y(m_t) \leq Th_{char} \cdot T \tag{8}$$

Similarly, T_{max} is the upper bound of the thermal tank and W_{max} represents the discharging rate constraint of the thermal tank.

2.3. Problem Statement

System State and Constraint: According to the components described in our system, we define the system state as a state vector Q_t:

$$Q_t \triangleq [L_e(t), L_w(t), R(t), C(t), B(t), T(t)] \tag{9}$$

We assume that Q_t is an i.i.d. process over time. Although some of the elements in Q_t can be arbitrarily correlated, the control decisions at each time slot only depends on current system state Q_t without any future system information.

Through jointly scheduling the power and heat energy storage, centralized power grid, the renewables, and co-generation, our system can realize the goal of minimizing the long-term time-averaged operating cost. In particular, the control vector at time slot t isdefined by:

$$U_t \triangleq [G_l(t), G_s(t), P_c(t), P_a(t), R_c(t), r(t), u(t), v(t)] \tag{10}$$

The total cost of our system includes the cost of power acquired from external power grid, the fuel consumption of the co-generation, and natural gas for generating heat, switching and sunk cost:

$$f(t) = C(t)[G_l(t) + G_s(t)] + C_f P_c(t)y(m_t) + C_g P_a(t) + C_u y(m_t) \tag{11}$$

We denote the real-time electricity price of power grid as $C(t)$, which is bounded by C_{min} and C_{max}. C_{min} and C_{max} is the minimum and maximum electricity price. So we have $C_{min} \leq C(t) \leq C_{max}$. It should be noticed that although $C(t)$ can also be a stochastic process, the statistics will not be depended in our algorithm. In this paper, we set the fuel price C_f and the price of natural gas C_g to be constants at each time slot. Actually, our algorithm is also available in the case that the fuel price and natural gas price are not fixed since our algorithm is based on the current system state which can be known at each time slot.

So far we can formulate our first optimization problem as follows:

$$\text{P1: } \min \lim_{T \to \infty} \frac{1}{T} \sum_{i=0}^{T-1} \mathbb{E}\{f(t)\} \tag{12}$$

subject to

$$[L_e(t)]^+ = G_1(t) + D(t) + (1 - r(t))\eta_{ce}P_c(t)y(m_t) \tag{13}$$

$$L_w(t) \le (1 - v(t))\eta_{ah}P_a(t) + (1 - u(t))\eta_{ch}P_c(t)y(m_t) + W(t) \tag{14}$$

$$B(t+1) = B(t) - \eta_d D(t) + \eta_c[R_c(t) + G_s(t) + r(t)\eta_{ce}P_c(t)y(m_t)] \tag{15}$$

$$T(t+1) = T(t) - \eta_{fi}W(t) + \eta_{ff}[u(t)\eta_{ch}P_c(t)y(m_t) + v(t)\eta_{ah}P_a(t)] \tag{16}$$

$$0 \le B(t) \le B_{max}, \ 0 \le T(t) \le T_{max} \tag{17}$$

$$0 \le G_s(t) + r(t)\eta_{ce}P_c(t)y(m_t) + R_c(t) \le C_{char} \tag{18}$$

$$0 \le v(t)\eta_{ah}P_a(t) + u(t)\eta_{ch}P_c(t)y(m_t) \le Th_{char} \tag{19}$$

$$0 \le u(t) \le 1, \ 0 \le v(t) \le 1 \tag{20}$$

$$0 \le D(t) \le D_{max}, \ 0 \le W(t) \le W_{max} \tag{21}$$

$$G_1(t), G_s(t), P_c(t), P_a(t), R_c(t), r(t), u(t), v(t) \ge 0 \tag{22}$$

At the beginning of each frame, the local generator make a decision on choosing the on/off statement by solving a mixed-integer stochastic optimization program with constraints. We then jointly decide other components $(G_1(t), G_s(t), P_c(t), P_a(t), R_c(t), r(t), u(t), v(t))$ in each time slot.

Solving P1 is challenging. In this paper, we aim to develop an online algorithm which requires no system statistics and is easy to implement.

3. The Co-Generation System Scheduling Algorithm

From the above, we know that P1 is a challenge to solve by the current algorithm due to the non-convex optimization. However, we have found a feasible method to work out a convex optimization problem already. Therefore, in this section, we will change P1 into a convex optimization problem. It is a real-time algorithm derived from the two-timescale Lyapunov optimization techniques [23].

3.1. Problem Relaxation

Stochastic optimization framework guarantees the balance of average energy consumption and average energy generation in the long term; however, it can not provide their hard bounds in any time slot. Thus, the problem above cannot be settled directly through stochastic optimization framework under those circumstances (17). To solve the problem, we try to take expectation on both sides of (15) and (16), which leads to P2 as follows:

$$\text{P2: } \min_{U_t} \lim_{T \to \infty} \frac{1}{T} \sum_{i=0}^{T-1} \mathbb{E}\{f(t)\} \tag{23}$$

$$\text{s.t. } \overline{D(t)} = \eta_c[\overline{R_c(t)} + \overline{G_s(t)} + \eta_{ce}\overline{r(t)P_c(t)y(m_t)}] \tag{24}$$

$$\overline{W(t)} = \eta_a[\eta_{ch}\overline{u(t)P_c(t)y(m_t)} + \eta_{ah}\overline{v(t)P_a(t)}] \tag{25}$$

$$(13), (14), (18), (19), (20), (21), (22).$$

After those operations, we finally obtain P2, which fits the stochastic optimization framework. P2 extends the limitation of Battery and Thermal tank storage. It no longer restricts the value of $B(t)$ and $T(t)$ in each time slot instead of restricting them in the whole process. Under the condition that the solutions must satisfy constraint (17) at each time slot, the framework is feasible to P1. As long as we define these two constants suitably, solutions to P2 can also be feasible solutions to P1.

3.2. Online Algorithm

To simplify the following discussion, the virtual queues $E(t)$ and $X(t)$ are respectively defined for the battery and thermal tank as follows:

$$E(t) = B(t) - \theta \tag{26}$$

$$X(t) = T(t) - \varepsilon \tag{27}$$

where θ and ε are two perturbation parameters, which are time-independent constants and will be specified later.

Then the queueing dynamics (15) and (16) can be transformed into:

$$E(t+1) = E(t) - \eta_d D(t) + \eta_c[R_c(t) + G_s(t) + \eta_{ce}r(t)P_c(t)y(m_t)] \tag{28}$$

$$X(t+1) = X(t) - \eta_\beta W(t) + \eta_\alpha[\eta_{ah}v(t)P_a(t) + \eta_{ch}u(t)P_c(t)y(m_t)] \tag{29}$$

In addition, the Lyapunov function is defined to be: $Q(t) = \frac{1}{2}[E(t)]^2 + \frac{1}{2}[X(t)]^2$. Then the T-slot conditional Lyapunov drift can be defined as follows:

$$\Delta(t) = E\{Q(t+T) - Q(t)|(E(t), X(t))\} \tag{30}$$

Consider any $\tau \in [t, ..., t+T-1]$, squaring both sides of (28) and (29). Considering the result in one time slot after carrying out sub calculations, we can obtain:

$$
\begin{aligned}
Q(t+1) - Q(t) = {} & 0.5\max\{\eta_d^2 D_{\max}^2, \eta_c^2[R_{c,\max} + G_{s,\max} + \eta_{ce}P_{c,\max}y(m_t)]^2\} \\
& - E(t)\{\eta_d D(t) - \eta_c[R_c(t) + G_s(t) + r(t)\eta_{ce}P_c(t)y(m_t)]\} \\
& + 0.5\max\{\eta_\beta^2 W_{\max}^2, \eta_\alpha^2[\eta_{ch}P_{c,\max}y(m_t) + \eta_{ah}P_{a,\max}]^2\} \\
& - X(t)\{\eta_\beta W(t) - \eta_\alpha[u(t)\eta_{ch}P_c(t)y(m_t) + v(t)\eta_{ah}P_a(t)]\}
\end{aligned}
\tag{31}
$$

In each time slot, the CHP consume the fuel while the thermal source consumes the gas. The maximum amount of them are $P_{c,\max}$ and $P_{a,\max}$ separately. Similarly, $R_{c,\max}$ and $G_{s,\max}$ denote the maximum charging power from the renewable energy resource and the external power grid, respectively. We define B as: $B = 0.5\max\{\eta_d^2 D_{\max}^2, \eta_c^2[R_{c,\max} + G_{s,\max} + \eta_{ce}P_{c,\max}y(m_t)]^2\} + 0.5\max\{\eta_\beta^2 W_{\max}^2, \eta_\alpha^2[\eta_{ch}P_{c,\max}y(m_t) + \eta_{ah}P_{a,\max}]^2\}$.

Summing (31) over $\tau \in [t, ..., t+T-1]$ and taking the expectation conditional on $E(t)$ and $X(t)$ yields:

$$
\begin{aligned}
\Delta(t) \leq {} & BT - E\{\sum_{\tau=t}^{t+T-1} E(\tau)[\eta_d D(\tau) - \eta_c(R_c(\tau) + G_s(\tau) + r(\tau)\eta_{ce}P_c(\tau)y(m_\tau))]\} \\
& + E\{\sum_{\tau=t}^{t+T-1} X(\tau)[\eta_\beta W(\tau) - \eta_\alpha[\eta_{ch}u(\tau)P_c(\tau)y(m_\tau) - \eta_{ah}v(t)P_a(\tau)]\}
\end{aligned}
\tag{32}
$$

For the purpose of keeping $E(t)$ and $X(t)$ stable under the stochastic optimization framework, we should minimize the right-hand side of (32). Beyond that, the goal of our control algorithm is to minimize the system energy cost. Accordingly, we set parameter V to denote the tradeoff between energy storage and consumption and the drift-plus-penalty function is defined as follows:

$$\Delta(t) + V\mathbb{E}\{f(t)\} \leq BT - \mathbb{E}\{\sum_{\tau=t}^{t+T-1} E(\tau)[\eta_d D(\tau) - \eta_c(R_c(\tau) + G_s(\tau) + r(\tau)\eta_{ce}P_c(\tau)y(m_\tau))]\}$$

$$+ \mathbb{E}\{\sum_{\tau=t}^{t+T-1} X(\tau)[\eta_\beta W(\tau) - \eta_\alpha[\eta_{ch}u(\tau)P_c(\tau)y(m_\tau) - \eta_{ah}v(t)P_a(\tau)]\}$$

$$+ V\mathbb{E}\{C(t)[G_l(t) + G_s(t) + C_g P_a(t)] + C_f P_c(t)y(m_t) + C_m y(m_t)\}$$

Replacing $G_l(t)$ in (33) use $G_l(t) = [L_e(t)]^+ - D(t) - (1 - r(t))\eta_{ce}P_c(t)y(m_t)$. In order to facilitate the algorithm, we conduct some manipulation and get the formula:

$$\Delta(t) + V\mathbb{E}\{f(t)\} \leq BT + V\mathbb{E}\{C(t)[L_e(t)]^+ | E(t)\} + V\mathbb{E}\{C_m y(m_t)\}$$

$$+ \mathbb{E}\{E(t)\eta_c R_c(t)|E(t)\} - \mathbb{E}\{D(t)[E(t)\eta_d + VC(t)]|E(t)\} - \mathbb{E}\{\eta_\beta W(t)X(t)|X(t)\}$$

$$+ \mathbb{E}\{G_s(t)[\eta_c E(t) + VC(t)]|E(t)\}$$

$$+ \mathbb{E}\{P_c(t)y(m_t)[r(t)\eta_{ce}\eta_c E(t) + \eta_\alpha\eta_{ch}u(t)X(t) - (1 - r(t))\eta_{ce}VC(t) - \eta_{ce}VC(t) + VC_f]\}$$

$$+ \mathbb{E}\{P_a(t)[\eta_\alpha\eta_{ah}v(t)X(t) + VC_g]|X(t)\}$$

The main concept of our control algorithm is minimizing the right-hand side of (34). In other words, by observing the system inputs, i.e., $C(t)$, $E(t)$, $X(t)$, $L_e(t)$ and $L_w(t)$ at each time slot in a frame, then the values of $G_l(t)$, $G_s(t)$, $r(t)$, $P_c(t)$, $P_a(t)$, $R_c(t)$, $D(t)$ can be determined.

Derived from the analysis above, an online algorithm can be developed by solving P3:

P3:   min $G_s(t)H_s(t) + P_c(t)H_c(t) + P_a(t)H_a(t) - D(t)H_d(t) - W(t)H_w(t) + R_c(t)E(t)$ (35)

          s.t. $G_l(t) + D(t) + (1 - r(t))\eta_{ce}P_c(t)y(m_t) = [L_e(t)]^+$ (36)

              $0 \leq G_s(t) + r(t)\eta_{ce}P_c(t)y(m_t) + R_c(t) \leq C_{char}$ (37)

              $0 \leq D(t) \leq D_{max}, 0 \leq W(t) \leq W_{max}$ (38)

              $(1 - v(t))\eta_{ah}P_a(t) + W(t) + (1 - u(t))\eta_{ch}P_c(t)y(m_t) \geq L_w(t)$ (39)

              $0 \leq u(t)\eta_{ch}P_c(t)y(m_t) + v(t)\eta_{ah}P_a(t) \leq Th_{char}$ (40)

              $0 \leq r(t) \leq 1, P_c(t), G_l(t), G_s(t), P_a(t) \geq 0$ (41)

        Here

$$H_{R_c}(t) = \eta_c E(t), \quad H_s(t) = \eta_c E(t) + VC(t) \tag{42}$$

$$H_c(t) = r(t)H_r(t) + u(t)H_u(t) + H_b(t) \tag{43}$$

$$H_r(t) = \eta_c\eta_{ce}E(t)y(m_t) + \eta_{ce}VC(t)y(m_t) \tag{44}$$

$$H_u(t) = \eta_{ch}\eta_\alpha X(t)y(m_t), \quad H_v(t) = \eta_\alpha\eta_{ah}X(t) \tag{45}$$

$$H_b(t) = VC_f - \eta_{ce}VC(t)y(m_t), \quad H_w(t) = \eta_\beta X(t) \tag{46}$$

$$H_a(t) = H_v(t)v(t) + VC_g, \quad H_d(t) = \eta_d E(t) + VC(t) \tag{47}$$

Observing P3, we can find the problem function includes the term $P_c(t)H_c(t)$ and $P_a(t)H_a(t)$ with $r(t)$ and $u(t)$ in $H_c(t)$ and $v(t)$ in $H_a(t)$. It follows that the Hessian matrix of the function is not positive semi-definite, which makes P3 a non-convex optimization problem and challenging to solve. However, with a further investigation of P3, $D(t)$ and $W(t)$ can be decoupled from $H_d(t)$ and $H_w(t)$.

At first, we take the terms $D(t)H_d(t)$ and $W(t)H_w(t)$ into account. If $H_d(t) < 0$, it is clear that $D(t) = 0$; otherwise, it can be easily obtained that $D(t) = \min\{D_{max}, L_e(t)\}$. Similarly, if $H_w(t) < 0$, we have $W(t) = 0$; otherwise, $W(t) = W_{max}$. Consequently, we can only concentrate on the key part of (36) which is listed as follows:

$$\min \ G_s(t)H_s(t) + P_c(t)H_c(t) + P_a(t)H_a(t) + R_c(t)E(t) \tag{48}$$
$$\text{s.t. } (37), (38) \text{ and } (40)$$

We discuss the solutions to minimize P3 when the local generator is on or off, respectively. We discuss the question under the circumstance that $y(m_t) = 1$ first.

First, we assume that both (37) and (40) are inactive. Analyzing those equations we can easily see the linear relationship between $H_a(t)$ and $v(t)$. Therefore, we can set $v(t)$ to be 0 or 1 to minimize $H_a(t)$. Similarly, $r(t)$ and $u(t)$ can also be decided to be 0 or 1 in order to minimize $H_c(t)$.

Secondly, supposing (37) to be active while (40) to be inactive, we have:

$$G_s(t) = C_{char} - r(t)\eta_{ce}P_c(t) - R_c(t) \tag{49}$$

Replacing $G_s(t)$ using (49) in (48), now the problem change into:

$$\min \ R_c(t)[E(t) - H_s(t)] + H_s(t)C_{char} + P_c(t)H_u(t)u(t) + P_c(t)H_b(t) \tag{50}$$

Then the minimum of the above equation can be achieved by setting $u(t)$ to be 0 or 1. Thirdly, supposing (37) to be inactive while (40) to be active, we have:

$$P_a(t) = \frac{Th_{char} - u(t)\eta_{ch}P_c(t)}{\eta_{ah}v(t)} \tag{51}$$

and we replace $P_a(t)$ using (51) in (48), then we have:

$$\min \ \frac{VC_g[Th_{char} - u(t)\eta_{ch}P_c(t)]}{\eta_{ah}v(t)} + G_s(t)H_s(t) + P_c(t)r(t)H_r(t) + P_c(t)H_b(t) + R_c(t)E(t) \tag{52}$$

Since $Th_{char} - u(t)\eta_{ch}P_c(t) \geq 0$, we can set both $u(t)$ and $v(t)$ to 1 to minimize (52). Finally, supposing (37) and (40) to be active, we can transform the problem into:

$$\min \ \frac{VC_g[Th_{char} - u(t)\eta_{ch}P_c(t)]}{\eta_{ah}v(t)} + P_c(t)H_b(t) + Th_{char}X(t)\eta_\alpha + R_c(t)E(t) \tag{53}$$
$$\text{s.t. } (37), (40), (41)$$

Since $T_{char} - \eta_{ch}u(t)P_c(t) \geq 0$, to minimize (53) we can set $v(t) = u(t) = 1$.

Then we discuss the circumstance that $y(m_t) = 0$.

Because $P_c(t)$ is related to $y(m_t)$, so when $y(m_t) = 0$, $P_c(t) = 0$. The problem is reduced to the equation as follows·

$$\min \ G_s(t)H_s(t) + P_a(t)H_a(t) + R_c(t)E(t) \tag{54}$$

Replacing $R_c(t)$ use (37). Setting $v(t) = 0$ or 1 can get the minimum value of $H_a(t)$. So the equation is more concise:

$$\min \ G_s(t)[H_s(t) - E(t)] + C_{char}E(t) \tag{55}$$

From this, what we need to discuss is the value of $G_s(t)$. If $(\eta_c - 1)E(t) + VC(t) \geq 0$, to minimize (55), $G_s(t)$ should be set to 0; If $(\eta_c - 1)E(t) + VC(t) < 0$, we have $G_s(t) = G_{s,max}$. The above analysis display the minimization is four different circumstances.

4. Performance Analysis

Theorem 1. *Define θ and ε to be:*

$$\theta = \frac{V \max\{C_{e,max}, C_f\}}{\eta_c} + T \min\{\eta_d D_{max}, L_{e,max}\} \tag{56}$$

$$\varepsilon = \frac{V C_g}{\eta_a \eta_{ah}} + T L_{w,max} \tag{57}$$

Then concluding from the process of minimizing P3 we can obtain the result:

$$0 \leq B(t) \leq \theta + TC_{char}, \forall t \in T \tag{58}$$

$$0 \leq T(t) \leq \varepsilon + Th_{char} \cdot T, \forall t \in T. \tag{59}$$

Proof. Using induction method, the upper and lower bounds of $B(t)$ and $T(t)$ can be proved under (56) and (57).

1. Firstly, we show the upper bounds. The main idea is to prove that the battery and thermal tank will not charging when there level exceed θ and ε, respectively.

    - Suppose $B(t) \leq \theta$. Since the electricity charged in a time slot will not be more than TC_{char}. Then $B(t+1) \leq \theta + TC_{char}$ holds, obviously.
    - Suppose $B(t) > \theta$, i.e., $E(t) > 0$. From (42), (44) and (47), it can be obtained that $H_s(t) > 0$, $H_r(t) > 0$, $H_d(t) > 0$ and $H_{R_c}(t) > 0$. According to P3, we must set $R_c(t) = 0$, $G_s(t) = 0$, $r(t) = 0$, $D(t) = \min\{\eta_d D_{max}, L_e(t)\}$ to minimize the problem function, i.e., the battery will not charge when its level rises over θ. Accordingly, $B(t+1) \leq B(t)$ holds when $B(t) > \theta$. With the conclusion when $B(t) \leq \theta$, we can know that $B(t) \leq \theta + TC_{char}, \forall t \in T$.
    - Suppose $T(t) \leq \varepsilon$. Similar to the battery, the heat charged in a time slot will not exceed $TC_{char} \cdot T$. Hence, $T(t+1) \leq \varepsilon + Th_{char} \cdot T$ holds then.
    - Suppose $T(t) > \varepsilon$, i.e., $X(t) > 0$. From (45) to (47), it can be obtained that $H_u(t) > 0$, $H_v(t) > 0$, $H_w(t) > 0$ and $H_a(t) > 0$. According to P3, we must set $P_a(t) = 0$, $v(t) = 0$ and $W(t) = \min\{\eta_\beta W_{max}, L_w(t)\}$ to minimize the problem function, i.e., the thermal tank will not charge when its level rises over ε. Accordingly, $T(t+1) \leq T(t)$ holds when $T(t) > \varepsilon$. With the conclusion when $T(t) \leq \varepsilon$, we know that $T(t) \leq \varepsilon + Th_{char} \cdot T, \forall t \in T$.

    The above proof presents the upper bounds of $B(t)$ and $T(t)$, i.e., the capacities of battery and thermal tank, respectively. To simplify the future investigation, we denote the capacities as B_{max} and T_{max}, respectively. Since the capacities are functions of V, the value of V can be changed to make a tradeoff between energy storage and cost. In contrast, the value of V can be obtained with a given battery pack or thermal tank capacity.

2. Secondly, we show the lower bounds using (56) and (57). To keep $B(t)$ and $T(t)$ from being negative, we only need to prevent the battery and thermal tank from discharging when they can not afford, i.e., when $B(t) < T \min\{\eta_d D_{max}, L_{e,max}\}$ and $T(t) < T L_{w,max}$

    - Suppose $B(t) \geq T \min\{\eta_d D_{max}, L_{e,max}\}$. Certainly, it follows that $B(t+1) > 0$.
    - Suppose $0 \leq B(t) < T \min\{\eta_d D_{max}, L_{e,max}\}$. From (42), (47) and (56), it can be obtained that $H_d(t) < 0$, $H_s(t) < 0$. According to P3, we have to set $D(t) = 0$, $G_s(t) \geq 0$, i.e., the battery will not discharge. Accordingly, $B(t) \leq B(t+1)$ holds then. Consequently, we can conclude that $0 \leq B(t), \forall t \in T$.

- Suppose $T(t) \geq TL_{w,max}$. Then $T(t+1) > 0$ follows apparently.
- Suppose $0 \leq T(t) < L_{w,max}$. From (46) and (57), it can be obtained that $H_w(t) < 0$. According to P3, we have to set $W(t) = 0$, i.e., the thermal tank will not discharge. Accordingly, $T(t) \leq T(t+1)$ holds then. Finally, we can conclude that $0 \leq T(t), \forall t \in T$.

Theorem 1 shows that the battery and thermal tank both have finite capacities under the proposed algorithm, which means solutions of P3 are feasible solutions to P1 as well. □

Theorem 2. *The gap between the optimal cost of P1 and the expected cost obtained by solving P3 is no more than $\frac{TB}{V}$, i.e.,*

$$\lim_{T \to \infty} \frac{1}{T} \sum_{i=0}^{T-1} \mathbb{E}\{f'(t)\} \leq P_1^* + \frac{TB}{V}, \tag{60}$$

where $f'(t)$ is the energy cost at time slot t under the proposed algorithm, and P_1^* is the optimal solution to the original problem P1.

5. Numerical Simulations

In this section, we use Matlab to evaluate the performance of the proposed algorithms by numerical simulations. We consider a hotel with battery and thermal tank as well as the CHP system. In our simulation, each time slot represents 15 min and each frame consists of 4 time slots (i.e., one hour). The parameter settings are partly listed in Table 1 and detailed in Section 5.1.

Table 1. Parameter settings in numerical simulations.

Parameter	Value	Unit
η_{ah}	0.8	
η_c	0.9	
η_d	1.1	
η_α	0.9	
η_β	1.1	
c_f	0.0035	$/kBtu
c_m	0.1	$/h
D_{max}	30	kWh/h
C_{char}	20	kWh/h
$G_{l,max}$	32	kWh/h
$G_{s,max}$	32	kWh/h
W_{max}	30	kBtu/h
$P_{c,max}$	50	kBtu/h
$P_{a,max}$	32	kBtu/h

5.1. Simulation Setup

Centralized Power Grid: We obtain the electricity price data of power grid from [24]. The data trace is shown in Figure 3. The maximal supply power $G_{l,max}$ and charging power $G_{s,max}$ obtained from power grid are both set to be 32 kWh/h.

External Gas Station: We assume that the natural gas price c_g varies across time and has a uniform distribution between (0.004, 0.010)$/kBtu. The maximal thermal output is set as $P_{a,max} = 32$ kBtu/h. The efficiency is set as $\eta_{ah} = 0.8$.

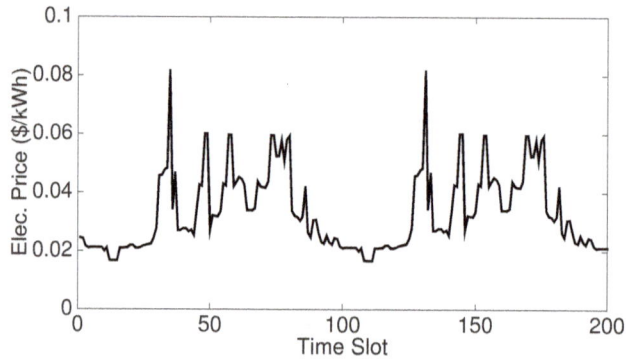

Figure 3. Data trace of electricity market prices.

CHP System: The maximal thermal output of CHP system is set as $P_{c,max} = 50$ kBtu/h. The overall CHP efficiency is assumed to be 80%, and the electricity conversion efficiency is in the range of 30–40%. The fuel cost of CHP is set as $c_f = 0.0035$ $/kBtu. We set the minimal on/off period of CHP to be 1 h, i.e., 4 time slots. The sunk cost for maintaining the system in its active state is set as $c_m = 0.1$ $/h.

Harvested Wind Power: The harvested wind power data is obtained from [25]. The data path is shown in Figure 4.

Battery and Thermal Tank Model: We set the maximal charging and discharging rates of the battery as $C_{char} = 20$ kWh/h and $D_{max} = 30$ kWh/h, while the charging and discharging efficiency is set as $\eta_c = 0.9$ and $\eta_d = 1.1$, respectively. Similarly, the heat storing and releasing efficiency of the thermal tank is also set as $\eta_{ff} = 0.9$ and $\eta_{fi} = 1.1$. The maximal heat output is set as 30 kBtu/h.

Electricity and Heat Demand: We use the real demand data provided by California Commercial End-Use Survey (CEUS) [26] in our simulation. The data traces in 50 h (i.e., 200 time slots) are shown in Figure 5.

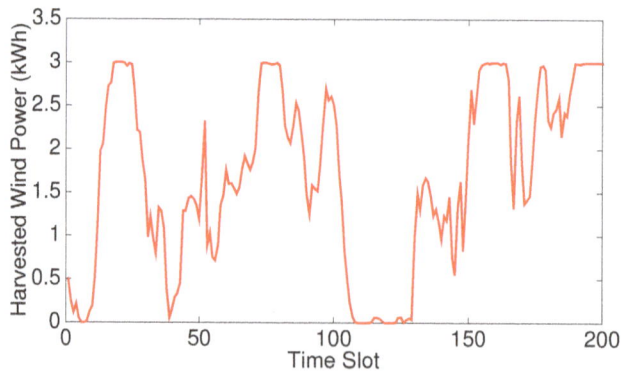

Figure 4. Data traces of harvested wind power.

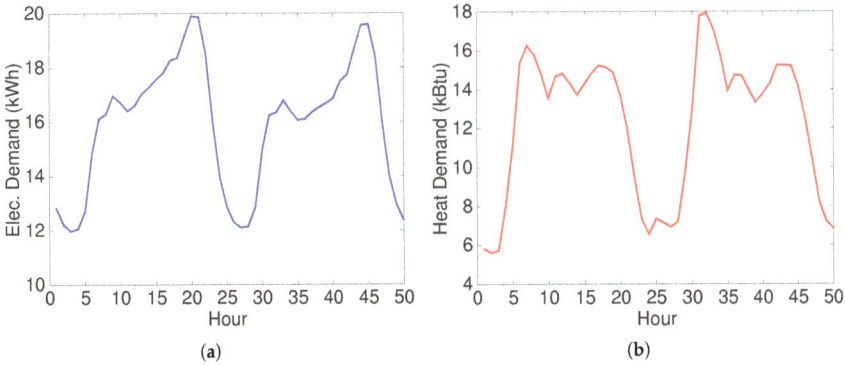

Figure 5. Demand data traces in 50 h (i.e., 200 time slots). (**a**) Electricity demand; (**b**) Heat demand.

5.2. Results of the Simulation

With the parameters above, we simulate for 200 time slots and each time slot stands for 15 min. We let $V = 5$.

Figure 6 shows the process where the CHP system adaptively makes on/off decisions in 200 time slots. We can see that the decisions are made every 4 time slots.

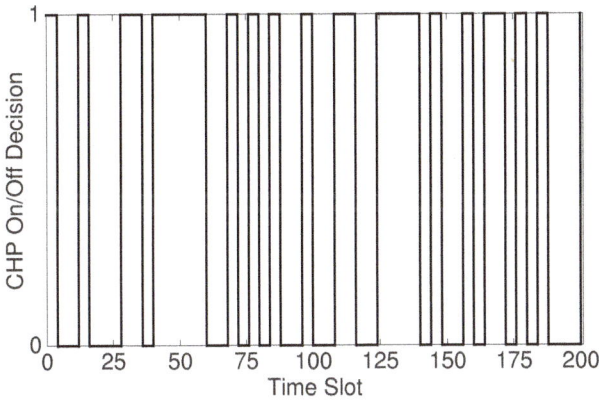

Figure 6. A sample path of on/off decisions in 200 time slots under $V = 5$.

The sample paths in Figure 7 depict electricity and heat supplies in the first 24 h (i.e., 96 time slots). Both the electricity and heat demands can be satisfied with hybrid sources.

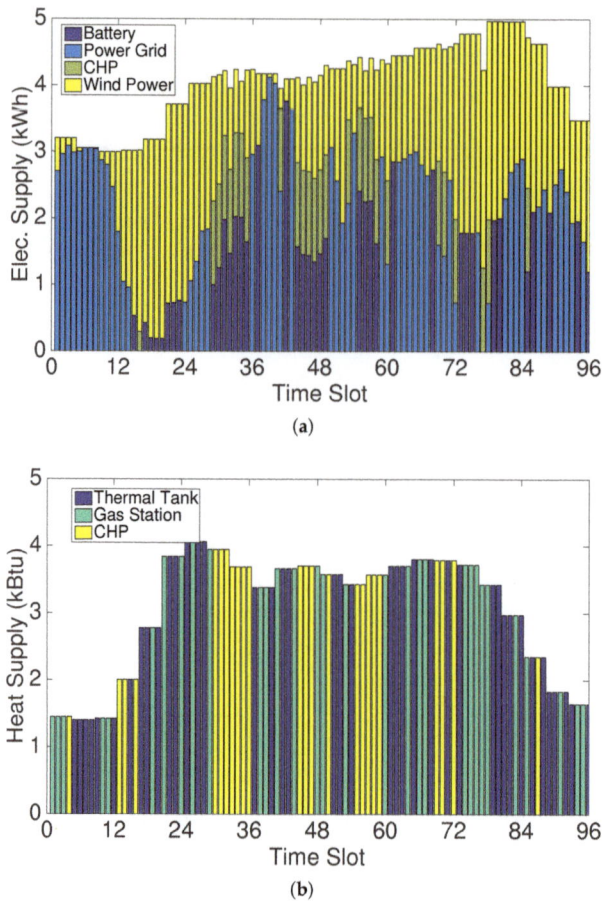

Figure 7. Sample paths of power supplies in the first 24 h (i.e., 96 time slots) under $V = 5$. (**a**) Electricity supply; (**b**) Heat supply.

Figure 8 specifies the charging/discharging behavior of the battery and thermal tank. As shown in Figures 4 and 5a, the electricity demand is larger than the harvested wind power at every time slot, there is no excessive renewable energy (wind power) left to charge into the battery. As a result, there is no wind power illustrated in Figure 7a. Actually, in our optimization problem and simulations, if there exists excessive wind power, it can be charged into the battery for future use under our designed algorithm.

Figure 9 shows the corresponding changes of battery level and thermal tank level in all the 200 time slots under $V = 5$. We can see that the tank level remains almost stationary due to the uniform distribution of the gas price, while the battery level fluctuates in reaction to the electricity price. However, the capacities of both battery and thermal tank are bounded.

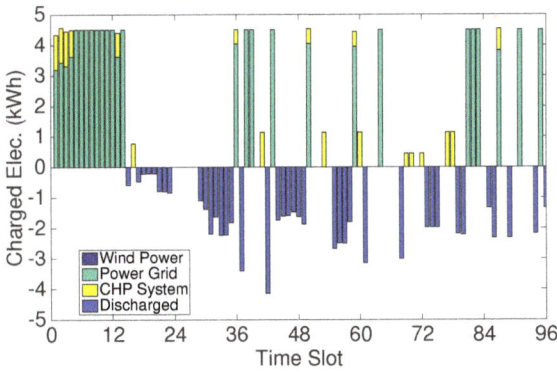

(a)

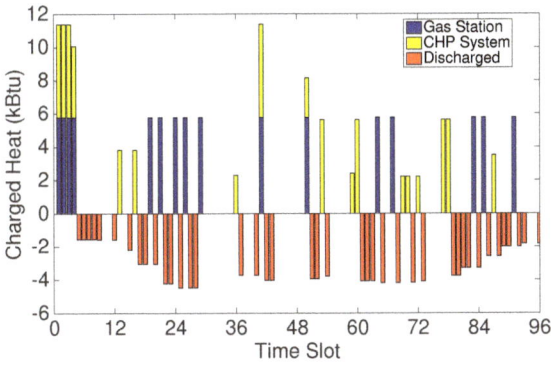

(b)

Figure 8. Behavior of charging and discharging in the first 24 h (i.e., 96 time slots) under $V = 5$. (**a**) Amount of electricity charged(discharged) into(from) battery; (**b**) Amount of heat stored (released) into (from) thermal tank.

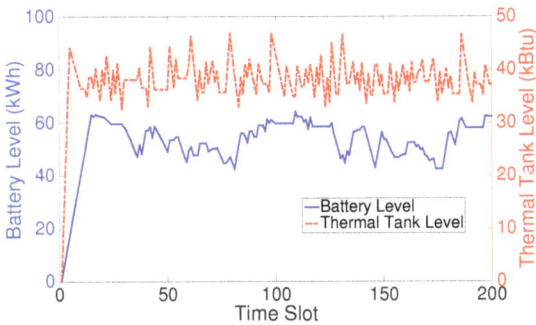

Figure 9. Sample paths of battery level and thermal tank power level under $V = 5$.

5.3. Performance vs. V and Charging/Discharging Efficiency

With further simulations under different values of η_α, η_β, η_c, η_d and V, we observe the impacts of the weight V and efficiency of charging/discharging.

As shown in Figure 10, the cost drops when charging and discharging get more efficient. Figure 11 shows that the capacity of battery and thermal tank is linear with V, which is also indicated in Theorem 1. The total cost curve shown in Figure 12, on the other hand, converges to the minimum with increasing V. Furthermore, by comparing with the situations where CHP is permanently on and off, the effectiveness of our adaptive on/off decision policy can be verified.

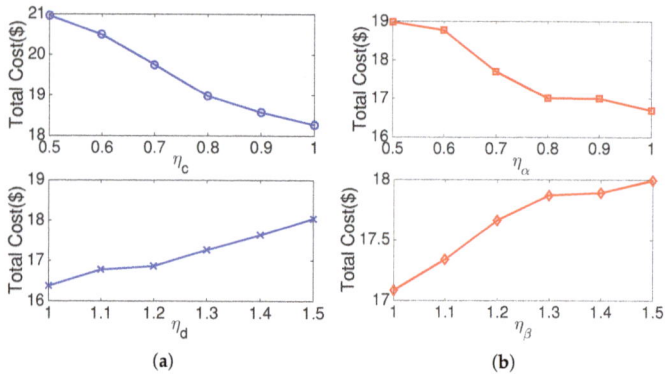

(a)

(b)

Figure 10. Impacts of the efficiency of charging and discharging. (a) Total cost vs. η_c and η_d; (b) Total cost vs. η_α and η_β.

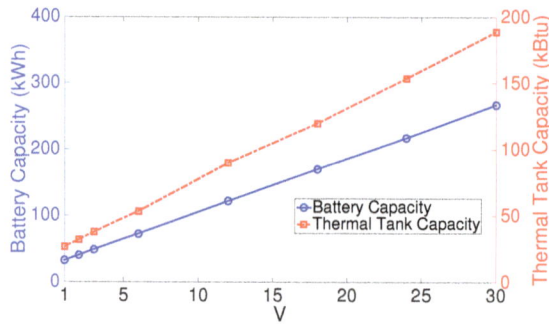

Figure 11. Capacities of battery and thermal tank vs. V.

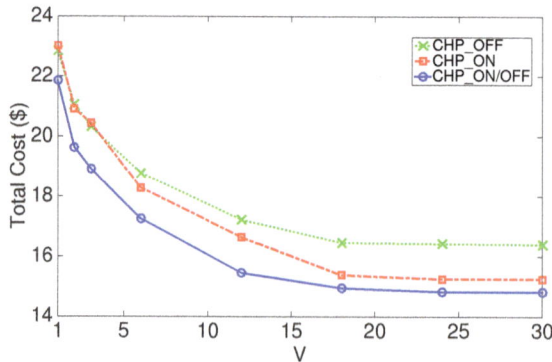

Figure 12. Total cost vs. V under different on/off policies of CHP.

6. Conclusions

In this paper, we studied the operating cost minimization problem for microgrids with CHP generation, energy storages, and renewable energy resources by using the Lyapunov approach. We designed an algorithm LYP that can achieve near-optimal performance by adjusting the value of V. According to a large amount of empirical evaluations, the microgrid operating cost can be reduced significantly through such an integration of centralized grid, renewable energy, power storage device, and co-generation.

Acknowledgments: This work is supported by the National Natural Science Foundation of China (Grant No. 61772130, 61301118); the International Science and Technology Cooperation Program of the Shanghai Science and Technology Commission (Grant No. 15220710600); the Innovation Program of the Shanghai Municipal Education Commission (Grant No. 14YZ130); and the Fundamental Research Funds for the Central Universities. This paper is an extended version of our conference paper: Optimal energy storage management for microgrids with ON/OFF co-generator: A two-time-scale approach [27].

Author Contributions: The work presented in this paper is a collaborative development by all of the authors. Guanglin Zhang and Yu Cao contributed to the idea of the incentive mechanisms and designed the algorithms. Yongheng Cao, Demin Li and Lin Wang developed and analyzed the experimental results. All authors contributed to the organization of the paper including writing and proofreading.

Conflicts of Interest: The authors declare no conflict of interest.

References

1. Marnay, C.; Firestone, R. Microgrids: An emerging paradigm for meeting building electricity and heat requirements efficiently and with appropriate energy quality. In *European Council for an Energy Efficient Economy Summer Study*; Lawrence Berkeley National Laboratory: Berkeley, CA, USA, 2007.
2. European Commission. 2020 Climate and Energy Package. 2017. Available online: https://ec.europa.eu/clima/policies/strategies/2020_en (accessed on 21 May 2017).
3. Senator Energy, Utilities and Communications Committee. Senate Bill X 1-2. 2011. Available online: http://www.leginfo.ca.gov/pub/11-12/bill/sen/sb_0001-0050/sbx1_2_cfa_20110214_141136_sen_comm.html (accessed on 21 May 2017).
4. Bitar, E.; Rajagopal, R.; Khargonekar, P.; Poolla, K. The role of co-located storage for wind power producers in conventional electricity markets. In Proceedings of the American Control Conference, San Francisco, CA, USA, 29 June–1 July 2011.
5. Varaiya, P.P.; Wu, F.F.; Bialek, J.W. Smart operation of smart grid: Risk-limiting dispatch. *IEEE Proc.* **2011**, *99*, 40–57.
6. Stadler, M.; Aki, H.; Lai, R.; Marnay, C.; Siddiqui, A. *Distributed Energy Resources On-Site Optimization for Commercial Buildings with Electric and Thermal Storage Technologies*; LBNL-293E; Lawrence Berkeley National Laboratory: Pacific Grove, CA, USA, 2008.
7. Guo, Y.; Gong, Y.; Fang, Y.; Khargonekar, P.P.; Geng, X. Energy and Network Aware Workload Management for Sustainable Data Centers with Thermal Storage. *IEEE Trans. Parallel Distrib. Syst.* **2014**, *25*, 2030–2042.
8. Koutsopoulos, I.; Hatzi, V.; Tassiulas, L. Optimal Energy Storage Control Policies for the Smart Power Grid. In Proceedings of the 2011 IEEE Third International Conference on Smart Grid Communications (SmartGridComm), Brussels, Belgium, 17–20 October 2011.
9. Van de ven, P.; Hegde, N.; Massoulie, L.; Salonidis, T. Optimal Control of Residential Energy Storage Under Price Fluctuations. In Proceedings of the 1st International Conference on Smart Grids, Green Communications and IT Energy-Aware Technologies 2011 (ENERGY 2011), Venice, Italy, 22–27 May 2011.
10. Urgaonkar, R.; Urgaonkar, B.; Neely, M.J.; Sivasubramaniam, A. Optimal Power Cost Management Using Stored Energy in Data Centers. In Proceedings of the ACM SIGMETRICS International Conference on Measurement and Modeling of Computer Systems, San Jose, CA, USA, 7–11 June 2011.
11. Guo, Y.; Ding, Z.; Fang, Y.; Wu, D. Cutting Down Electricity Cost in Internet Data Centers by Using Energy Storage. In Proceedings of the 2011 IEEE Global Communications Conference (GLOBECOM 2011), Houston, TX, USA, 5–9 December 2011.

12. Mishra, A.; Irwin, D.; Shenoy, P.; Kurose, J.; Zhu, T. SmartCharge: Cutting the Electricity Bill in Smart Homes with Energy Storage. In Proceedings of the 3rd International Conference on Energy-Efficient Computing and Networking (e-Energy), Madrid, Spain, 9–11 May 2012.

13. Li, J.; Xiong, R.; Yang, Q.; Liang, F.; Zhang, M.; Yuan, W. Design/test of a hybrid energy storage system for primary frequency control using a dynamic droop method in an isolated microgrid power system. *Appl. Energy* **2017**, *201*, 257–269.

14. Chau, C.; Zhang, G.; Chen, M. Cost Minimizing Online Algorithms for Energy Storage Management with Worst-Case Guarantee. *IEEE Trans. Smart Grid* **2016**, *7*, 2691–2702.

15. Lu, L.; Tu, J.; Chau, C.; Chen, M.; Lin, X. Online Energy Generation Scheduling for Microgrids with Intermittent Energy Sources and Co-Generation. In Proceedings of the International Conference on Measurement and Modeling of Computer Systems (SIGMETRICS '13), Pittsburgh, PA, USA, 17–21 June 2013.

16. Zhou, K.; Cai, L.; Pan, J. Optimal Combined Heat and Power System Scheduling in Smart Grid. In Proceedings of the IEEE Conference on Computer Communications Workshops, Toronto, ON, Canada, 27 April–2 May 2014.

17. ICF International Inc. *CHP Policy Analysis and 2011–2030 Market Assessment*; February 2012. Available online: http://www.energy.ca.gov/2012publications/CEC-200-2012-002/CEC-200-2012-002.pdf (accessed on 21 May 2017).

18. Li, J.; Yang, Q.; Robinson, F.; Liang, F.; Zhang, M.; Yuan, W. Design and test of a new droop control algorithm for a SMES/battery hybrid energy storage system. *Energy* **2017**, *118*, 1110-1122.

19. Rooijers, F.; Amerongen, R. Static economic dispatch for cogeneration systems. *IEEE Trans. Power Syst.* **1994**, *9*, 1392–1398.

20. Tao, G.; Henwood, M.; van Ooijen, M. An algorithm for combined heat and power economic dispatch. *IEEE Trans. Power Syst.* **1996**, *11*, 5453360.

21. Li, J.; Wang, X.; Zhang, Z.; le Blond, S.; Yang, Q.; Zhang, M.; Yuan, W. Analysis of a new design of the hybrid energy storage system used in the residential m-CHP systems. *Appl. Energy* **2017**, *187*, 169–179.

22. Georgiadis, L.; Neely, M.J.; Tassiulas, L. *Resource Allocation and Cross-Layer Control in Wireless Networks*; Foundations and Trends in Networking: Hanover, MA, USA, 2006; Volume 1, pp. 1–144.

23. Huang, L.; Walrand, J.; Ramchandran, K. Optimal Demand Response with Energy Storage Management. In Proceedings of the 2012 IEEE Third International Conference on Smart Grid Communications (SmartGridComm), Tainan, Taiwan, 5–8 November 2012.

24. ISO New England. Available online: https://www.iso-ne.com/isoexpress/ (accessed on 21 May 2017).

25. National Renewable Energy Laboratory. Available online: http://wind.nrel.gov (accessed on 21 May 2017).

26. California Commercial End-Use Survey. Available online: http://capabilities.itron.com/CeusWeb (accessed on 21 May 2017).

27. Shen, Y.; Ou, X.; Xu, J.; Zhang, G.; Wang, L.; Li, D. Optimal energy storage management for microgrids with on/off co-generator: A two-time-scale approach. In Proceedings of the 2015 IEEE Global Conference on Signal and Information Processing (GlobalSIP), Orlando, FL, USA, 14–16 December 2015; pp. 997–1001.

energies

MDPI

Article

Energy Flexibility Management Based on Predictive Dispatch Model of Domestic Energy Management System

Amin Shokri Gazafroudi [1,*], Francisco Prieto-Castrillo [1,2,3], Tiago Pinto [1], Javier Prieto [1,4], Juan Manuel Corchado [1,5] and Javier Bajo [1]

[1] BISITE Research Group, University of Salamanca, Edificio I+D+i, 37008 Salamanca, Spain; franciscop@usal.es (F.P.-C); tpinto@usal.es (T.P.); javierp@usal.es (J.P.); corchado@usal.es (J.M.C.); jbajope@usal.es (J.B.)

[2] MediaLab, Massachusetts Institute of Technology, Cambridge, MA 02139-4307, USA

[3] Harvard T.H. Chan School of Public Health, Harvard University, Boston, MA 02115, USA

[4] StageMotion, R&D Department, C/Orfebres 10, 34005 Palencia, Spain

[5] Osaka Institute of Technology, Asahi-ku Ohmiya, Osaka 535-8585, Japan

[*] Correspondence: shokri@usal.es

Academic Editor: Pedro Faria
Received: 3 August 2017; Accepted: 5 September 2017; Published: 13 September 2017

Abstract: This paper proposes a predictive dispatch model to manage energy flexibility in the domestic energy system. Electric Vehicles (EV), batteries and shiftable loads are devices that provide energy flexibility in the proposed system. The proposed energy management problem consists of two stages: day-ahead and real time. A hybrid method is defined for the first time in this paper to model the uncertainty of the PV power generation based on its power prediction. In the day-ahead stage, the uncertainty is modeled by interval bands. On the other hand, the uncertainty of PV power generation is modeled through a stochastic scenario-based method in the real-time stage. The performance of the proposed hybrid Interval-Stochastic (InterStoch) method is compared with the Modified Stochastic Predicted Band (MSPB) method. Moreover, the impacts of energy flexibility and the demand response program on the expected profit and transacted electrical energy of the system are assessed in the case study presented in this paper.

Keywords: decision-making under uncertainty; domestic energy management system; energy flexibility; interval optimization; stochastic programming

1. Introduction

1.1. Aims and Motivation

In the last decade, power systems have faced new challenges due to the increment of the distributed renewable energy resources. Renewable energy resources decrease the greenhouse gas emissions and costs related to electricity production [1]. However, the integration of these intermittent energy resources leads to energy management problems based on the scale of the energy system [2]. At a smaller scale, Domestic Energy Management Systems (DEMSs) enable the residential customers to manage their loads in order to minimize the electricity cost. Generally, there are two approaches for energy management of the DEMSs. These approaches consist of centralized and decentralized systems. Based on the approach of the system, different structures of the controlling and communicating systems are required [3].

1.2. Literature Review

Various research has been presented for optimal energy scheduling at the scale of the smart homes and smart grids that can be classified based on their goals, strategies, utilized technologies and software. In [4,5], the authors discussed the necessities of using computational intelligence in the DEMSs and a review of energy management systems based on multi-agent systems, respectively. On the other hand, in [6], the present and future perspectives regarding EVs and their operation modes in smart grids and smart homes have been discussed. In [7], the authors proposed a method to schedule the local energy resources optimally. Minimizing the loss of energy and purchasing electricity cost were the main goals of the authors. In [8], the authors defined the DEM problem in connection with local energy nodes. Furthermore, in [8], homes have a two-way communication with the market and can manage energy locally. In [9], a domestic demand response has been implemented in a distribution grid. The real-time price is the main goal of the demand response program based on the direct load control in [9]. Furthermore, the uncertainty of the price and load has been considered in [9]. Price prediction has been used instead of communication between homes in the distribution network.

In [10], a DEMS has been developed for day-ahead energy scheduling considering hourly pricing and the peak power constraint based on the demand response programs. The authors of [11] proposed a method to manage the energy of EV and energy storage systems according to the dynamic pricing, peak power limitation and demand response programs. The proposed Domestic Energy Management (DEM) problem has been modeled by mixed-integer linear programming. In [12], the authors propose a decentralize strategy for optimal energy scheduling under the large penetration of EVs. This interaction has been considered between consumers and the aggregator in [12]. This way, end-users send their optimum demand decisions and reschedule their demands based on the signals of the aggregator. In [13], a rescheduling DEMS has been presented to make the optimum decisions through the day and avoid the negative impact of price uncertainty.

The authors of [14] presented a multi-time scale DEM problem that includes EV and different types of electrical loads. Furthermore, the authors introduced the improved optimization algorithm to solve the DEM problem. In [15], a chance constraint model has been presented to optimize the performance of the domestic devices. The improved particle swarm optimization method has been used to optimize the problem based on the proposed demand response program. In [16], the authors introduced the Stackelberg game-based method to maximize the profits of the costumers and retailers simultaneously. In the proposed home model of [16], electrical loads of customers are elastic based on EVs. Furthermore, the price-based DR has been implemented in [16]. In [17], the DEM problem has been solved by stochastic dynamic programming considering EV. The authors proved that the EV is one source of uncertainty in the system due to the EV's plug-in time, plug-out time and charge demand for mobility. Hence, the plug-state of the EV has been modeled through a Markov chain in [17]. In [18], a DEMS has been presented as part of an organization-based multi-agent system. Besides, the uncertainty of distributed energy resources has been considered through an Modified Stochastic Predicted Band (MSPB) method used to model the DEM problem.

1.3. Contributions

In the literature, several relevant advances have been accomplished in the DEMS domain. These mostly refer to the study and analysis of the several resources' impact on the management process, namely flexible loads, EVs, batteries and generation. The interaction with the electricity market and the participation in demand response programs has also been explored, and this research is leading to promising outcomes, but modeling the electricity market at the local level has not been of interest to the authors in the previous works. However, the local electricity market has been introduced in [18], but the impact of the flexibility has not been evaluated in [18]. The research dealing with the uncertainty associated with the several resources is, however, still at an initial stage. Although some relevant works can be found in the literature, there are still many loose ends due to the difficulty in correctly identifying, measuring, modeling and representing the different sources of uncertainty, so that these can be correctly considered by DEMS scheduling, dispatch and management models.

In order to overcome the limitations in the field, this paper proposes a two-stage predictive dispatch model to manage energy flexibility in the domestic energy system. EV, battery and shiftable loads are in charge of providing the energy flexibility. A novel hybrid (InterStoch) method is defined for the first time in this paper to model the uncertainty of the PV power generation. In the first stage, the day-ahead stage, the uncertainty is modeled by interval bands. However, a stochastic scenario-based method is used to consider the uncertainty of PV power generation in the second stage, the real-time stage. Finally, the performance of the proposed hybrid Interval-Stochastic (InterStoch) method is compared with the MSPB method that was introduced in [18,19].

The rest of this paper is organized as follows. Section 2 introduces the proposed hybrid interval-stochastic method. Then, the domestic energy management problem is described in Section 3. Section 4 provides the simulation results. Finally, Section 5 summarizes the conclusions.

2. Interval-Stochastic Method

2.1. Data

In this paper, the predicted data from [18] have been used. For simplicity, only the uncertainty of PV power generation is considered. As shown in Table 1, the predicted data in each time step consist of the central forecasting and up/down deviation. Hence, the predicted data are limited to the upper/lower band based on the central forecasting and up/down deviation. It is noticeable that this paper concentrates only on modeling the uncertainty due to PV power prediction in the system. Hence, the forecasting system is not explained in this paper. The presented data of Table 1 are the inputs of the energy management system. Therefore, the energy management system makes optimum decisions through the InterStoch method.

Table 1. Predicted data of uncertain variables [18].

t	$P_{pv_t}^{pred}$ (kW)	σ_{pv}^{down} (kW)	σ_{pv}^{up} (kW)	$\theta_{out_t}^{pred}$ (°C)	$L_{mrs_t}^{pred}$ (kW)
1	0	0.00	0.00	5.5	0.005
2	0	0.00	0.00	5.5	0.005
3	0	0.00	0.00	5.2	0.005
4	0	0.00	0.00	5.2	0.005
5	0	0.00	0.00	4.8	0.005
6	0	0.00	0.00	5.5	0.005
7	0.10	0.01	0.02	6.5	0.005
8	0.20	0.02	0.04	7.5	0.005
9	0.42	0.03	0.07	9.8	0.005
10	0.76	0.08	0.26	10	0.005
11	1.1	0.12	0.23	11	0.005
12	1.32	0.13	0.26	12	0.005
13	1.91	0.10	0.19	12	0.005
14	0.85	0.02	0.04	12	0.005
15	0.29	0.02	0.04	11	0.005
16	0.31	0.02	0.03	10	0.005
17	0.06	0.01	0.01	9	0.005
18	0	0.00	0.00	8.5	0.005
19	0	0.00	0.00	8	0.005
20	0	0.00	0.00	7.5	1.218
21	0	0.00	0.00	7	0.262
22	0	0.00	0.00	6.5	0.14
23	0	0.00	0.00	6.2	0.127
24	0	0.00	0.00	6	0.005

2.2. Interval Model

In the day-ahead stage, PV system power generation is limited between bands according to the forecasting deviations. The minimum band represents the deviation below the central forecasting, and the maximum band represents the deviation above the central forecasting. $P^{da}_{pv_t}$ intends to converge to the maximum/minimum band in the best/worst case. Therefore, Equation (1) can be divided into Equations (2) and (3) in the best and worst cases, respectively. This way, an auxiliary parameter is added in these equations as a slack parameter for the decision-maker. This parameter is denoted as the Optimistic Coefficient (OC), α, which is between zero and one, and had been defined for the first time in [19]. Hence, $P^{da}_{pv_t}$ converges to the best/worst case when the decision-maker has the pessimistic/conservative perspective by adding α to Equations (2) and (3) and summing over them, as seen in Equation (4). Then, Equations (4) and (5) are obtained through simplification of Equations (1)–(3).

$$[H]P^{pred}_{pv_t} - \sigma^{down}_{pv_t} \leq P^{da}_{pv_t} \leq P^{pred}_{pv_t} + \sigma^{up}_{pv_t} \tag{1}$$

$$P^{pred}_{pv_t} \leq P^{da}_{pv_t} \leq P^{pred}_{pv_t} + \sigma^{up}_{pv_t} : OC = 1 \tag{2}$$

$$P^{pred}_{pv_t} - \sigma^{down}_{pv_t} \leq P^{da}_{pv_t} \leq P^{pred}_{pv_t} : OC = 0 \tag{3}$$

$$P^{pred}_{pv_t}\alpha_{pv} - (P^{pred}_{pv_t} - \sigma^{down}_{pv_t})(1 - \alpha_{pv}) \leq P^{da}_{pv_t} \tag{4}$$
$$\leq (P^{pred}_{pv_t} + \sigma^{up}_{pv_t})\alpha_{pv} + P^{pred}_{pv_t}(1 - \alpha_{pv})$$

$$P^{pred}_{pv_t} - \sigma^{down}_{pv_t}(1 - \alpha_{pv}) \leq P^{da}_{pv_t} \leq P^{pred}_{pv_t} + \sigma^{up}_{pv_t}\alpha_{pv} \tag{5}$$

2.3. Stochastic Model

In the real-time stage, stochastic programming is used to model the uncertainty of the PV power. Therefore, scenarios with their corresponding probabilities are defined in this section. This way, the prediction mean and deviation are defined as metric parameters by Equations (6) and (7), respectively. These are used to generate the scenarios of the PV power in the real-time stage. In this step, three scenarios are defined to model the uncertainty of the PV system's power generation. The first scenario, the up scenario, describes data that have a deviation above the central forecasting. The second scenario, the down scenario, represents data that have a deviation below the central forecasting. Then, the third scenario describes the central forecasting data. The amounts of these scenarios are determined through Equations (8)–(10). Moreover, the corresponding probabilities are obtained according to Equations (11)–(13).

$$P^{mean}_{pv_t} = P^{pred}_{pv_t} + \frac{\sigma^{up}_{pv_t} - \sigma^{down}_{pv_t}}{2} \tag{6}$$

$$\Delta_{pv_t} = \frac{\sigma^{up}_{pv_t} + \sigma^{down}_{pv_t}}{2} \tag{7}$$

$$P^{rt}_{pv_t}(\omega = \omega_1) = P^{pred}_{pv_t} + \sigma^{up}_{pv_t} \tag{8}$$

$$P^{rt}_{pv_t}(\omega = \omega_2) = P^{pred}_{pv_t} - \sigma^{down}_{pv_t} \tag{9}$$

$$P^{rt}_{pv_t}(\omega = \omega_3) = P^{pred}_{pv_t} \tag{10}$$

$$\pi(\omega = \omega_1) = Prob(P^{pred}_{pv_t} + \sigma^{up}_{pv_t} > P^{mean}_{pv_t} + \Delta_{pv_t}) \tag{11}$$

$$\pi(\omega = \omega_2) = Prob(P^{pred}_{pv_t} - \sigma^{down}_{pv_t} < P^{mean}_{pv_t} - \Delta_{pv_t}) \tag{12}$$

$$\pi(\omega = \omega_3) = 1 - \pi(\omega = \omega_1) - \pi(\omega = \omega_2) \tag{13}$$

3. Domestic Energy Management Problem

We consider that each smart home can participate in two different types of Local Electricity Market (LEM), not the wholesale market [18]. These LEMs are called day-ahead and real-time markets. In practice, the proposed LEMs can be operated by distribution system operator or retailers. Hence, the distribution system operator or retailers are responsible for providing the local electricity market framework for their agents that are in their region or have contracts to transact energy with them. Besides, it is considered that smart homes are price-takers in the LEM, and they can buy electricity from the local electricity market based on the Time of Use (ToU) tariff. Furthermore, it is assumed that the sold/bought electricity prices to/from the local electricity market are different. The domestic energy management problem is modeled as a two-stage problem. The first stage is called the day-ahead stage, and the second stage is called the real-time stage. Here, the Expected Profit (EP) is defined by an Objective Function (OF) to maximize the profit of energy services. In Equation (14), EP is the sum of the day-ahead EP, EP^{da}, and the real-time EP, EP^{rt}, which are OFs of the day-ahead and real-time stages, respectively.

$$EP = EP^{da} + EP^{rt} \tag{14}$$

3.1. Day-Ahead Stage

The objective function of the domestic energy management system in the day-ahead local electricity market is defined in the Day-Ahead (DA) stage. The purpose is to make the best decisions in each of the time periods during the day d. However, the DA stage obtains optimum decisions for the system in day d-1. Hence, the objective function for the DA stage is represented in (15):

$$EP^{da} = \sum_{t=1}^{N_t} (\lambda_t' P_{pv,out_t}^{da} + \sum_k \gamma_k \lambda_t' P_{dis,out_t}^{da}(k) - \lambda_t P_{net_t}^{da}) \tag{15}$$

EP^{da} consists of three parts. The first and second parts represent the revenue of selling the electrical energy produced by PV and Energy Storage Systems (ESSs) to the local market. The third part states the costs of buying the electrical energy from the local market. It should be mentioned that participation factor, γ_k, is a binary parameter that is defined for the first time in this paper in order to consider the participation of the ESSs in the DA stage. If the participation factor is equal to zero, ESSs are used to trade energy only in the real-time LEM. In other words, homes can utilize the full capacity of the ESSs in the day-ahead market if the participation factor equals one. The constraints of the DA stage are:

$$P_{net_t}^{da} + P_{pv,in_t}^{da} + \sum_k \gamma_k P_{dis,in_t}^{da}(k) = \sum_{j=1}^{N_j} L_{j_t}^{da} + \gamma_k P_{ch_t}^{da}(k) \tag{16}$$

$$-S_{max} \leq P_{net_t}^{da} - P_{pv,out_t}^{da} - \sum_k \gamma_k P_{dis,out_t}^{da}(k) \leq S_{max} \tag{17}$$

Equation (16) establishes the power balance equation due to the power outputs of the PV, P_{pv,in_t}^{da}, and ESSs, $P_{dis,in_t}^{da}(k)$, injected into the home, the grid power input, $P_{net_t}^{da}$, electrical loads, $L_{j_t}^{da}$, and the charged power of ESSs, $P_{ch_t}^{da}(k)$. In this paper, power loss is not considered for simplicity. Equation (17) represents the power flow limitation through the distribution line, which ends at the building. S_{max} expresses the maximum power capacity of the distribution line that links the smart home with the distribution power network. Besides, there are some limitations corresponding to all appliances. Only the maximum and minimum limitations of the energy produced/consumed are defined in each device at this stage because the uncertainty is not considered in the DA stage.

$$P_{pv_t}^{da} = P_{pv,in_t}^{da} + P_{pv,out_t}^{da} \tag{18}$$

$$P_{pv_t}^{pred} - \sigma_{pv_t}^{down}(1 - \alpha_{pv}) \leq P_{pv_t}^{da} \leq P_{pv_t}^{pred} + \sigma_{pv_t}^{up}\alpha_{pv} \tag{19}$$

$$L_{j_t}^{da} = L_{j_t}^{pred} \tag{20}$$

$$\sum_{j=1}^{N_j} L_{j_t}^{da} = L_{sh_t}^{da} + L_{swh_t}^{da} + L_{pp_t}^{da} + L_{mrs_t}^{da} \tag{21}$$

The total power generation of the PV is stated in (18). Equation (19) represents the power output limitations of the PV system. Besides, Equation (20) represents the total electrical power consumed.

Energy Storage Systems

ESSs can be utilized economically based on the charge and discharge strategies in the DEM problem. Mobility patterns and storage characteristics of the ESSs are different factors that should be considered in modeling the ESSs. However, the mobility pattern is only related to the EVs.

$$C_t^{da}(k) = C_{t-1}^{da}(k) + P_{ch_t}^{da}(k)\eta_{B2V} - P_{dis_t}^{da}(\omega)/\eta_{V2B}, t \geq 2 \tag{22}$$

$$C_t^{da}(\omega) = C_i, t = 1$$

$$P_{ev}^{min} \leq C_t^{da}(k) \leq P_{ev}^{max} \tag{23}$$

$$-w^{min} \leq C_t^{da}(k) - C_{t-1}^{da}(k) \leq w^{max}, t \geq 2 \tag{24}$$

$$-w^{min} \leq C_t^{da}(k) - C_i(k) \leq w^{max}, t = 1$$

$$0 \leq P_{dis_t}^{da}(k) \leq w^{max}u_t^{da} \tag{25}$$

$$0 \leq P_{ch_t}^{da}(k) \leq w^{min}(1 - u_t^{da}) \tag{26}$$

$$P_{dis_t}^{da}(k) = P_{dis,in_t}^{da}(k) + P_{dis,out_t}^{da}(k) \tag{27}$$

3.2. Real-Time Stage

In this stage, the objective function of the home due to participating in the RTLEM is defined. In addition, the uncertainties of decision-making variables are considered through a stochastic scenario-based method. These variables are determined based on the outputs of the first stage and the prediction engine. It is noticeable that the traded energy of the homes in the real-time market is different from their traded energy in the day-ahead market because of the PV power generation uncertainty. In other words, the traded energy of smart homes in real time can be positive or negative due to the prediction error of the PV power generation. The expected profit of the real-time stage, EP^{rt}, is represented as:

$$
\begin{aligned}
EP^{rt} = {} & \sum_{t=1}^{N_t} \sum_{\omega=1}^{N_\Omega} \pi(\omega)(\lambda_t(P_{pv,out_t}^{rt}(\omega) - P_{pv,out_t}^{da}) \\
& + \sum_k(\lambda_t(P_{dis,out_t}^{rt}(k,\omega) - \gamma_k P_{dis,out_t}^{da}(k)) - \lambda_t(P_{ch_t}^{rt}(\omega) - \gamma_k P_{ch_t}^{da}(k))) \\
& - \sum_{j=1}^{N_j} VOLL_j L_{j_t}^{shed}(\omega) - V_{pv}^s S_{pv_t}(\omega))
\end{aligned}
\tag{28}
$$

EP^{rt} consisting of five parts. The first part represents the revenue for selling energy produced by PV to the real-time local electricity market. The total cost of electrical energy that is bought from the BLEMis represented in the second part. The third part expresses the profit due to selling the stored electrical energy of ESSs to the local market. The Value of Loss Load (VOLL) cost, $VOLL_j$, is stated in the fourth part. Finally, the spillage cost of the PV system is represented in the last part. As seen in (28), it is proposed that if the PV power generation in the real-time stage, $P_{pv,out_t}^{rt}(\omega)$, is more than the PV power generation in the DA stage, the DEMS can only sell its extra power at the net price, λ, that is less than the price that is established for the purchase of the power generated by the PV on the day-ahead local market, λ'. Hence, the DEMS can increase its expected revenue if it has better day-ahead prediction accuracy of its PV power generation.

$$P_{net_t}^{rt}(\omega) + P_{pv,in_t}^{rt}(\omega) + \sum_k P_{dis,in_t}^{rt}(k,\omega) = \sum_{j=1}^{N_j} (L_{j_t}^{rt}(\omega) - L_{j_t}^{shed}(\omega))$$
$$+ \sum_k P_{ch_t}^{rt}(k,\omega) \tag{29}$$

$$-S_{max} \le P_{net_t}^{rt}(\omega) - (P_{pv,out_t}^{rt}(\omega) + \sum_k P_{dis,out_t}^{rt}(k,\omega)) \le S_{max} \tag{30}$$

In the balancing stage, Equation (29) is the power balance equation, and (30) shows the power flow limitation in a distribution line. Besides, there are specific definitions for all appliances in the building energy system whose uncertainties are considered in the balancing stage.

3.2.1. PV System

The power output of PV in the real-time stage, $P_{pv_t}^{rt}$, is obtained based on (31).

$$P_{pv_t}^{rt}(\omega) = P_{pv,p_t}^{rt}(\omega) - S_{pv_t}(\omega) \tag{31}$$
$$P_{pv_t}^{rt}(\omega) = P_{pv,in_t}^{rt}(\omega) + P_{pv,out_t}^{rt}(\omega) \tag{32}$$
$$0 \le S_{pv_t}(\omega) \le P_{pv,p_t}^{rt}(\omega) \tag{33}$$

Here, $P_{pv,p_t}^{rt}(\omega)$ is the potential power generation of PV in real time, and $S_{pv_t}(\omega)$ is the spillage power of the PV system. Equation (32) represents that the total power output of PV equals its power output consumed in the home, $P_{pv,in_t}^{rt}(\omega)$, and the amount of power generation that is sold to the real-time local market, $P_{pv,out_t}^{rt}(\omega)$. The PV spillage is the amount of power that is spilled in period t. This amount is positive or equal to zero and is limited to the actual power generation of PV as represented in (33).

3.2.2. Energy Storage Systems

ESSs can be utilized economically based on the charge and discharge strategies in the domestic energy management problem.

$$C_t^{rt}(k,\omega) = C_{t-1}^{rt}(k,\omega) + P_{ch_t}^{rt}(k,\omega)\eta_{B2V} - P_{dis_t}^{rt}(k,\omega)/\eta_{V2B}, t \ge 2 \tag{34}$$
$$C_t^{rt}(k,\omega) = C_i, t = 1$$
$$P_{ev}^{min} \le C_t^{rt}(k,\omega) \le P_{ev}^{max} \tag{35}$$
$$-w^{min} \le C_t^{rt}(k,\omega) - C_{t-1}^{rt}(k,\omega) \le w^{max}, t \ge 2 \tag{36}$$
$$-w^{min} \le C_t^{rt}(k,\omega) - C_i \le w^{max}, t = 1$$
$$0 \le P_{dis_t}^{rt}(k,\omega) \le w^{max}u_t^{rt} \tag{37}$$
$$0 \le P_{ch_t}^{rt}(k,\omega) \le w^{min}(1 - u_t^{rt}) \tag{38}$$
$$P_{dis_t}^{rt}(k,\omega) = P_{dis,in_t}^{rt}(k,\omega) \mid P_{dis,out_t}^{rt}(k,\omega) \tag{39}$$

The power generation of ESSs, $P_{dis_t}^{rt}(\omega)$, is expressed in (39). Equation (34) represents the state of charge balance equation in an ESS, where C_i is the initial state of charge in the ESS. Maximum and minimum limitations of the ESSs' state of charge are represented in Equation (35). Ramping constraints of ESSs are represented in Equation (36). Moreover, Equations (37) and (38) express the constraints of ESS in the discharge and charge states, respectively.

3.3. Electrical Loads

Electrical loads include loads that can be controllable and/or shiftable. In this paper, three types of loads are modeled: the space heater, L_{sh_t}, which is a controllable load, the storage water heater, L_{swh_t}, which is a shiftable load, and the must-run services, L_{mrs_t}, which are non-controllable-shiftable loads. Equations (40) and (41) define total electrical load and total load shedding, respectively. These loads are described in the following.

$$\sum_{j=1}^{N_j} L_{j_t}^{rt}(\omega) = L_{sh_t}^{rt}(\omega) + L_{swh_t}^{rt}(\omega) + L_{pp_t}^{rt}(\omega) + L_{mrs_t}^{rt}(\omega) \tag{40}$$

$$\sum_{j=1}^{N_j} L_{j_t}^{shed}(\omega) = L_{sh_t}^{shed}(\omega) + L_{swh_t}^{shed}(\omega) + L_{pp_t}^{shed}(\omega) + L_{mrs_t}^{shed}(\omega) \tag{41}$$

3.3.1. Space Heater

The space heater provides the indoor temperature at the desired temperature. Equation (42) represents the relation between the indoor temperature and its power consumption. In Equation (42), θ_0 is the initial indoor temperature, which is assumed to be equal to the desired temperature. Equation (43) expresses that indoor temperature is limited to 1 °C more or less than the desired temperature. Furthermore, the maximum and minimum bands of the space heater load are represented in (44). In addition, the load shedding constraint of the space heater is represented in (45).

$$\theta_{in_t+1}(\omega) = \theta_{in_t}(\omega)e^{-1/RC} + L_{sh_t}^{rt}(\omega)R(1 - e^{-1/RC}) \tag{42}$$
$$+ \theta_{out_t}^{pred}(1 - e^{-1/RC}), t \geq 2$$

$$\theta_{in_t}^{rt}(\omega) = \theta_0 = \theta_{des}, t = 1$$

$$-1 \leq \theta_{in_t}^{rt}(\omega) - \theta_{des} \leq 1 \tag{43}$$

$$L_{sh}^{min} \leq L_{sh_t}^{rt}(\omega) \leq L_{sh}^{max} \tag{44}$$

$$0 \leq L_{sh_t}^{shed}(\omega) \leq L_{sh_t}^{rt}(\omega) \tag{45}$$

3.3.2. Storage Water Heater

The storage water heater stores the heat in the water tank. The maximum and minimum limitations of the storage water heater's load and energy consumption are represented in (46) and (47), respectively. The load shedding constraint of the storage water heater is expressed in (48).

$$L_{swh}^{min} \leq L_{swh_t}^{rt}(\omega) \leq L_{swh}^{max} \tag{46}$$

$$\sum_{t=1}^{N_t} L_{swh_t}^{rt}(\omega) = U_{swh} \tag{47}$$

$$0 \leq L_{swh_t}^{shed}(\omega) \leq L_{swh_t}^{rt}(\omega) \tag{48}$$

3.3.3. Pool Pump

The maximum running hours of the pool pump equal T_{on} hours per day. Equation (49) represents the limitations of the pool pump power consumption in each hour. Equation (50) represents the maximum hour constraint that that pool pump can be turned on. Moreover, the load-shedding constraint related to the pool pump is represented in (51).

$$L_{pp}^{min}z_t(\omega) \le L_{pp}^{rt}(\omega) \le L_{pp_t}^{max}z_t(\omega) \tag{49}$$

$$\sum_{t=1}^{N_t} z_t(\omega) \le T_{on} \tag{50}$$

$$0 \le L_{pp_t}^{shed}(\omega) \le L_{pp_t}^{rt}(\omega) \tag{51}$$

3.3.4. Must-Run Services

Must-run services are defined as loads that should be provided quickly. In this paper, it is assumed that there is no uncertainty due to the prediction of must-run services as represented in Equation (52). Furthermore, the load shedding constraint is represented by (53).

$$L_{mrs_t}^{rt}(\omega) = L_{mrs_t}^{pred} \tag{52}$$

$$0 \le L_{mrs_t}^{shed}(\omega) \le L_{mrs_t}^{rt}(\omega) \tag{53}$$

4. Simulation Results

4.1. Case Study

To evaluate the performance of the proposed DEMS, the modified test system from [19] is used for which the wind micro-turbine has been omitted from the test system in this paper. The maximum power produced by the PV system is 2 kW. The battery can store between 0.48 and 2.4 kWh, and the maximum charging/discharging rates are 400 W. Besides, the charging and discharging efficiencies are 90%. The maximum heating power of the Space Heater (SH) equals 2 kW to maintain the temperature of the house within ±1 of the desired temperature (23 °C). The thermal resistance of the building shell equals 18 °C/kW, and Cequals 0.525 kWh/°C . The energy capacity of the Storage Water Heater (SWH) is 10.46 kWh, which has a 2-kW heating element. The rated power of the Pool Pump (PP) is 1.1 kW, and it can run for a maximum of 6 h during the day. The program implemented is solved in GAMS 23.7 [20]. Table 2 gives the price data of the system. Moreover, VOLL and the spillage costs of PV-battery power generation are shown in Table 3.

Table 2. ToU Price data of the system.

Time (hour)	Price ($/MW)	
	λ_i	λ_{net}
23–7	2.2	0.0814
8–14	2.2	0.1408
15–20	2.2	0.3564
21–22	2.2	0.1408

Table 3. Value of Loss Load (VOLL) and spillage costs. SH, Space Heater; SWH, Storage Water Heater; PP, Pool Pump.

Time (hour)	VOLL ($/MW)				Spillage Cost ($/MW)
	SH	SWH	PP	MRS	PV
22–7	1	1	−0.5	2.2	4
8–21	1	1	0.25	2.2	4

4.2. Impact of Energy Flexibility

In this section, the energy flexibility of the proposed DEMS is assessed. Hence, four scenarios are defined to analyze the performance of the system. In Scenario 1, neither the battery nor the EV are defined in the day-ahead stage of the energy management problem ($\gamma_{battery} = \gamma_{EV} = 0$). In Scenario 2, only the battery is considered in the day-ahead stage ($\gamma_{battery} = 1$, $\gamma_{EV} = 0$). However, only the EV is considered in the day-ahead stage in Scenario 3 ($\gamma_{battery} = 0$, $\gamma_{EV} = 1$). In Scenario 4, both (battery and EV) are modeled in the day-ahead stage ($\gamma_{battery} = \gamma_{EV} = 1$).

The impact of ESSs on the total, day-ahead and real-time expected profits of the system is shown in Figure 1. Furthermore, the influence of the optimistic coefficient, α, is evaluated in Figure 1. From this figure, it is clear that an increment of α increases the total and day-ahead expected profits because α can directly affect the power produced by the PV system through interval bands in the day-ahead stage. Hence, α increases the power generated from the PV panels in the day-ahead stage and the day-ahead expected profit. However, α has a negative impact on the amounts of the real-time expected profit. Moreover, the expected profit of the system is maximum in Scenario 4. In other words, increasing the energy flexibility of the system increases the total, day-ahead and real-time expected profits of the system. Hence, the maximum and minimum amounts of the expected profit are in Scenarios 4 and 1, respectively. Furthermore, the expected profit in Scenario 3 is more than Scenario 2 because the ramping rate of the EV is more than the battery. Therefore, the EV can provide more energy flexibility than the battery in this proposed system.

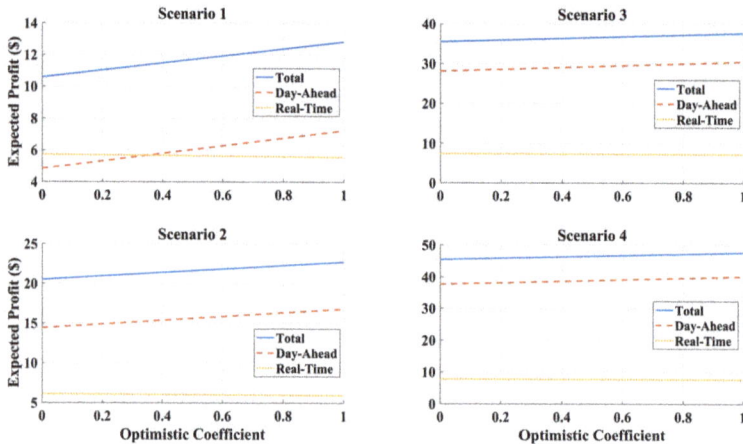

Figure 1. Impact of energy flexibility on the amounts of total, day-ahead and real-time expected profits.

4.3. Impact of Prediction Accuracy

The prediction accuracy due to the PV power generation and its influence on the total expected profit is analyzed in this section. It is noticeable that the prediction accuracy of the outdoor temperature of the home and must-run services is considered to be 100% in this paper in order to simplify the model. Besides, it is considered that the battery and EV are modeled in the day-ahead stage in this case. As mentioned before, α increases the amount of total expected profit of the system.

According to Figure 2, the impact of the prediction accuracy on the total expected profit is evaluated based on the optimistic coefficient. Furthermore, an increase in the prediction accuracy has a smooth negative effect on the expected profit. In other words, an increment in the prediction accuracy causes a decrease of the managed power of the PV in the proposed DEMS. Hence, this decreases the expected profit of the system. According to this assessment, the maximum amount of the total expected profit of the system is where α and the prediction accuracy equal one and zero, respectively.

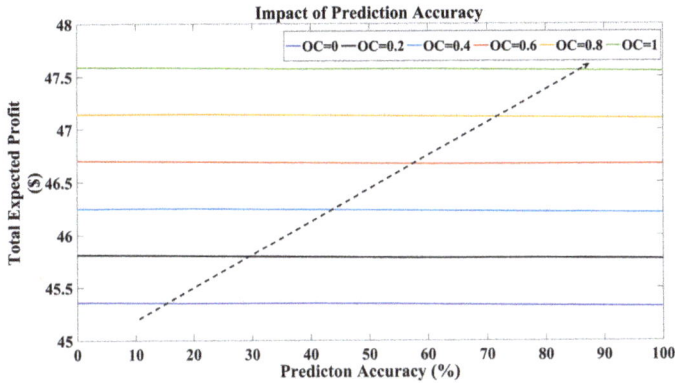

Figure 2. Impact of prediction accuracy on the total expected profit of the system. OC, Optimistic Coefficient.

4.4. Impact of Demand Response

In this section, the effect of the Demand Response Program (DRP) on the EPs and the home's electrical energy that is sold/bought to/from the local electricity market is assessed in four scenarios: with DRP, with only flexible VOLL, with only the ToU price and without DRP. Here, DRP consists of the flexible VOLL and ToU price.

As seen in Table 4, DRP causes a positive effect on the amount of total expected profit of the DEMS. In other words, while EP^{da} is increased when DRP is not considered in the system, EP^{rt} is decreased because electrical loads are not flexible when DRP is not considered in the DEMS. Furthermore, The sold/bought electrical energy of the DEMS considering DRP is more/less than without considering DRP because it makes DEMS able to shift the electrical load in the time horizon of the energy management problem and reduce the loads under some conditions. However, the impact of the flexible VOLL and ToU price are not the same. Although both of them increase the sold electrical energy, the total expected profit considering only flexible VOLL is more than considering only the ToU price. This is because of the positive effect of the flexible VOLL program in the real-time stage of the DEM problem.

Table 4. Impact of demand response program on the amount of expected profit of the system and sold/bought electrical energy to/from the local electricity market. DRP, Demand Response Program.

Demand Response Scenarios	$\alpha = 1$				
	EP_{total}	EP_{da}	EP_{rt}	E_{sold}	E_{bought}
With DRP (Flexible VOLL + ToU)	47.571	40.003	7.568	18.605	43.033
With Only Flexible VOLL	47.775	42.409	5.365	14.406	37.995
With Only ToU Price	42.071	40.003	2.068	15.236	49.432
Without DRP	42.275	40.409	−0.135	13.847	47.842

4.5. Impact of Uncertainty Modeling

In this section, the modeling of uncertainty is evaluated through a comparison of the InterStoch method and MSPB. Although the InterStoch method has been defined in this paper, MSPB has been defined in [18,19]. For simplicity, only the battery has been considered, and $\gamma_{battery}$ is equal to zero in this section. The amounts of total, day-ahead and real-time expected profits are compared in optimistic and conservative cases based on the InterStoch and MSPB methods. As seen in Table 5, the optimistic case of both methods is where α equals one. However, the pessimistic case based on the InterStoch and MSPB methods is where α equals zero and 0.4, respectively, as seen in Table 6. Tables 5 and 6 show that the difference between the amounts of the expected profits in the optimistic and conservative cases based on the InterStoch method is less than the MSPB method. Besides, Figure 3 shows the impact of α on the total expected profit in both methods. Figure 3 also illustrates that the worst case of the DEMS based on the InterStoch method is where α equals zero, and there is a linear pattern between the increment of the optimistic coefficient and the total expected profit when uncertainty is modeled by the InterStoch method. This point makes the system easier to analyze and more reliable, as it is able to further mitigate the uncertainty, dealing with it in away that its impact on the expected results is highly reduced. Moreover, the amount of the total expected profit in the worst case of the InterStoch is less than its amount in the worst case of the MSPB method. Hence, the InterStoch method is more robust than the MSPB method to model uncertainty in the proposed domestic energy management problem.

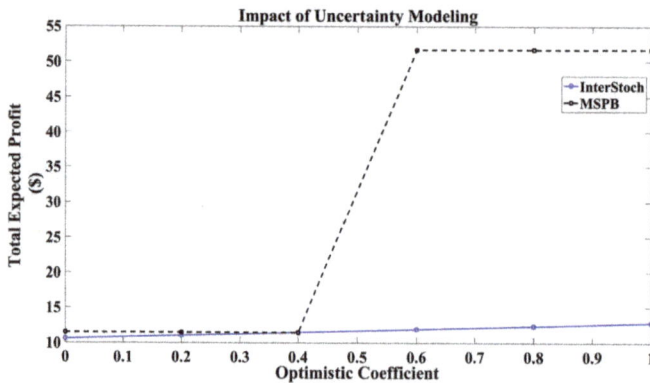

Figure 3. Impact of uncertainty modeling on the total expected profit of the system.

Table 5. Impact of uncertainty modeling on day-ahead, real time and total expected profits under the optimistic case. InterStoch, Interval-Stochastic; MSPB, Modified Stochastic Predicted Band.

Expected Profit ($)	InterStoch ($\alpha = 1$)		MSPB ($\alpha = 1$)	
	With Uncertainty	Without Uncertainty	With Uncertainty	Without Uncertainty
EP_{total}	12.798	10.549	51.707	51.618
EP_{da}	7.234	4.836	49.232	49.232
EP_{rt}	5.564	5.713	2.475	2.386

Table 6. Impact of uncertainty modeling on day-ahead, real time and total expected profits under the conservative case.

Expected Profit ($)	InterStoch ($\alpha = 0$)		MSPB ($\alpha = 0.4$)	
	With Uncertainty	Without Uncertainty	With Uncertainty	Without Uncertainty
EP_{total}	10.569	10.549	11.449	51.618
EP_{da}	4.836	4.836	4.836	49.232
EP_{rt}	5.733	5.713	6.613	2.386

5. Conclusions

In this paper, the energy flexibility management of the home electricity system based on the predictive dispatch model has been introduced. Furthermore, the InterStoch hybrid method to model the uncertainty of the PV power generation has been defined for the first time in this paper. The proposed method consists of two stages. In the first stage, the day-ahead domestic energy management problem has been modeled by an interval method to consider the uncertainty due to the prediction error of PV power generation. However, a real-time problem has been represented based on the stochastic method to consider the uncertainty. The performance of the proposed domestic energy management problem has been evaluated based on a comparison between the proposed hybrid (InterStoch) and MSPB methods. Furthermore, the impact of the proposed energy flexibility model, prediction accuracy and demand response program on the expected profit and transacted electrical energy of the system and the reliability of the results has been assessed. From the simulation, it is concluded that:

- Increasing the energy flexibility increases the total, day-ahead and real-time expected profits of the system.
- The EV can provide more energy flexibility than the battery in the proposed system.
- The increment of α increases the PV power produced in the day-ahead stage and day-ahead expected profit. However, α has a negative impact on the amounts of the real-time expected profit.
- The increment of the prediction accuracy has a smooth negative impact on the expected profit.
- For the considered case study, the demand response program has a positive effect on the amount of the DEMS's total expected profit. Furthermore, the demand response program decreases the domestic electrical energy load.
- The amount of the total expected profit in the worst case of InterStoch is less than its amount in the worst case of the MSPB method. Hence, the InterStoch method is more robust than the MSPB method to model uncertainty in the proposed domestic energy management problem.

Finally, it should be mentioned that the uncertainty of electrical load, the EV mobility pattern and market prices have not been modeled in our proposed DEMS. Our future work will consist of modeling the uncertainty related to the EV and must-run services and to evaluate their impacts on the transacted energy of the homes for the local electricity market.

Acknowledgments: This work has been supported by the European Commission H2020 MSCA-RISE-2014: Marie Sklodowska-Curie project DREAM-GO Enabling Demand Response for short and real-time Efficient And Market Based Smart Grid Operation—An intelligent and real-time simulation approach Ref. 641794, Grant Agreement No. 703689 (Project ADAPT) and Project SURF: Intelligent System for integrated and sustainable management of urban fleets TIN2015-65515-C4-3-R that has been supported by the Spanish Ministry, Ministerio de Economía y Competitividad and FEDER funds. Moreover, Amin Shokri Gazafroudi acknowledge the support by the Ministry of Education of the Junta de Castilla y León and the European Social Fund through a grant from predoctoral recruitment of research personnel associated with the research project "Arquitectura multiagente para la gestión eficaz de redes de energía a través del uso de técnicas de intelligencia artificial" of the University of Salamanca.

Author Contributions: Amin Shokri Gazafroudi developed the hybrid interval-stochastic method and model the residential energy management problem. Francisco Prieto-Castrillo developed the mathematical model to realize the stochastic scenarios of the system. Tiago Pinto implemented the energy management system in the test system. The remaining co-authors addressed key ideas for the project development.

Conflicts of Interest: The authors declare no conflict of interest regarding the publication of this paper.

Nomenclature

Indices

t	Index of time periods
j	Index of electrical loads
k	Index of energy storage systems
ω	Index of PV power scenarios

Variables

EP	Expected profit
EP^{da}	Day-ahead expected profit
EP^{rt}	Real-time expected profit
$P^{da}_{pv_t}$	Day-ahead total power generation for the PV system in period t
P^{da}_{pv,out_t}	Day-ahead power generation for the PV system that is injected to the power grid in period t
P^{da}_{pv,in_t}	Day-ahead power generation for the PV system that is injected to the home in period t
$P^{da}_{dis_t}(k)$	Day-ahead total discharged power for energy storage system k in period t
$P^{da}_{dis,out_t}(k)$	Day-ahead discharged power for energy storage system k that is injected to the power grid in period t
$P^{da}_{dis,in_t}(k)$	Day-ahead discharged power for energy storage system k that is injected to the home in period t
$P^{da}_{ch_t}(k)$	Day-ahead charged power for energy storage system k that is injected to the home in period t
$P^{da}_{net_t}$	Day-ahead power generation that is bought from the local electricity market in period t
$L^{da}_{j_t}$	Day-ahead electrical load j in period t
$L^{da}_{sh_t}$	Day-ahead electrical load of the space heater in period t
$L^{da}_{swh_t}$	Day-ahead electrical load of the storage water heater in period t
$L^{da}_{pp_t}$	Day-ahead electrical load of the pool pump in period t
$L^{da}_{mrs_t}$	Day-ahead electrical load of the must-run services in period t
$C^{da}_t(k)$	Day-ahead state of charge for energy storage system k in period t
u^{da}_t	Day-ahead discharging commitment binary variable for energy storage system k in period t
$P^{rt}_{pv_t}(\omega)$	Real-time total power generation for the PV system in period t and in scenario ω
$P^{rt}_{pv,out_t}(\omega)$	Real-time power generation for the PV system that is injected to the power grid in period t and in scenario ω
$P^{rt}_{pv,in_t}(\omega)$	Real-time power generation for the PV system that is injected to the home in period t and in scenario ω
$P^{rt}_{dis_t}(k,\omega)$	Real-time total discharged power for energy storage system k in period t and in scenario ω
$P^{rt}_{dis,out_t}(k,\omega)$	Real-time discharged power for energy storage system k that is injected to the power grid in period t and in scenario ω
$P^{rt}_{dis,in_t}(k,\omega)$	Real-time discharged power for energy storage system k that is injected to the home in period t and in scenario ω
$P^{rt}_{ch_t}(k,\omega)$	Real-time charged power for energy storage system k that is injected to the home in period t and in scenario ω
$P^{rt}_{net_t}(\omega)$	Real-time power generation that is bought from local electricity market in period t and in scenario ω
$L^{rt}_{j_t}$	Real-time electrical load j in period t and in scenario ω
$L^{shed}_{j_t}(\omega)$	Load shedding for load j in period t and in scenario ω
$S_{pv_t}(\omega)$	Spillage amount for PV in period t and in scenario ω
$P^{rt}_{pv,p_t}(\omega)$	Potential power generation for PV in real time in period t and in scenario ω
$L^{rt}_{sh_t}(\omega)$	Real-time electrical load of the space heater in period t and in scenario ω
$L^{rt}_{swh_t}(\omega)$	Real-time electrical load of the storage water heater in period t and in scenario ω
$L^{rt}_{pp_t}(\omega)$	Real-time electrical load of the pool pump in period t and in scenario ω
$L^{rt}_{mrs_t}(\omega)$	Real-time electrical load of the must-run services in period t and in scenario ω
$C^{rt}_t(k,\omega)$	Real-time state of charge for energy storage system k in period t and in scenario ω
$u^{rt}_t(\omega)$	Real-time discharging commitment binary variable for energy storage system k in period t and in scenario ω
$L^{shed}_{sh_t}(\omega)$	Load shedding for the space heater in period t and in scenario ω
$L^{shed}_{swh_t}(\omega)$	Load shedding for the storage water heater in period t and in scenario ω
$L^{shed}_{pp_t}(\omega)$	Load shedding for the pool pump in period t and in scenario ω
$L^{shed}_{mrs_t}(\omega)$	Load shedding for the must-run services in period t and in scenario ω
$\theta_{in_t}(\omega)$	Indoor temperature in period t and in scenario ω
$z_t(\omega)$	Commitment binary variable for the pool pump k in period t and in scenario ω

Parameters

$P_{pv_t}^{pred}$	Central forecasting of the PV power generation in period t
$\sigma_{pv_t}^{down}$	Down deviation of the PV power prediction in period t
$\sigma_{pv_t}^{up}$	Up deviation of the PV power prediction in period t
α_{pv}	Optimistic coefficient related to the PV power prediction
$P_{pv_t}^{mean}$	Mean of the PV power prediction in period t
Δ_{pv_t}	Mean deviation of the PV power prediction in period t
$\pi(\omega)$	Probability of the PV power generation in scenario ω
$\lambda_t^{'}$	Sold electricity price to the local electricity market in period t
λ_{net_t}	Bought electricity price from the local electricity market in period t
γ_k	Participation factor for energy storage system k
S_{max}	Maximum power capacity for the line
$L_{j_t}^{pred}$	Predicted electrical load j in period t
η_{B2V}	Charging efficiency for energy storage systems j
η_{V2B}	Discharging efficiency for energy storage systems j
C_i	Initial state of charge for energy storage systems
w^{max}	Maximum charging/discharging for energy storage systems
w^{min}	Minimum charging/discharging for energy storage systems
$VOLL_j$	Value of loss load for electrical load j
V_{pv}^s	Spillage cost for the PV system
θ_0	Initial indoor temperature
θ_{des}	Desired indoor temperature
$\theta_{out_t}^{pred}$	Predicted outdoor temperature
L_{sh}^{max}	Maximum electrical consumption for the space heater
L_{sh}^{min}	Minimum electrical consumption for the space heater
R	Thermal resistance of the building shell
L_{swh}^{max}	Maximum electrical consumption for the storage water heater
L_{swh}^{min}	Minimum electrical consumption for the storage water heater
U_{swh}	Energy consumption for the storage water heater
L_{pp}^{max}	Maximum electrical consumption for the pool pump
L_{pp}^{min}	Minimum electrical consumption for the pool pump
T_{on}	Maximum running hours for the pool pump
$L_{mrs_t}^{pred}$	Predicted electrical load of the must-run services in period t

References

1. Abrishambaf, O.; Gomes, L.; Faria, P.; Afonso, J.L.; Vale, Z. Real-time simulation of renewable energy transactions in microgrid context using real hardware resources. In Proceedings of the 2016 IEEE/PES Transmission and Distribution Conference and Exposition (T&D), Dallas, TX, USA, 3–5 May 2016; pp. 1–5.

2. Vale, Z.; Morais, H.; Faria, P.; Ramos, C. Distribution system operation supported by contextual energy resource management based on intelligent SCADA. *Renew. Energy* **2013**, *52*, 143–153.

3. Abrishambaf, O.; Ghazvini, M.A.F.; Gomes, L.; Faria, P.; Vale, Z.; Corchado, J.M. Application of a Home Energy Management System for Incentive-Based Demand Response Program Implementation. In Proceedings of the 2016 27th International Workshop on Database and Expert Systems Applications (DEXA), Porto, Portugal, 5–8 September 2016; pp. 153–157.

4. Manic, M.; Wijayasekara, D.; Amarasinghe, K.; Rodriguez-Andina, J.J. Building Energy Management Systems The Age of Intelligent and Adaptive Buildings. *IEEE Ind. Electron. Mag.* **2016**, *10*, 25–39.

5. Shokri Gazafroudi, A.; de Paz, J.F.; Prieto-Castrillo, F.; Villarrubia, G.; Talari, S.; Shafie-khah, M.; Catalão, J.P.S. A Review of Multi-agent Based Energy Management Systems. In Proceedings of the ISAmI 2017: Ambient Intelligence—Software and Applications—8th International Symposium on Ambient Intelligence (ISAmI 2017), Porto, Portugal, 21–23 June 2017; pp. 203–209.

6. Monteiro, V.; Pinto, J.G.; Afonso, J.L. Operation Modes for the Electric Vehicle in Smart Grids and Smart Homes: Present and Proposed Modes. *IEEE Trans. Veh. Technol.* **2016**, *65*, 1007–1020.

7.   Zhao, C.; Dong, S.; Li, F.; Song, Y. Optimal Home Energy Management System with Mixed Types of Loads. *CSEE J. Power Energy Syst.* **2015**, *1*, 1–11.

8.   Pratt, A.; Krishnamurthy, D.; Ruth, M.; Wu, H.; Lunacek, M.; Vaynshenk, P. Transactive Home Energy Management Systems: The Impact of Their Proliferation on the Electric Grid. *IEEE Electrif. Mag.* **2016**, *4*, 8–14.

9.   Wang, Z.; Paranjape, R. Optimal Residential Demand Response for Multiple Heterogeneous Homes with Real-Time Price Prediction in a Multiagent Framework. *IEEE Trans. Smart Grid* **2017**, *8*, 1173–1184.

10. Paterakis, N.G.; Erdinç, O.; Bakirtzis, A.G.; Catalão, J.P.S. Optimal Household Appliances Scheduling under Day-Ahead Pricing and Load-Shaping Demand Response Strategies. *IEEE Trans. Ind. Inform.* **2015**, *11*, 1509–1519.

11. Erdinc, O.; Paterakis, N.G.; Mendes, T.D.P.; Bakirtzis, A.G.; Catalão, J.P.S. Smart Household Operation Considering Bi-Directional EV and ESS Utilization by Real-Time Pricing-Based DR. *IEEE Trans. Smart Grid* **2015**, *6*, 1281–1291.

12. Sarker, M.R.; Ortega-Vazquez, M.A.; Kirschen, D.S. Optimal Coordination and Scheduling of Demand Response via Monetary Incentives. *IEEE Trans. Smart Grid* **2015**, *6*, 1341–1352.

13. Althaher, S.; Mancarella, P.; Mutale, J. Automated Demand Response From Home Energy Management System Under Dynamic Pricing and Power and Comfort Constraints. *IEEE Trans. Smart Grid* **2015**, *6*, 1874–1883.

14. Althaher, S.; Mancarella, P.; Mutale, J. Equivalence of Multi-Time Scale Optimization for Home Energy Management Considering User Discomfort Preference. *IEEE Trans. Smart Grid* **2017**, *8*, 1876–1887.

15. Huang, Y.; Wang, L.; Guo, W.; Kang, Q.; Wu, Q. Chance Constrained Optimization in a Home Energy Management System. *IEEE Trans. Smart Grid* **2016**, *PP*, 1-1.

16. Yoon, S.; Choi, Y.; Park, J.; Bahk, S. Stackelberg Game based Demand Response for At-Home Electric Vehicle Charging. *IEEE Trans. Veh. Technol.* **2016**, *65*, 4172–4184.

17. Wu, X.; Hu, X.; Yi, X.; Moura, S. Stochastic Optimal Energy Management of Smart Home with PEV Energy Storage. *IEEE Trans. Smart Grid* **2016**, *PP*, 1-1.

18. Shokri Gazafroudi, A.; Pinto, T.; Prieto-Castrillo, F.; Prieto, J.; Corchado, J.M.; Jozi, A.; Vale, Z.; Venayagamoorthy, G.K. Organization-based Multi-Agent Structure of the Smart Home Electricity System. In Proceedings of the 2017 IEEE Congress on Evolutionary Computation (CEC), San Sebastian, Spain, 5–8 June 2017.

19. Shokri Gazafroudi, A.; Prieto-Castrillo, F.; Corchado, J.M. Residential Energy Management Using a Novel Interval Optimization Method. In Proceedings of the 4th International Conference on Control, Decision and Information Technology 2017 (CoDIT 2017), Barcelona, Spain, 5–8 April 2017.

20. GAMS Release 2.50. A User's Guide. GAMS Development Corporation, 1999. Available online: http://www.gams.com (accessed on 2 August 2017).

energies

MDPI

Article

Demand Response Unit Commitment Problem Solution for Maximizing Generating Companies' Profit

K. Selvakumar *, K. Vijayakumar and C. S. Boopathi

Department of EEE, SRM University, Chennai 603203, India; kvijay_srm@rediffmail.com (K.V.); csbsrm@gmail.com (C.S.B.)
* Correspondence: selvakse@gmail.com; Tel.: +91-984-338-9383

Received: 11 August 2017; Accepted: 19 September 2017; Published: 22 September 2017

Abstract: Over the recent years there has been an immense growth in load consumption due to which, Load Management (LM) has become more significant. Energy providers around the world apply different load management concepts and techniques to improve the load profile. In order to reduce the stress over the load management, Demand Response Unit Commitment (DRUC), a new concept, has been implemented in this paper. The main feature of this concept is that both the energy providers and consumers must participate in order to get mutual benefits hence maximizing each of their profits. In this paper we discuss the time-based Demand Response Program since there is no penalty observed in this program. When the Demand Response was combined with Unit Commitment and compiled it was observed that a satisfactory solution resulted, which is proved to be mutually beneficial for both Generating Companies (GENCOs) and their customers. Here, we have used a Cat Swarm Optimization (CSO) technique to find the solution for the DRUC problem. The results are obtained using CSO technique for UC problem with and without DR program. This is compared with the results obtained using other conventional methods. The test system considered for the study is IEEE39 bus system.

Keywords: Unit Commitment (UC); Demand Response (DR); Demand Response Unit Commitment (DRUC); Cat Swarm Optimization (CSO)

---

1. Introduction

With the improvements in the power sector field over the decades, there has also been a vast increase in load consumption due to heavy demand. Sometimes the load required is very high due to multiple consumers requiring power at the same time [1]. Due to this issue, GENCOs are sometimes not able to meet the customer demands, hence making them unsatisfied or prompting them to terminate their contracts. Some of the growing issues associated with power system operation include limited supply of system resources that in turn forces the operators to operate their systems at their maximum capacity, resulting in regular price hikes in the electricity market [2]. All the aforementioned limitations motivate us to search for and explore novel ways to increase the efficiency of resource utilization in power operations. As one of these new ways, Demand Response (DR) has recently become a major concept in power system operation. The use of Demand Response management in power systems enables the operators to efficiently utilize their resources as well as the power system operation. The use of Demand Response Programs (DRPs) in power system operation increases the profit of customers as well as the operators. It also encourages customers' participation in the Demand Response Program (DRP) by rewarding them with incentives, if they agree to reduce their load demands during the peak hours of the day [3].

As per the Federal Energy Regulatory Commission (FERC), Demand Response can be defined as "Changes in electric usage by end-use customers from their normal consumption patterns in response to changes in the price of electricity over time, or to incentive payments designed to induce lower electricity use at times of high wholesale market prices or when system reliability is jeopardized" [4]. This is quite a different concept from energy efficiency that involves using less power for the same task. Demand response is also a component of smart energy demand that includes energy efficiency, home and building energy management, distributed renewable resources and electric vehicle charging [5]. The implementation of DRP in power system operation reduces the load stress on the equipment, hence ensuring a maximum efficiency and power. According to the Federal Energy Regulatory Commissions (FERC) report on demand response programs implemented in the US electricity markets from 2006 [6,7] DRP is broadly divided into two major categories:

(a)   Time-Based Rate Programs (TBRP):

Time-Based Rate Programs (TBRPs) are programs that involve changes in the forecasted price that varies with the time of day, so the consumer can change or reduce their load usage for the respective hours accordingly. TBRPs are subcategorized into three programs, namely time of use, critical peak pricing and real time pricing programs. In time of use programs the main aim is to reduce the demand (peak periods) by increasing the prices at the high demand hour causing customers to shift or reduce their loads and lowering the prices where load management (off peak) use is possible. This attracts and encourages the customers to use load during off-peak hours. It is a basic type, where the rates of load per unit consumption vary in different time blocks. The rates during peaks are high and during off-peak periods are low [8]. Critical peak pricing rates consist of a pre-specified high load usage price imposed on Time of Use rates. These rates are applied for a short period of days or hours of a year. In real time pricing programs, the consumers are faced with hourly varying prices that reflect the real price of load in the market at that time. Customers under this program are informed in advanced about the prices on a day before or an hour before [9].

(b)   Incentive-Based Programs (IBP):

IBPs are all based on paying or receiving a small amount in the form of penalties/incentives. IBPs are sub categorized into (i) Direct Load Control (ii) Emergency DR Program (iii) Demand Bidding/Buyback program (iv) Ancillary Services Program (v) Interruptible/Curtailable Service and (vi) Capacity Market program.

Direct Load Control involves programs where the loads are remotely controlled by the GENCOs, so they can be remotely committed or decommitted during peak hours in order to reduce the load stress. Some of the remotely controlled loads may include air conditioners, pumps and water heaters. Emergency DR Programs (EDR) require customers to curtail their loads during system emergencies [10]. The customers are in turn rewarded with incentives for curtailing their loads. In both Direct Load Control (DLC) and EDR programs, the customers are not penalized, if they fail to achieve the objectives, because they are involved in voluntary programs. In Demand Bidding/Buyback programs, the customers are encouraged to curtail load at a rate by which they are satisfied or how much load they are willing to curtail at the given price. In Ancillary Services Programs customers are made to bid and challenge their load curtailment values in markets as operating reserves [11].

Interruptible/Curtailable Service programs are the programs where the enrolled customers are asked to curtail their loads during the peak demand hours of a day in order to reduce load stress. They are in turn paid certain incentives to do so. If they fail to curtail the desired amount of load, they are penalized. In Capacity Market Programs, the customers are willing to perform pre-informed load curtailments for certain incentive rewards. Failing to do so will cause a penalization. Implementation of DR along UC not only reduces the load stress during peak hours, but it also increases the profits of GENCOs and makes the system more reliable. DR helps UC by shaving off loads during peak hours using various methods and thus causing an increase in profits and making systems more reliable and robust [12,13].

In order to implement demand response in smart grids, we should be able to coordinate large number of distributed resources using sensors, communication protocols and actuators. In addition, the increased presence of different renewable generation drives a much larger need for officials to procure more ancillary services in order to balance the grid [14,15]. Demand response is also provided by industrial customers. Industrial manufacturing plants' magnitude of power consumption is very large compared to commercial and residential loads [16]. Demand response implementation was imposed in the United States by FERC Order No. 745 in March 2011 [4]. Reduction of loads during peak hours decreases the need for installing new units. According to the demand response smart grid coalition, around 10–20% of electricity costs are due to peak demand in the United States [17]. It was found in the California electricity crisis in 2000–2001 that lowering the demand by mere 5% would have resulted in a 50% of price reduction during peak hours [18]. A suit was filed regarding legality of order 745 by many affected parties, including the State of California [19]. From December 2009, the UK national grid has contracted to provide DR of 839 MW (35%) [20]. The mathematical formulation of the Market Clearing Model based on DRP was implemented in Singapore to improve the wholesale market profit [21]. The analysis on various power sectors of Germany was improved with wind power prediction [22].

The impact of UC and DRUC problems was studied by a dynamic approach on an IEEE 10 unit system [23]. Zhang et al. proposed how renewable energy resources can play a vital role in the future power system. How it can be used along with DR and electric vehicles in a UC problem to utilize wind power efficiently by using fuzzy chance constraints has also been studied [24]. The wind uncertainty can be overcome using ancillary services from Pumped Hydro Energy System (PHES) and DR and simultaneous scheduling of PHES and DR along with wind uncertainties has been attempted by solving an LR-based probabilistic UC [25].

The IBP based multi-objective energy management system is proposed in order to optimize micro grids by PSO [26]. Kwag et al. discussed virtual generation and the various costs reduction by using DR [27].The growing load factors in the Spanish electric energy system causing higher loads and increased cost and its reduction by demand shifting and curtailment were examined in [28]. The UC model is presented for accessing the reserve requirements resulting from large scale integration of renewable energy sources and deferrable demand in power systems and the alternative DR paradigms are discussed for accessing the benefits of demand flexibility in [29]. A robust optimization technique with wind power to derive an optimal UC was developed in [30]. Based on the explosion of fireworks in the sky, a unit commitment problem in a deregulated environment was modeled and the GENCOs' profits were maximized [31]. An economic model of responsive loads is derived based upon price elasticity of demand and customers benefit function in [7]. Govardhan proposed a linear load economic model for solving the demand response unit commitment problem by using an Artificial Bee Colony algorithm [32]. The critical kick-back effect has been applied to a DR program for maximizing the profit in peak hours in a day in [33].

2. Demand Response Unit Commitment Problem Formulation

Traditional Unit Commitment (TUC) is the process of scheduling power generation, without violating the systems or units operational constraints. The traditional unit commitment problem objective function focuses on minimization of generation cost along with fuel costs and startup costs [32,34–37]. In this paper, the demand response based unit commitment problem is modeled and the main objective of the demand response unit commitment problem is used to maximize the profits of the GENCO using a Time-Based Demand Response Program (TBDRP) [38,39].

The objective function is as follows:

$$\text{Max } P_R = [TR_V - TO_{\text{COST}}] \tag{1}$$

where, P_R—is the total profit of the GENCOs and Demand Response Service Provider (DRSP) combined, TR_V—is the total revenue calculated from the GENCOs and DRSP, TO_{COST}—is the total operating cost of the GENCOs and DRSP combined:

$$TR_v = \left[\sum_{t=1}^{T} \sum_{i=1}^{N} P_{gen}^{i,t} \, S_{price}^{i} \, U_{stat}^{i,t} \right] + \left[\sum_{t=1}^{T} \sum_{di=1}^{dN} P_{gen}^{di,t} \, S_{price}^{di} \, U_{stat}^{di,t} \right] \tag{2}$$

$$TO_{cost} = \left[\sum_{t=1}^{T} \sum_{i=1}^{N} F_{cost}^{i} \left(P_{gen}^{i,t} \right) U_{stat}^{i,t} + SU_{cost} \right] + \left[\sum_{t=1}^{T} \sum_{di=1}^{dN} F_{cost}^{di} \left(P_{gen}^{di,t} \right) U_{stat}^{di,t} + SU_{cost} \right] \tag{3}$$

$$F_{cost}^{i} = a^{i} \left(P_{gen}^{i} \right)^{2} + b^{i} \left(P_{gen}^{i} \right) + c^{i} \tag{4}$$

2.1. Mathematical Modelling of DRUC

The main objective of GENCOs in a deregulated environment is to maximize their profits, their objective being minimizing the cost of energy supplied to the consumers. Hence the traditional unit commitment is modeled with demand response program. The market clearing price in the demand response program is calculated from the DRSP supply curve coefficients and based on customer's willingness to participate in a Demand Response Program. The demand response market clearing price is formulated by the following equation:

$$DR_{price} = \theta^{di} DR_{gen}^{di} + \delta^{di} \left(1 - \mu^{di} \right) ; \; (di = 1 \cdots dN) \tag{5}$$

Here, μ^{di} is the customer's willingness to participate in a *DR* program. Its value is between 0 and 1, and the higher the willingness of customers, the less is the *DR* cost. θ^{di} and δ^{di} are DRSP coefficients for all customers [40]. Rewriting the above equation as:

$$DR_{price} = \theta^{di} DR_{gen}^{di} + \Delta\delta^{di} \tag{6}$$

where:

$$\Delta\delta^{di} = \delta^{di} \left(1 - \mu^{di} \right) \tag{7}$$

Rearranging the above equation, we get:

$$DR_{gen}^{di} = \frac{DR_{price} - \Delta\delta^{di}}{\theta^{di}} , \; i = 1 \cdots dN \tag{8}$$

Equality must be maintained between the sold and purchased value of *DR*, and using this constraint the following equation is modeled:

$$DR_{req} = \sum_{di=1}^{dN} DR_{gen}^{di} = \sum_{di=1}^{dN} \frac{DR_{price} - \Delta\delta^{di}}{\theta^{di}} \tag{9}$$

$$DR_{price} = \frac{DR_{req} + \sum_{di=1}^{dN} \frac{\Delta\delta^{di}}{\theta^{di}}}{\sum_{di=1}^{dN} \frac{1}{\theta^{di}}} \tag{10}$$

The higher the willingness of customers to participate in DPR, the less will be the value of $\Delta\delta^{di}$. Similarly, the value of $\Delta\delta^{di}$ increases as customer willingness decreases. The profit maximization equation for DRSPs is defined as:

$$PDR_{f}^{di} = DR_{price} DR_{gen}^{di} - DRO_{cost}^{di} ; \; di = 1 \cdots dN \tag{11}$$

Substituting DR^{di}_{gen} and DRO^{di}_{cost} in Equation (11), we get:

$$PDR^{di}_f = DR_{price} \times \left(\frac{DR_{price} - \Delta\delta^{di}}{\theta^{di}} \right) - \left[\theta m^{di} \times \left(\frac{DR_{price} - \Delta\delta^{di}}{\theta^{di}} \right)^2 + \delta m^{di} \times \left(\frac{DR_{price} - \Delta\delta^{di}}{\theta^{di}} \right) + \phi m^{di} \right] \; ; di = 1 \dots dN \quad (12)$$

where, the coefficients θm^{di}, δmdi and ϕm^{di} are referred to the customers' supply curve cost coefficients. θ^{di} is always considered equal to θm^i. Taking derivation of the profit function with respect to $\Delta\delta^{di}$ we get:

$$
\begin{bmatrix} \Delta\delta^1(k) \\ \Delta\delta^2(k) \\ \vdots \\ \Delta\delta^{dN}(k) \end{bmatrix} =
\begin{bmatrix}
0 & \frac{\theta^1}{\theta^2} \times \frac{1}{(\theta^1)^2 K^2 - 1} & \cdot & \frac{\theta^1}{\theta^{dN}} \times \frac{1}{(\theta^1)^2 K^2 - 1} \\
\frac{\theta^2}{\theta^1} \times \frac{1}{(\theta^2)^2 K^2 - 1} & 0 & \cdot & \frac{\theta^2}{\theta^{dN}} \times \frac{1}{(\theta^2)^2 K^2 - 1} \\
\vdots & \vdots & \vdots & \vdots \\
\frac{\theta^{dN}}{\theta^1} \times \frac{1}{(\theta^{dN})^2 K^2 - 1} & \frac{\theta^{dN}}{\theta^2} \times \frac{1}{(\theta^{dN})^2 K^2 - 1} & \cdots & 0
\end{bmatrix}
\times
\begin{bmatrix} \Delta\delta^1(k) \\ \Delta\delta^2(k) \\ \vdots \\ \Delta\delta^{dN}(k) \end{bmatrix}
+
\begin{bmatrix}
\frac{\theta^1 K}{\theta^1 K+1} & 0 & \cdots & 0 & \frac{\theta^1}{(\theta^1)^2 K+1} \\
0 & \frac{\theta^2 K}{\theta^2 K+1} & \cdots & 0 & \frac{\theta^2}{(\theta^2)^2 K+1} \\
\vdots & \vdots & \vdots & \vdots & \vdots \\
0 & 0 & \cdot & \frac{\theta^{dN} K}{\theta^{dN} K+1} & \frac{\theta^{dN}}{(\theta^{dN})^2 K+1}
\end{bmatrix}
\begin{bmatrix} \Delta\delta m^1(k) \\ \Delta\delta m^2(k) \\ \vdots \\ \Delta\delta m^{dN}(k) \\ DR_{req} \end{bmatrix}
$$

and:

$$DR_{price} = \begin{bmatrix} \frac{1}{\theta^1 K} & \frac{1}{\theta^2 K} & \cdots & \frac{1}{\theta^{dN} K} \end{bmatrix} \begin{bmatrix} \Delta\delta^1(k) \\ \Delta\delta^2(k) \\ \vdots \\ \Delta\delta^{dN}(k) \end{bmatrix} + \left(\frac{1}{K} \times DR_{req} \right) \quad (13)$$

Equating Equations (10) and (13) we get:

$$DR_{price} = \frac{DR_{req} + \frac{\Delta\delta^{di}}{\theta^{di}} + \sum\limits_{di \neq dj}^{dN} \frac{\Delta\delta^{dj}}{\theta^{dj}}}{\sum\limits_{di=1}^{dN} \frac{1}{\theta^{di}}} \quad (14)$$

Rearranging the above equation, we get:

$$\Delta\delta^{di} = \sum\limits_{di \neq dj}^{dN} \left(\frac{\theta^{di}}{\theta^{dj}} \times \frac{1}{\left(\theta^{di}\right)^2 K - 1} \times \Delta\delta^{dj} \right) + \left(\frac{\theta^{di}}{\left(\theta^{di}\right)^2 K - 1} \times DR_{req} \right) + \left(\frac{\theta^{di} K}{\theta^{di} K - 1} \times \delta m^{di} \right) \quad (15)$$

where:

$$K = \sum\limits_{di=1}^{dN} \frac{1}{\theta^{di}} \quad$$

2.2. Traditional Unit Commitment Constraints

2.2.1. Equality Constraint

$$\sum\limits_{i=1}^{N} \left(P^{i,t}_{gen} \, U^{i,t}_{stat} \right) = P^t_{dem} \; ; (t = 1 \quad T) \quad (16)$$

2.2.2. Inequality Constraint

$$P^{i,min}_{gen} \leq P^{i,t}_{gen} \leq P^{i,max}_{gen} \; ; (i = 1 \cdots N) \quad (17)$$

2.2.3. Ramp up Rate

$$P^{i,t \, max}_{gen} = min \left(P^{i \, max}_{gen}, P^{i(t-1)}_{gen} + \psi R^i_{up} \right) \quad (18)$$

2.2.4. Ramp down Rate

$$P_{\text{gen}}^{i,t\ \min} = \max\left(P_{\text{gen}}^{i\ \min}, P_{\text{gen}}^{i(t-1)} - \psi R_{\text{down}}^{i}\right) \tag{19}$$

2.2.5. Minimum up Time

$$\text{ON}^{i} \geq M_{\text{up}}^{i} \tag{20}$$

2.2.6. Minimum down Time

$$\text{OFF}^{i} \geq M_{\text{down}}^{i} \tag{21}$$

2.2.7. Reserve Constraints

$$0 \leq R_{\text{gen}}^{i,t} \leq \left(P_{\text{gen}}^{i,\max}\ P_{\text{gen}}^{i,\min}\right) \tag{22}$$

$$P_{\text{gen}}^{i,\min} \leq \left(P_{\text{gen}}^{i,t} + R_{\text{gen}}^{i,t}\right) U_{\text{stat}}^{i,t} \leq P_{\text{gen}}^{i,\max} \tag{23}$$

$$\sum_{i=1}^{N} R_{\text{gen}}^{i,t}\ U_{\text{stat}}^{i,t} \leq R_{\text{gen}}^{i,t\ \max} \tag{24}$$

2.2.8. Spinning Reserve

$$S_{\text{res}}^{i} = \sum_{i=1}^{N} \left(P_{\text{gen}}^{i\ \max} - P_{\text{gen}}^{i}\right) \tag{25}$$

2.2.9. Startup Cost of Units

$$SU_{\text{cost}}^{i} = \begin{cases} H_{\text{cost}}^{i} & , \text{if } M_{\text{down}}^{i} \leq \text{OFF}^{i} \leq CS_{\text{time}}^{i} \\ C_{\text{cost}}^{i} & , \text{if } \text{OFF}^{i} \geq CS_{\text{time}}^{i} \end{cases} \tag{26}$$

$$CS_{\text{time}}^{i} = CS_{\text{Hour}}^{i} + M_{\text{down}}^{i} \tag{27}$$

2.3. Demand Response Unit Commitment Constraints

2.3.1. Equality Constraint

$$\sum_{i=1}^{N} \left(P_{\text{gen}}^{i,t}\ U_{\text{stat}}^{i,t}\right) + \sum_{di=1}^{dN} \left(P_{\text{gen}}^{di,t}\ U_{\text{stat}}^{di,t}\right) = P_{\text{dem}}^{t}\ ; (t = 1 \cdots T) \tag{28}$$

2.3.2. Minimum up Time

$$\text{ON}^{di} \geq M_{\text{up}}^{di} \tag{29}$$

2.3.3. Minimum down Time

$$\text{OFF}^{di} \geq M_{\text{down}}^{di} \tag{30}$$

2.3.4. Ramp up Rate

$$P_{gen}^{di,t \; max} = min\left(P_{gen}^{di \; max}, P_{gen}^{di(t-1)} + \psi R_{up}^{di}\right) \tag{31}$$

2.3.5. Ramp down Rate

$$P_{gen}^{di,t \; min} = max\left(P_{gen}^{di \; min}, P_{gen}^{di(t-1)} - \psi R_{down}^{di}\right) \tag{32}$$

3. Cat Swarm Optimization (CSO)

CSO optimization overcomes the limitations of PSO and DE that they are influenced by parameters and stagnation problem [41]. CSO is a meta-heuristic evolutionary optimization technique that intimates the natural behavior of felines. Cats have a strong curiosity towards objects that move. The cat group has superior hunting skills. Although it may be seen as always being at rest and they may seem to move slowly, they are always alert and aware of their surroundings [42]. Upon sensing the presence of prey, they chase it very quickly thereby spending a large amount of energy. These mentioned two characteristics, that is, the slow movement resting and sudden chase with high speed are described as seeking and tracking modes [43,44]. Each of these modes can be separately modeled mathematically.

3.1. Seeking Mode

There are four essential factors used in seeking, these factors are described as:

(a) Seeking Memory Pool (SMP): number of copies of a cat produced.
(b) Seeking Range of selected Dimension (SRD): difference between the new and old in the dimension selected for mutation.
(c) Counts of Dimensions to Change (CDC): number of dimensions to be mutated.
(d) Mixture Ratio (MR): to state that most of the time spent by the cats is resting and observing.

Steps executed in seeking mode:

(1) Randomly select MR fraction of population as seeking cats: rest of them as tracing cats.
(2) SMP copies of the ith seeking cat is created.
(3) Update the position of each copy based on CDC by randomly adding or subtracting SRD fraction.
(4) Evaluate error fitness values of copies.
(5) Best candidate is picked from all copies and placed at ith seeking cat.
(6) Repeat Step 2 until all seeking cats are involved.

3.2. Tracing Mode

This mode corresponds to the local search technique of an optimization problem. This method involves the cats tracing a target while spending a huge amount of energy. The rapid chase of cats is mathematically modeled as follows:

Define the position and velocity of ith cat in the D-dimensional space as:

$$X^i = \left(X_1^i, X_2^i, \ldots\ldots, X_D^i\right) \tag{33}$$

and:

$$V^i = \left(V_1^i, V_2^i, \ldots\ldots, V_D^i\right) \tag{34}$$

The global best position of a cat is represented as:

$$G_{best} = \left(G_{best}^1, G_{best}^2, \ldots\ldots, G_{best}^D\right) \tag{35}$$

Updated equations are:

$$V_D^i = w \times V_D^i + C \times r\left(G_{best}^D - X_D^i\right) \tag{36}$$

and:

$$X_D^i = X_D^i + V_D^i \tag{37}$$

The proposed method algorithm is given by the following steps:

Step 1: Create N number of population.

Step 2: Initialize time $t = 0$ and $i = 0$.

Step 3: Find the overall cost and revenue for TUC and DRSP from the data provided using iterations and store the values and evaluate the profit for TUC and DRSP using the formula $P_f = R_v - T_{cost}$.

Step 4: Check if all units are over and whether the cat is in seeking mode based on MR value

Step 5: If yes, Seeking Mode.

Create SMP copies and update position based on CDC, then take best value from SMP copies.

Step 6: If no, then Tracing mode.

Update position and velocity by using the equations:

$$V_D^i = w \times V_D^i + C \times r\left(G_{best}^D - X_D^i\right)$$

$$X_D^i = X_D^i + V_D^i$$

And save the highest profit unit.

Step 7: Check if all cats are updated, if yes, then proceed or else go back to Step 4.

Step 8: Check if maximum iteration is over, if yes, then stop and display the result, else go back to Step 2.

4. Result and Discussion

In this paper, we have used IEEE 39 bus system with conventional 10-units for a scheduling period of 24 h. The data for the load demand curve of the 10 unit systems is listed in Table 1. The operator data are listed in Table 2. The load data for the 10 unit 39 bus system is shown in Table 3. The forecasted price values for 24 h in a 10 bus system are shown in Table 4 and plotted in Figure 1. Six separate DRSPs are considered here, each generating load at a capacity of 50 MW. The load data curve value for these DRSPs is given in Table 5. The curve for forecasted price along with load demand variation for 24 h is plotted in Figure 2. Here it is noted that the price value during peak hours is high compared to the non-peak hours.

Table 1. Load curve data for the 10 unit IEEE 39 bus system.

Unit No.	a_i	b_i	c_i
U-1	1000	16.19	0.00048
U-2	970	17.26	0.00031
U-3	700	16.6	0.002
U-4	680	16.5	0.00211
U-5	450	19.7	0.00398
U-6	370	22.26	0.00712
U-7	480	27.74	0.00079
U-8	660	25.92	0.00413
U-9	665	27.27	0.0022
U-10	670	27.799	0.00173

Table 2. Operator data for the 10 unit IEEE 39 bus system.

Unit No.	P_{gen}^{max} (MW)	P_{gen}^{min} (MW)	M_{up}^i (h)	M_{down}^i (h)	C_{cost}^i ($)	H_{cost}^i ($)	CS_{time}^i (h)	U_{stat}^i (h)
U-1	455	150	8	8	4500	9000	5	8
U-2	455	150	8	8	5000	10,000	5	8
U-3	130	20	5	5	550	1100	4	−5
U-4	130	20	5	5	560	1120	4	−5
U-5	162	25	6	6	900	1800	4	−6
U-6	80	20	3	3	170	340	2	−3
U-7	85	25	3	3	260	520	2	−3
U-8	55	10	1	1	30	60	0	−1
U-9	55	10	1	1	30	60	0	−1
U-10	55	10	1	1	30	60	0	−1

Table 3. Load demand for 24 h.

Time (h)	1	2	3	4	5	6	7	8	9	10	11	12
Load Demand (MW)	700	750	850	950	1000	1100	1150	1200	1300	1400	1450	1500
Time (h)	13	14	15	16	17	18	19	20	21	22	23	24
Load Demand (MW)	1400	1300	1200	1050	1000	1100	1200	1400	1300	1100	900	800

Table 4. Forecasted price values for 24 h.

Time (h)	1	2	3	4	5	6	7	8	9	10	11	12
Price ($)	22.15	22	23.1	22.65	23.25	22.95	22.5	22.15	22.8	29.35	30.15	31.65
Time (h)	13	14	15	16	17	18	19	20	21	22	23	24
Price ($)	24.6	24.5	22.5	22.3	22.25	22.05	22.2	22.65	23.1	22.95	22.75	22.55

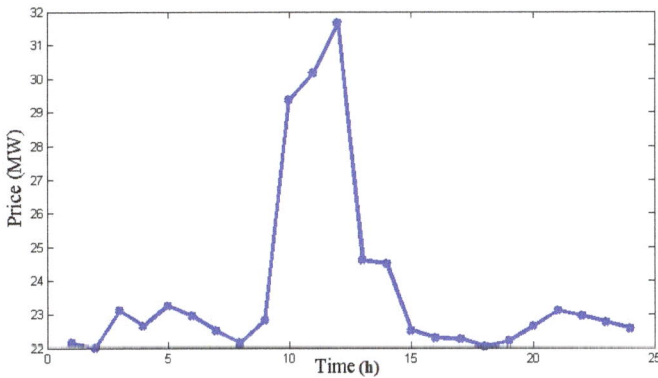

Figure 1. Forecasted price curve for 24 h.

Table 5. Load curve data for DRSP's.

-	θ^{di}	δ^{di}	φ^{di}
DRSP1	0.07	70	240
DRSP2	0.095	100	200
DRSP3	0.09	85	220
DRSP4	0.09	110	200
DRSP5	0.08	105	220
DRSP6	0.075	120	190

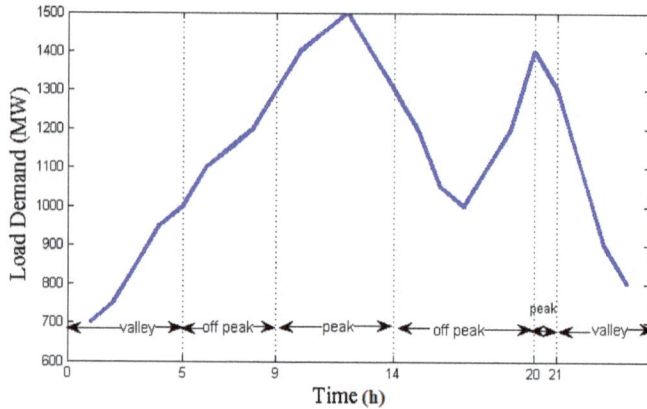

Figure 2. Valley, off peak and peak load unit operating system.

5. Simulation Results

The CSO formulation and solution methodology was implemented in MATLAB (2015, The MathWorks, Natick, MA, USA) and executed on a core i5 (2.6 GHz) personal computer equipped with 4 GB RAM. The proposed methodology that has been tested on a 10 unit generating system to solve TUC and DRUC problem is shown in Tables 6–8. The parameters assumed here are as follows; population size = 50, max iterations cycles = 100, SMP = 5, CDC = 0.6, SRD = 2, MR = 0.1, inertia weight w = 0.4 and acceleration constant C = 1.5 [41].

Table 6. Output data for base case using Traditional Unit Commitment (TUC).

Hour	1	2	3	4	5	6	7	8	9	10	Reserve (MW)	F_{cost} ($)	SC ($)	TO_{cost} ($)	R_v ($)	P_R ($)
1	455	245	0	0	0	0	0	0	0	0	210	13,683.13	0	13,683.13	15,505	1821.87
2	455	295	0	0	0	0	0	0	0	0	160	14,554.50	0	14,554.50	16,500	1945.50
3	455	370	0	0	25	0	0	0	0	0	222	16,809.45	900	17,709.45	19,635	1925.55
4	455	455	0	0	40	0	0	0	0	0	122	18,597.67	0	18,597.67	21,517.5	2919.83
5	455	390	0	130	25	0	0	0	0	0	202	20,020.02	560	20,580.02	23,250	2669.98
6	455	360	130	130	25	0	0	0	0	0	232	22,387.04	1100	23,487.04	25,245	1757.96
7	455	410	130	130	25	0	0	0	0	0	182	23,261.98	0	23,261.98	25,875	2613.02
8	455	455	130	130	30	0	0	0	0	0	132	24,150.34	0	24,150.34	26,580	2429.66
9	455	455	130	130	85	20	25	0	0	0	197	27,251.06	860	28,111.06	29,640	1528.94
10	455	455	130	130	162	33	25	10	0	0	152	30,057.55	60	30,117.55	41,090	10,972.45
11	455	455	130	130	162	73	25	10	10	0	157	31,916.06	60	31,976.06	43,717.5	11,741.44
12	455	455	130	130	162	80	25	43	10	10	162	33,890.16	60	33,950.16	47,475	13,524.84
13	455	455	130	130	162	33	25	10	0	0	152	30,057.55	0	30,057.55	34,440	4382.45
14	455	455	130	130	85	20	25	0	0	0	197	27,251.06	0	27,251.06	31,850	4598.94
15	455	455	130	130	30	0	0	0	0	0	132	24,150.34	0	24,150.34	27,000	2849.66
16	455	310	130	130	25	0	0	0	0	0	282	21,513.66	0	21,513.66	23,415	1901.34
17	455	260	130	130	25	0	0	0	0	0	332	20,641.82	0	20,641.82	22,250	1608.18
18	455	360	130	130	25	0	0	0	0	0	232	22,387.04	0	22,387.04	24,255	1867.96
19	455	455	130	130	30	0	0	0	0	0	132	24,150.34	0	24,150.34	26,640	2489.66
20	455	455	130	130	162	33	25	10	0	0	152	30,057.55	490	30,547.55	31,710	1162.45
21	455	455	130	130	85	20	25	0	0	0	197	27,251.06	0	27,251.06	30,030	2778.94
22	455	455	0	0	145	20	25	0	0	0	137	22,735.52	0	22,735.52	25,245	2509.48
23	455	425	0	0	0	20	0	0	0	0	90	17,645.36	0	17,645.36	20,475	2829.64
24	455	345	0	0	0	0	0	0	0	0	110	15,427.42	0	15,427.42	18,040	2612.58
				TOTAL COST ($)								55,9847.7	4090	563,937.7	651,380	87,442.31

Two cases are considered for solving unit commitment problem.

5.1. Case 1: Base Case

In this case, TUC is formulated using CSO programming for the 10 unit generating system considering the initial loads. The output obtained for this is shown in Table 6. Here the total revenue generated is $651,380 and the total operating cost calculated is $563,937.7. The profit obtained with this TUC is $87,442.31, which is 13.42% as shown in Table 9. In Table 9 the average TUC profit of the proposed CSO method gives better results compared to the LR [45], BCGA, ICGA [46], BFA [47] and ICA [48] methods.

5.2. Case 2: Base Case Established Using DR

In this case, we have used a real time-based demand response program to reduce load during the peak hours of the day. The peak hours can be seen in Figure 2 where the various valleys, off peak and peak load hours are plotted. 20% of load is reduced only in those particular hours and a TUC problem is executed and the output is shown in Table 7. The total revenue generated is $593,389.50 and the total operating cost calculated is $507,954.30. The profit obtained here is $85,435.21 as shown in Table 9. In the output Table 7, it is seen that the generators 8, 9 and 10 are not committed thereby reducing the total operating cost. The various generator running hours are depicted in Figure 3. From Figure 3, it is observed that the TUC methodology uses the entire generators in its distribution hence causing rise in cost. Whereas in DRUC the last three units are idle and don't take part in generation hence reducing the overall cost.

Table 7. Output data for base case established with Demand Response (DR) using Demand Response Unit Commitment (DRUC).

Hour	1	2	3	4	5	6	7	8	9	10	Reserve (MW)	F_{cost} ($)	SC ($)	TO_{cost} ($)	R_v ($)	P_R ($)
1	455	245	0	0	0	0	0	0	0	0	210	13,683.13	0	13,683.13	15,505	1821.87
2	455	295	0	0	0	0	0	0	0	0	160	14,554.49	0	14,554.5	16,500	1945.5
3	455	370	0	0	25	0	0	0	0	0	222	16,809.45	900	17,709.45	19,635	1925.55
4	455	455	0	0	40	0	0	0	0	0	122	18,597.67	0	18,597.67	21,517.5	2919.83
5	455	390	0	130	25	0	0	0	0	0	202	20,020.02	560	20,580.02	23,250	2669.98
6	455	360	130	130	25	0	0	0	0	0	232	22,387.04	1100	23,487.04	25,245	1757.95
7	455	410	130	130	25	0	0	0	0	0	182	23,261.98	0	23,261.98	25,875	2613.02
8	455	455	130	130	30	0	0	0	0	0	132	24,150.34	0	24,150.34	26,580	2429.66
9	405	360	130	120	25	0	0	0	0	0	292	21,386.63	0	21,386.63	23,712	2325.37
10	455	380	130	130	25	0	0	0	0	0	212	22,736.83	0	22,736.83	32,872	10,135.17
11	455	395	130	130	25	0	25	0	0	0	257	24,173.33	520	24,693.33	34,974	10,280.67
12	455	435	130	130	25	0	25	0	0	0	217	24,874.02	0	24,874.02	37,980	13,105.98
13	455	355	130	130	25	0	25	0	0	0	297	23,473.63	0	23,473.63	27,552	4078.37
14	410	355	130	120	25	0	0	0	0	0	292	21,382.13	0	21,382.13	25,480	4097.87
15	455	435	130	130	50	0	0	0	0	0	132	24,199.99	0	24,199.99	27,000	2800.01
16	415	350	130	130	25	0	0	0	0	0	282	21,547.94	0	21,547.94	23,415	1867.06
17	455	260	130	130	25	0	0	0	0	0	332	20,641.82	0	20,641.82	22,250	1608.18
18	455	350	130	130	35	0	0	0	0	0	232	22,411.63	0	22,411.63	24,255	1843.37
19	455	445	130	130	40	0	0	0	0	0	132	24,174.74	0	24,174.74	26,640	2465.26
20	455	380	130	130	25	0	0	0	0	0	212	22,736.83	0	22,736.83	25,368	2631.17
21	415	360	110	110	25	20	0	0	0	0	372	21,859.06	340	22,199.06	24,024	1824.94
22	455	445	0	120	35	45	0	0	0	0	182	22,398.79	0	22,398.79	25,245	2846.21
23	455	425	0	0	0	20	0	0	0	0	90	17,645.36	0	17,645.36	20,475	2829.64
24	455	345	0	0	0	0	0	0	0	0	110	15,427.42	0	15,427.42	18,040	2612.58
				TOTAL COST ($)								504,534.29	3420	507,954.3	593,389.5	85,435.21

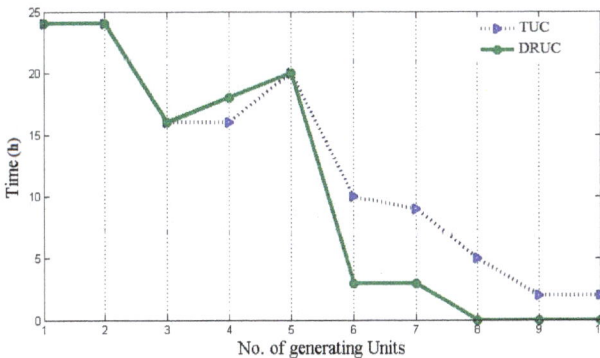

Figure 3. Operating hours of generating units in TUC and DRUC.

Along with this TUC programming, the six separate DRSPs that were installed are now used for generation. These generators generate the 20% load that was reduced from the initial case for their corresponding hours, respectively. The output of these generators are shown in Table 8. The same forecasted price given in table 4 is used to calculate the revenue. The revenue generated for these hours is $57,990.5, and the total operating cost is $40,512.5 for DRSP. The total revenue obtained when combined with the DRUC and DRSP is $651,380. This is same as that of our base case hence proving the same value of price is considered in our proposed case too as shown in Table 9. The total operating cost is $548,466.80. This is lower than our base case hence increasing the profit to $102,913.20 (15.8%), thus giving a profit rise of 2.37% shown in Table 9. Even an amount as low as a dollar saved per day will sum up to be much greater amount at the end of a year, although upon comparison, the amount doesn't seem to be much higher, but considering long term generation, it will make a huge difference.

Table 8. Output data of DRSP's during peak hours.

Hour	DRSP 1	DRSP 2	DRSP 3	DRSP 4	DRSP 5	DRSP 6	Reserve (MW)	F_{dcost} ($)	R_v ($)	PDR_f^{di} ($)
9	50	50	50	50	50	10	40	4810	5928	1118
10	50	50	50	50	50	30	20	5110	8218	3108
11	50	50	50	50	50	40	10	5282.5	8743.5	3461
12	50	50	50	50	50	50	0	5470	9495	4025
13	50	50	50	50	50	30	20	5110	6888	1778
14	50	50	50	50	50	10	40	4810	6370	1560
20	50	50	50	50	50	30	20	5110	6342	1232
21	50	50	50	50	50	10	40	4810	6006	1196
TOTAL COST ($)								40,512.5	57,990.5	17,478

Table 9. Various data comparisons.

-	F_{cost} ($)	SC ($)	TO_{cost} ($)	R_v ($)	P_R ($)	% Rise
TUC (LR) [45]	-	-	565,825	651,380	85,555	13.13
TUC (BCGA) [46]	-	-	567,367	651,380	84,013	12.89
TUC (ICGA) [46]	-	-	566,404	651,380	84,976	13.05
TUC (BFA) [47]	-	-	565,872	651,380	85,508	13.13
TUC (ICA) [48]	-	-	563,938	651,380	87,442	13.42
TUC (CSO)	559,847.7	4090	563,937.7	651,380	87,442.31	13.42
DRUC	504,534.3	3420	507,954.3	593,389.5	85,435.21	14.40
DRSP	-	-	40,512.5	57,990.5	17,478	30.14
DRUC + DRSP	-	-	548,466.8	651,380	102,913.2	15.80
% Difference	Total cost variation = 2.74%			Total profit variation = 2.37%		

The cost comparison for the base case and the base case with DR and DRSPs is shown in Figure 4. The base case is observed to have the maximum cost while the base case with DR has less due to the reduced load during the peak hours. The final case being the total cost combined of base case with DR and DRSPs. It is noticed that the profit is maximum when DRUC is scheduled rather than TUC. The various running data for the separate DGs installed is shown in Figure 5. It is observed that all the units are committed for peak hours and generating 50 MW each, except for the last unit, that is varied throughout peak hours in order to equalize the demand power.

The load demand versus time curve is depicted in Figure 6. The peak time load has been reduced using DRUC when compared with TUC. The reduced load value is generated by the DRSP's which is shown in Figure 5. The profit calculated for base case, base with DR and DRSPs are shown in Figure 7.

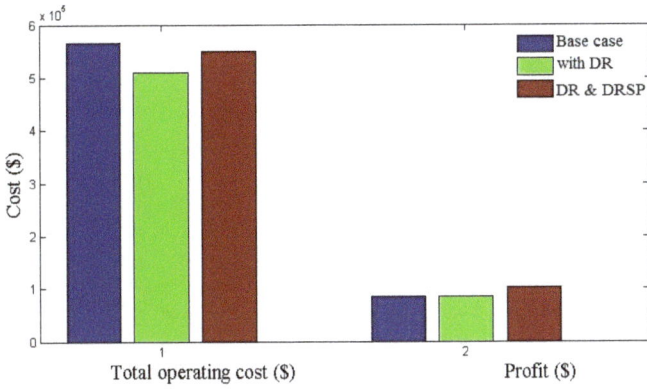

Figure 4. Cost comparison for TUC, TUC using DR and DRUC.

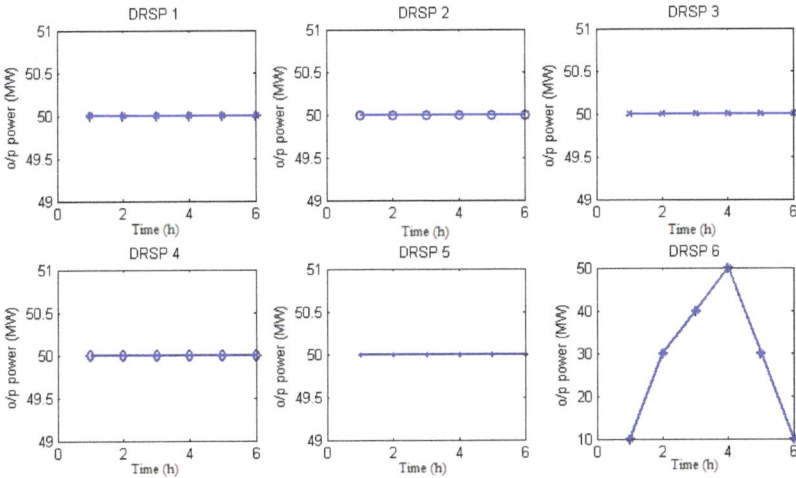

Figure 5. Unit running data for DRSP switched on during peak hours.

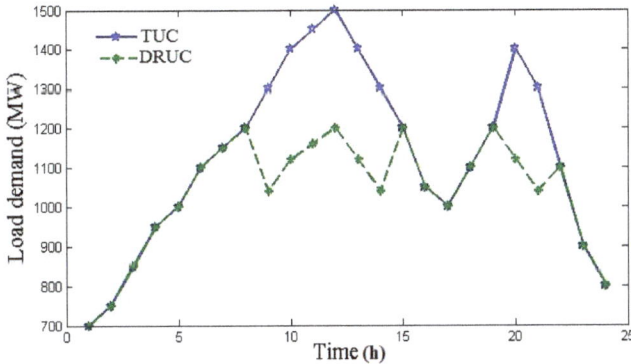

Figure 6. Load demand vs. time curve for TUC and DRUC.

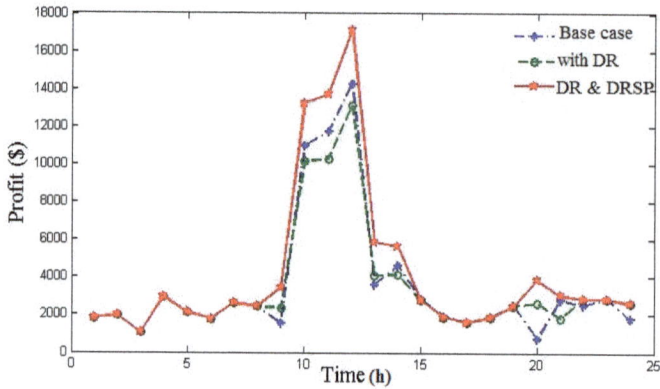

Figure 7. Profit for TUC, TUC using DR and DRUC.

It is noted that the profit is more using DRUC when compared with TUC. Also the curve when only the base case with DR excluding DRSPs is plotted and depicted. The various revenues calculated in TUC and DRUC are shown in Figure 8.

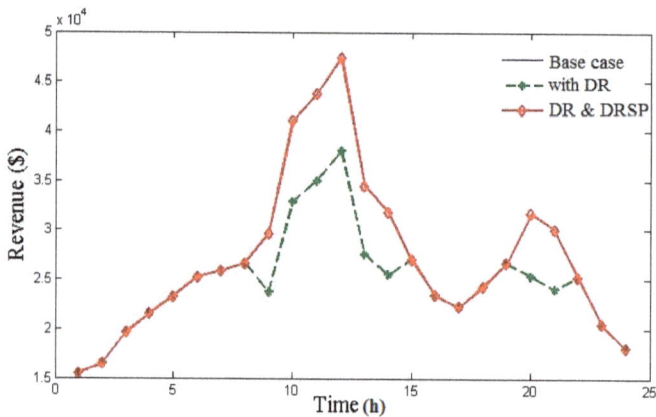

Figure 8. Revenue calculated in TUC, TUC using DR and DRUC.

It can be seen that the revenue when DR is established is reduced. This reduced revenue is calculated for DRSPs using the same spot price values. It should be noted that the revenue for both cases are one and the same. The total operating cost for the base case, base case using DR without DRSPs and base case using DR with DRSPs are shown in Figure 9. It is observed that the overall cost is reduced in DRUC when compared with TUC.

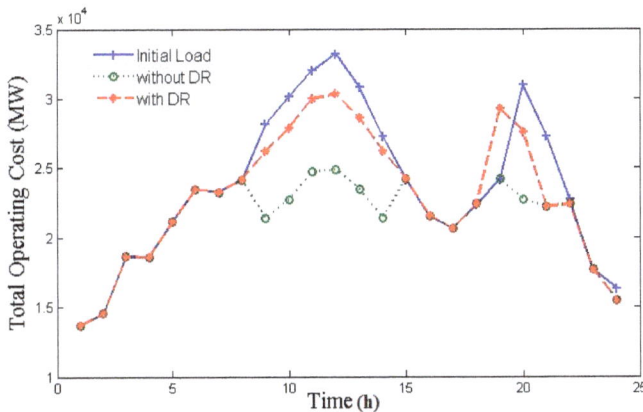

Figure 9. Operating cost for TUC, TUC using DR and DRUC.

6. Conclusions

In this paper, a demand response-based unit commitment model is solved using the Cat Swarm Optimization technique. A real time-based demand response program is used here to reduce load stress during peak hours and reduce the overall cost of the generation system. Also, six Demand Response Service Providers are used to compensate for the reduced load values. It is observed that using demand response unit commitment maximizes the profit for both GENCOs and the Demand Response Service Providers. Even though the load is reduced during peak periods the GENCOs gain higher percentage of profit. The consumer gains profit by installing DRSPs that supply the shaved-off loads during peak hours thereby decreasing the overall cost and maximizing the profit. From the simulation studies, although the revenue remains the same in TUC as in DRUC, it is observed that by implementing DRUC in generation systems, there is an overall decrease of around 2.74% in total cost and an increased profit gain of around 2.37%. Also it is proved that using DRSP the profit of the consumer is increased by reduction in the fuel cost. The proposed algorithm gives better results when compared to other optimization methods.

Author Contributions: K. Selvakumar and K. Vijayakumar are formulated the problem and obtained the solution. K. Selvakumar and C. S. Boopathi analyzed the results with other methods and wrote the paper.

Conflicts of Interest: The authors declare no conflict of interest.

Nomenclature

Constants

N	Total number of units
Ψ	Unit commitment time step (60 min)
T	Dispatch period in hours
dN	Total number of DRSP units
$\Theta^{di}, \delta^{di}, \varphi^{di}$	Supply curve coefficients of DRSP generating units
$\theta m^{di}, \delta m^{di}, \varphi m^{di}$	Customer's supply curve cost coefficients
a^i, b^i, c^i	Supply curve coefficients of IEEE 10 generating units
μ^{di}	Customer willingness coefficient
R^i_{up}	Ramp up rate of unit i
R^i_{down}	Ramp down rate of unit i
M^i_{up}	Minimum up time limit of unit i
M^i_{down}	Minimum down time limit of unit i
H^i_{cost}	Hot start cost of unit i

C^i_{cost}	Cold start cost of unit i
CS^i_{time}	Cold start hour of unit i

Variables

i	Index of generator unit
di	Index of DRSP generator unit
P_R	Total profit of the GENCO's and DRSP combined
TR_V	Total revenue calculated from GENCO's and DRSP
TO_{cost}	Total operating cost of GENCO's and DRSP combined
F^i_{cost}	Fuel cost of generator unit
F^{di}_{cost}	Fuel cost of DRSP generating unit
DR_{price}	Demand response clearing price
DR^{di}_{gen}	DRSP generator output per hour
PDR^{di}_f	Total profit of DRSP
DR_{req}	Required power output from DRSP generating units
$p^{i,t}_{gen}$	Power generator output of ith unit at tth hour
$p^{di,t}_{gen}$	DRSP output of dith unit at tth hour
$U^{i,t}_{stat}$	Unit status of ith unit at tth hour
P^t_{dem}	Total power demand at hour t
$p^{i,min}_{gen}$	Minimum generation output power of ith unit
$p^{i,t\ min}_{gen}$	Minimum generation output power of ith unit at tth hour
$p^{i,max}_{gen}$	Maximum generation output power of ith unit
$p^{i,t\ max}_{gen}$	Maximum generation output power of ith unit at tth hour
$p^{i(t-1)}_{gen}$	Power generated in the previous hour
ON^i	Number of hours the unit was committed
OFF^i	Number of hours the unit was not committed
$R^{i,t}_{gen}$	Reserve generation of unit i at tth hour
S^i_{res}	Spinning reserve of unit i
S^i_{price}	Forecasted spot price of unit i
S^{di}_{price}	Forecasted spot price of DRSP generating units di
SU^i_{cost}	Startup cost of unit i

References

1. Muzhikyan, A.; Farid, A.M.; Youcef-Toumi, K. Relative merits of load following reserves & energy storage market integration towards power system imbalances. *Int. J. Electr. Power Energy Syst.* **2016**, *74*, 222–229.
2. Paterakis, N.G.; Erdinç, O.; Catalão, J.P. An overview of Demand Response: Key-elements and international experience. *Renew. Sustain. Energy Rev.* **2017**, *69*, 871–891. [CrossRef]
3. Kyoho, R.; Goya, T.; Mengyan, W.; Senjyu, T.; Yona, A.; Funabashi, T.; Kim, C.H. Thermal unit commitment with demand response to optimize battery storage capacity. In Proceedings of the 2013 IEEE 10th International Conference on Power Electronics and Drive Systems (PEDS), Kitakyushu, Japan, 22–25 April 2013; pp. 1207–1212.
4. Federal Energy Regulatory Commission (FERC). *Federal Energy Regulatory Commission Report to Congress: Implementation Proposal for the National Action Plan on Demand Response*; FERC: Washington, DC, USA, 2011.
5. Federal Energy Regulatory Commission (FERC). *News Release: FERC Approves Market-Based Demand Response Compensation Rule*; FERC: Washington, DC, USA, 2011.
6. *Benefits of Demand Response in Electricity Markets and Recommendations for Achieving Them United States DOE Report to the Congress*; U.S. Department of Energy: Washington, DC, USA, 2006.
7. Abdollahi, A.; Moghaddam, M.P.; Rashidinejad, M.; Sheikh-El-Eslami, M.K. Investigation of economic & environmental—Driven demand response measures incorporating UC. *IEEE Trans. Smart Grid* **2012**, *3*, 12–25.
8. Rastegar, M.; Fotuhi-Firuzabad, M.; Choi, J. Investigating the impacts of different price-based demand response programs on home load management. *J. Electr. Eng. Technol.* **2014**, *9*, 1125–1131. [CrossRef]
9. Pierluigi, S. Demand response and smart grids—A survey. *Renew. Sustain. Energy Rev.* **2014**, *30*, 461–478.

10. Howlader, H.O.R.; Matayoshi, H.; Senjyu, T. Distributed generation integrated with thermal unit commitment considering demand response for energy storage optimization of smart grid. *Renew. Energy* **2016**, *99*, 107–117. [CrossRef]

11. Babar, M.; Tp, I.A.; Alammar, E.A. The consumer rationality assumption in incentive based demand response program via reduction bidding. *J. Electr. Eng. Technol.* **2015**, *10*, 64–74. [CrossRef]

12. Ghazvini, M.A.F.; Soares, J.; Horta, N.; Neves, R.; Castro, R.; Vale, Z. A multi-objective model for scheduling of short-term incentive-based demand response programs offered by electricity retailers. *Appl. Energy* **2015**, *151*, 102–118. [CrossRef]

13. Aalami, H.A.; Parsa Moghaddam, M.; Yousefi, G.R. Demand response modeling considering interruptible/curtailable loads and capacity market programs. *Appl. Energy* **2010**, *80*, 426–435. [CrossRef]

14. Motalleb, M.; Thornton, M.; Reihani, E.; Ghorbani, R. A nascent market for contingency reserve services using demand response. *Appl. Energy* **2016**, *179*, 985–995. [CrossRef]

15. Motalleb, M.; Thornton, M.; Reihani, E.; Ghorbani, R. Providing frequency regulation reserve services using demand response scheduling. *Energy Convers. Manag.* **2016**, *124*, 439–452. [CrossRef]

16. Zhang, X.; Hug, G.; Kolter, Z.; Harjunkoski, I. Industrial demand response by steel plants with spinning reserve provision. *N. Am. Power Symp. (NAPS)* **2015**, 1–6. [CrossRef]

17. *How Smart Is the Smart Grid*; NPR: Washington, DC, USA, 2010.

18. The Power to Choose—Enhancing Demand Response in Liberalised Electricity Markets Findings of IEA Demand Response Project, Presentation. 2003. Available online: http://www.schneider-electric.us/documents/solutions1/demand-response-solutions/powertochoose_2003.pdf (accessed on 21 September 2017).

19. Electric Power Supply Association. *Demand Response Compensation in Organized Wholesale Energy Markets*; Docket No. RM10-17-000, Request for Clarification or, in the Alternative, Request for Rehearing of the Public Utilities Commission of the State of California; Electric Power Supply Association: Washington, DC, USA, 2011.

20. Grunewald, P.; Torriti, J. Demand response from the non-domestic sector: Early UK experiences and future opportunities. *Energy Policy* **2013**, *61*, 423–429. [CrossRef]

21. Zhou, S.; Shu, Z.; Gao, Y.; Gooi, H.B.; Chen, S.; Tan, K. Demand response program in Singapore's wholesale electricity market. *Electr. Power Syst. Res.* **2017**, *142*, 279–289. [CrossRef]

22. Klobasa, M. Analysis of demand response and wind integration in Germany's electricity market. *IET Renew. Power Gener.* **2010**, *4*, 55–63. [CrossRef]

23. Arasteh, H.R.; Moghaddam, M.P.; Sheikh-El-Eslami, M.K.; Abdollahi, A. Integrating commercial demand response resources with unit commitment. *Electr. Power Energy Syst.* **2013**, *51*, 153–161. [CrossRef]

24. Zhang, N.; Hu, Z.; Han, X.; Zhang, J.; Zhou, Y. A fuzzy chance-constrained program for unit commitment problem considering demand response, electric vehicle and wind power. *Electr. Power Energy Syst.* **2015**, *65*, 201–209. [CrossRef]

25. Kiran, B.D.H.; Kumari, M.S. Demand response and pumped hydro storage scheduling for balancing wind power uncertainties: A probabilistic unit commitment approach. *Electr. Power Energy Syst.* **2016**, *81*, 114–122. [CrossRef]

26. Aghajani, G.R.; Shayanfar, H.A.; Shayeghi, H. Presenting a multi-objective generation scheduling model for pricing demand response rate in micro-grid energy management. *Energy Convers. Manag.* **2015**, *106*, 308–321. [CrossRef]

27. Kwag, H.G.; Kim, J.O. Optimal combined scheduling of generation and demand response with demand resource constraints. *Appl. Energy* **2012**, *96*, 161–170. [CrossRef]

28. Dietrich, K.; Latorre, J.M.; Olmos, L.; Ramos, A. Demand response in an isolated system with high wind integration. *IEEE Trans. Power Syst.* **2012**, *27*, 20–29. [CrossRef]

29. Ikeda, Y.; Ikegami, T.; Kataoka, K.; Ogimoto, K. A unit commitment model with demand response for the integration of renewable energies. In Proceedings of the 2012 IEEE Power and Energy Society General Meeting, San Diego, CA, USA, 22–26 July 2012.

30. Chaoyue, Z.; Jianhui, W.; Watson, J.P.; Yongpei, G. Multi-stage robust unit commitment considering wind and demand response uncertainties. *Power Syst. IEEE Trans.* **2013**, *28*, 2708–2717.

31. Reddy, K.S.; Panwar, L.K.; Kumar, R.; Panigrahi, B.K. Binary fireworks algorithm for profit based unit commitment (PBUC) problem. *Electr. Power Energy Syst.* **2016**, *83*, 270–282. [CrossRef]

32. Govardhan, M.; Master, F.; Roy, R. Economic analysis of different demand response programs on unit commitment. In Proceedings of the 2014 IEEE Region 10 Conference—TENCON 2014, Bangkok, Thailand, 22–25 October 2014; pp. 1–6.

33. Han, X.; You, S.; Bindner, H. Critical kick-back mitigation through improved design of demand response. *Appl. Therm. Eng.* **2017**, *114*, 1507–1514. [CrossRef]

34. Wang, Q.; Wang, J.; Guan, Y. Stochastic unit commitment with uncertain demand response. *IEEE Trans. Power Syst.* **2013**, *28*, 562–563. [CrossRef]

35. Liu, G.; Tomsovic, K. Robust unit commitment considering uncertain demand response. *Electr. Power Syst. Res.* **2015**, *119*, 126–137. [CrossRef]

36. Magnago, F.H.; Alemany, J.; Lin, J. Impact of demand response resources on unit commitment and dispatch in a day-ahead electricity market. *Electr. Power Energy Syst.* **2015**, *68*, 142–149. [CrossRef]

37. Morales-España, G.; Ramírez-Elizondo, L.; Hobbs, B.F. Hidden power system inflexibilities imposed by traditional unit commitment formulations. *Appl. Energy* **2017**, *191*, 223–238. [CrossRef]

38. Reddy, G.V.S.; Ganesh, V.; Rao, C.S. Implementation of clustering based unit commitment employing imperialistic competition algorithm. *Electr. Power Energy Syst.* **2016**, *82*, 621–628. [CrossRef]

39. Venkatesan, K.; Selvakumar, G.; Rajan, C. EP based PSO method for solving multi area unit commitment problem with import and export constraints. *J. Electr. Eng. Technol.* **2014**, *9*, 415–422. [CrossRef]

40. Sahraei-Ardakani, M.; Rahimi-Kian, A. A dynamic replicator model of the players' bids in an oligopolistic electricity market. *Electr. Power Syst. Res.* **2009**, *79*, 781–788. [CrossRef]

41. Saha, S.K.; Ghoshal, S.P.; Kar, R.; Mandal, D. Cat Swarm Optimization algorithm for optimal linear phase FIR filter design. *ISA Trans.* **2013**, *52*, 781–794. [CrossRef] [PubMed]

42. Chu, S.-C.; Tsai, P.; Pan, J.-S. *Cat Swarm Optimization*; Springer-Verlag: Berlin/Heidelberg, Germany, 2006.

43. Panda, G.; Pradhan, P.M.; Majhi, B. IIR system identification using Cat Swarm Optimization. *Expert Syst. Appl.* **2011**, *38*, 12671–12683. [CrossRef]

44. Chu, S.-C.; Tsai, F.-W. Computational Intelligence based on the behavior of Cats. *Int. J. Innov. Comput. Inf. Control* **2007**, *3*, 163–173.

45. Kazarlis, S.A.; Bakirtzis, A.G.; Petridis, V. A genetic algorithm solution to the unit commitment problem. *IEEE Trans. Power Syst.* **1996**, *11*, 83–92. [CrossRef]

46. Damousis, I.G.; Bakirtzis, A.G.; Dokopoulos, P.S. A solution to the unit-commitment problem using integer-coded genetic algorithm. *IEEE Trans. Power Syst.* **2004**, *19*, 1165–1172. [CrossRef]

47. Eslamian, M.; Hosseinian, S.H.; Vahidi, B. Bacterial foraging-based solution to the unit-commitment problem. *IEEE Trans. Power Syst.* **2009**, *24*, 1478–1488. [CrossRef]

48. Hadji, M.M.; Vahidi, B. A solution to the unit commitment problem using imperialistic competition algorithm. *IEEE Trans. Power Syst.* **2012**, *27*, 117–124. [CrossRef]

Article

A Decentralized Multi-Agent-Based Approach for Low Voltage Microgrid Restoration

Ebrahim Rokrok [1], Miadreza Shafie-khah [1], Pierluigi Siano [2,*] and João P. S. Catalão [1,3,4,*]

1   Centre for Mechanical and Aerospace Science and Technologies (C-MAST), University of Beira Interior,
     R. Fonte do Lameiro, 6201-001 Covilhã, Portugal; ebrahim.rokrok@gmail.com (E.R.);
     miadreza@ubi.pt (M.S.-k.)
2   Department of Industrial Engineering, University of Salerno, 84084 Fisciano (SA), Italy
3   Institute for Systems and Computer Engineering, Technology and Science (INESC TEC),
     Faculty of Engineering, University of Porto, 4200-465 Porto, Portugal
4   Institute for Systems and Computer Engineering, Research and Development (INESC-ID),
     Instituto Superior Técnico, University of Lisbon, Av. Rovisco Pais, 1049-001 Lisbon, Portugal
*   Correspondence: psiano@unisa.it (P.S.); catalao@ubi.pt (J.P.S.C.);
     Tel.: +39-089-961-1111 (P.S.); +351-22-508-1850 (J.P.S.C.)

Academic Editor: Pedro Faria
Received: 11 August 2017; Accepted: 18 September 2017; Published: 27 September 2017

Abstract: Although a well-organized power system is less subject to blackouts, the existence of a proper restoration plan is nevertheless still essential. The goal of a restoration plan is to bring the power system back to its normal operating conditions in the shortest time after a blackout occurs and to minimize the impact of the blackout on society. This paper presents a decentralized multi-agent system (MAS)-based restoration method for a low voltage (LV) microgrid (MG). In the proposed method, the MG local controllers are assigned to the specific agents who interact with each other to achieve a common decision in the restoration procedure. The evaluation of the proposed decentralized technique using a benchmark low-voltage MG network demonstrates the effectiveness of the proposed restoration plan.

Keywords: average consensus algorithm (ACA); black start; local controller; microgrid (MG); multi-agent system (MAS); power system restoration (PSR)

---

1. Introduction

1.1. Motivations

Nowadays, the limited operating margins of the power systems have increased the risk of power blackouts and system collapse. In recent years, several major blackouts have occurred around the world. For instance, the blackout that occurred on 14 August 2003 in North America, caused an immense loss and the power system restoration (PSR) lasted nearly two weeks [1]. Also, a European power outage affected 15 million people on 4 November 2006, and it lasted up to 2 h. The blackout that occurred on 31 July 2012, in north India deenergized 50 GW of loads and affected 670 million people. The 2009 Brazil and Paraguay blackout was a power outage that occurred in many sections of Brazil and for a short time affected the entirety of Paraguay. The Fukushima nuclear power plant was faced with a series of equipment failures after the earthquake and tsunami on 11 March 2011, and a significant amount of radioactive materials were also released into ground and ocean waters [2]. When a blackout occurs, the main priority is to restore the power system in a proper manner so that the maximum load is restored as soon as possible considering the operating conditions and the system security.

During the past years, various aspects of the PSR problem have been studied, and its theories and methods are largely mature [3–7]. In conventional power systems, the restoration process begins from the transmission system by starting up those power plants which provide the black start capability in the shortest time. This allows the supply of a large part of the consumers near the power plants and to energize the transmission network [4]. The start of the restoration process from the transmission system causes many consumers in the distribution system to be supplied in the final stages of the restoration process, so the reliability of the system is decreased.

With the restructuring of the power grids toward smart grids which are based on the smart energy infrastructure consisting of microgrids (MGs) and distributed energy resources, the possibility of restoration of a large part of the loads at the distribution level along with restoration of the power plants and transmission network is provided. Therefore, the combination of the distributed energy resources and the flexible demands in the form of MGs can facilitate the implementation of local self-healing methods and accelerate the restoration process. In this sense, the PSR process can be carried out using a top-down procedure starting from the high voltage (HV) transmission system along with a bottom-up approach starting from the low voltage (LV) distribution system by using the capabilities of the LV MGs. The LV MGs and the HV transmission system will be synchronized and connect together at the medium voltage (MV) distribution level [8–11].

The motivation of this paper was to design and develop a decentralized multi-agent-based approach for restoration of a MG after a general blackout. The proposed decentralized approach provides an adequate restoration sequence to maximize the amount of restored loads.

1.2. Literature Review

The aims of restoration are to enable the power system to return to its normal conditions rapidly and securely, to minimize the losses and the restoration time, and to alleviate the adverse effects on the society after an outage. Many methods and technologies are employed for preparing the restoration schemes to address the abovementioned goals. Although the nature of the outages is unique, certain common guidelines exist to help operators restore and rebuild a stable power system after an outage [12]. The PSR can be categorized based on several different criteria as follows:

(a) Different parts of the power system: PSR needs to be carried out in different types of power systems and at different levels. In [13], the restoration of the transmission system with the goal of finding an appropriate sequence of actions to minimize the size of the blackout over time is presented. To solve the restoration ordering problem (ROP), the DC model and the linear programming approximation of AC (LPAC) power flow are used, and it is shown that the DC model is not sufficiently accurate to solve the ROP. In contrast, the LPAC power flow model is sufficiently accurate to obtain the restoration plans. In [14], the PSR is stated as a multi-objective, multi-variable, and multi-constrained nonlinear optimization problem and a multi-objective model based on the combination of the multi-agent technology and Tabu search method (TSM) is proposed for the restoration of the transmission system. Some of the studies investigate the restoration of the distribution system. In [15], by using the genetic algorithm (GA), the switching operation is minimized during the restoration process. It also reduces the required calculations time. The capabilities of the distributed generations (DGs) in distribution systems are used in [16] to minimize the restoration time and maximize the amount of restored loads.

(b) Outage range: Some researchers focus on the condition in which only a small part of the power system is deenergized [13] while other researchers focus on the restoration procedure after a total blackout [17].

(c) Sub problems: Much researchers have focused on the different sub problems in PSR such as generator start-up sequence [18], standing voltage phase angles [19], and selecting suitable islands to restart [20].

(d)   Modeling: There is a trade-off between speed and accuracy of PSR analysis; capturing the behavior of the real system reduces the computation time and the implementation complexity. Both static power flow calculations [17] and dynamic electromechanical models [8,21,22] are used.

Many recent reports focus on the using the capabilities of the MGs as the new effective solution for PSR at the distribution system level. In [8], the feasibility of MG restoration after a blackout is investigated using dynamic modeling. The microgrid central controller (MGCC) is responsible for making the restoration decisions such as starting-up the black start units, energizing the feeders, and restoring the loads and non-black start units. Using the information received periodically from the local controllers about generation and consumption levels, the centralized control system of the MG makes the restoration decisions. Such dynamic studies for MG restoration with the centralized approach are widely used in the literature. In [22], the restoration of the distribution system is investigated in the presence of multi-MGCC. Similar to [8], it is supposed that the sequence of the restoration actions is determined by the centralized control system. In [23] and [11], the dynamic studies of the centralized restoration process are performed for the MGs implemented in the northern region of Launceston in Tasmania, and Illinois Institute of Technology (IIT), respectively.

The centralized control schemes are low cost and easy to design, however, they suffer from single-point-failure. Furthermore, they are not adaptive to the changes of the power network structure. For instance, when new loads or generators are installed, the centralized control schemes may need to be redesigned. To avoid these shortcomings, the decentralized control scheme is introduced. One of the most popular decentralized control solutions is the multi-agent system (MAS). The MAS has the advantage of surviving single-point-failure, and it can do the decentralized data processing which, in turn, leads to task distribution and faster decision-making process [24].

MASs need to share the information process among the agents. The problem of communication of the agents can be solved by the average consensus algorithm (ACA). ACA shares the information among the agents in a distributed way to achieve an agreement on a common decision. This algorithm is widely employed in different areas including the collective behavior of swarms [25], random networks [26,27], formation flight control of multi-unmanned aerial vehicle (UAV) system [28], cooperative control of satellites [29], networks of cameras [30], and coordination and control of mobile robots [31].

1.3. Contributions

This paper introduces a decentralized multi-agent-based approach for restoration of an LV MG. In the proposed scheme, the MG local controllers are assigned to specific agents. The agents only know their own local information and communicate with their neighboring agents to access the required global information. The communication among the agents for sharing the local information and accessing to the global information is based on ACA. After completing the sharing information process, all agents take a common decision based on the discovered information to determine which load or generation unit should be chosen and connected for maximizing the amount of the restored load in the shortest possible time.

The centralized restoration scheme uses the information about the last generation-consumption scenarios of the MG to determine the sequence of restoration actions. The information is periodically sent by the MG local controllers to the MGCC. After a blackout, the MGCC performs service restoration based on the latest updated information stored in a database and determines all restoration actions. Since the database information is gathered during a certain period of time before the blackout, in the case of changing in the load or generation scenario or lack of preparedness for the restoration of a generation unit or load during the restoration process, the restoration scheme will be in trouble [8,22]. Thus, there is a need for a decentralized restoration approach that uses the online information of the generation/consumption of the MG and determines a proper sequence of restoration actions to restore the maximum possible amount of loads in the shortest possible time. The proposed decentralized multi-agent based restoration approach uses the online information of the generation/consumption of the MG during the restoration process and determines a proper sequence of restoration actions.

1.4. Paper Organization

The rest of the paper is organized as follows: in Section 2, the centralized control structure for the black start of the MG is briefly explained. Section 3 gives the dynamic modeling of an LV MG. The employed dynamical models are compatible with the type of study. In Section 4, the proposed decentralized multi-agent based approach for the restoration of an LV MG is explained. The proposed scheme is simulated in MATLAB-Simulink (R2013b (8.2.0.701)), The MathWorks, Inc., Natick, MA, USA) environment, and the study results are presented in Section 5. Finally, Section 6 provides the conclusions.

2. Microgrid Control Structure for Black Start

Figure 1 shows the structure of an inverter-based LV MG. The presence of a synchronous generator in an inverter-based LV MG is not common [8]. Generally, an LV MG includes the loads, microsources (photovoltaic (PV), wind energy conversion system (WECS), fuel cell, and microturbine), and storage devices (battery energy storage systems (BESSs) and flywheels).

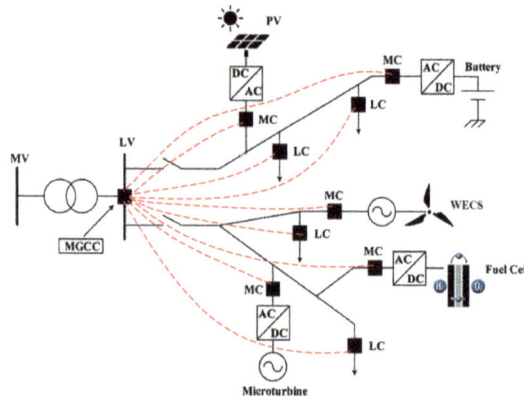

Figure 1. Typical low voltage (LV) microgrid (MG). MGCC: microgrid central controller; MC: microsource controller; LC: load controller; WECS: wind energy conversion system; PV: photovoltaic.

The safe, economic and stable operation of an MG in both grid connected and islanded mode depends on the existence of a proper control system [11,32–34]. An LV MG can be controlled centrally by the MGCC installed at the LV side of MV/LV substation. The load controller (LC) and microsource controller (MC) are local controllers that control the loads and microsources, respectively, and exchange the required information (such as set-points and load/consumption situations) with the MGCC through a narrow-band communication link. LC controls the loads using the local load shedding schemes in emergency conditions while MC controls the active and reactive power of microsources [35].

Under normal operation, the MG is connected to the MV network. However, in order to deal with the islanded mode and black start of the MG following a blackout, an emergency operation mode should be provided. If a blackout occurs, the restoration process time needs to be reduced as much as possible. The restoration plan is defined step by step. The main steps are building the LV network, connecting the microsources, connecting the controllable loads, controlling the voltage and frequency, and synchronizing the MG with the MV network, when it is available [8,22].

In the centralized restoration scheme, the MG black start is guided by the MGCC. The information about generation-consumption scenarios of the MG is periodically sent by the MCs and LCs to MGCC using the communication links and they will be stored in a database. After the blackout occurrence, based on the information available in the database, the MGCC determines a sequence of restoration actions and send the proper control commands to the local controllers [8,22].

To implement such a centralized restoration approach, all of the MCs and LCs must have a direct communication with the MGCC. In the case of communication failure for each local controller, the centralized restoration scheme faces with some problems. That's why it is said that the centralized control schemes easily suffer from single point failure. Moreover, the centralized restoration scheme uses the information available in its database to determine the sequence of restoration actions. Since the database information is gathered during a certain period of time before the blackout, in the case of changing in the load or generation scenario or lack of preparedness for the restoration of a generation unit or load during the restoration process, the restoration scheme will be in trouble. Thus, there is a need for a decentralized restoration approach that uses the online information of the generation/consumption of the MG and determines a proper sequence of restoration actions to restore the maximum possible amount of loads in the shortest possible time.

This paper proposes a decentralized multi-agent based approach for the MG restoration in which the online data related to the generation and consumption are used to determine a proper sequence of restoration actions. In the following sections, first, the MG components will be dynamically modeled. Then, according to the mathematical discussion on the distributed averaging problem, the proposed multi-agent based method will be presented.

3. Modeling of the Microgrid Components

Generally, MGs include some components such as microsources, storage devices, and loads. In order to study the dynamic behavior of the MG during the restoration process, it is essential to provide a proper dynamic model for each component that is compatible with the type of the study [36]. In the following subsections, various components of the MG are modeled.

3.1. Microsource Modeling

There are several dynamic models for microsources in the literature. To model the PV cell, this paper uses the single-diode model or five-parameter model [13]. This model provides an adequate trade-off between simplicity and accuracy. A PV system is commercially available in the form of modules in which there is a number of series cells. The modules are connected in series to make a string with an appropriate voltage level. While, to increase the current rating, the strings are connected in parallel and form an array. In [37], the model of the PV array based on the five parameter model is found. In this paper, it is assumed that the PV arrays work at the maximum power point.

The WESS model used in this study is based on the constant speed wind turbine that is available in [38]. Likewise, regarding the period of study, only the average value of the wind speed is considered (i.e., the wind speed is constant).

To model the dynamic behavior of the microturbine, the gas turbine (GAST) model [39] is used. There are two types of the microturbine including single-shaft microturbine (high-speed) and split-shaft microturbine (low-speed). In the single-shaft microturbine, the turbine speed range is from 50,000 to 120,000 rpm. So, this type of microturbine requires an AC/DC/AC converter for connecting to the grid. The split-shaft microturbine uses a power turbine that is rotated at 3600 rpm and can be connected to a conventional induction generator using a gearbox [39].

Reference [40] provides a basic dynamic model for a solid-oxide fuel cell (SOFC) that is used in this paper. This model has some assumption to achieve an integrated dynamic model for using in the power systems simulations.

3.2. Converter Modeling

There are two kinds of control mode for operating the converters: (1) grid-forming mode and (2) grid-following mode [41]. The grid-forming converters emulate the behavior of a synchronous generator and provide the voltage and frequency references for the MG. The grid-forming converter acts as a voltage source and controls its output voltage and frequency using the droop control. When the MG works in the islanded mode, at least one converter must operate as a grid-forming converter.

DGs must meet two requirements to have the black start capability: (i) equipped with storage devices (batteries or super-capacitors) in the DC link of their inverter and (ii) operation of their inverter in the grid-forming control mode. These DGs are capable of restarting without any external power source, energize the network, supply a part of loads, and provide remote cranking power for the other DGs with the grid-following inverter control system [42]. Figure 2 shows the control structure of a droop-based grid-forming converter.

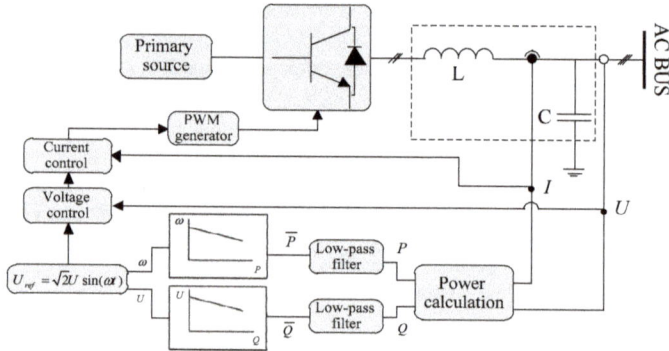

Figure 2. Structure of a grid-forming converter control system. PWM: Pulse width modulation.

The grid-following converters are mainly designed to deliver a pre-determined power to an energized grid. If there is no synchronous generator or grid-forming converter in the MG, the grid-following converter cannot operate. Figure 3 shows the control structure of a grid-following converter [43].

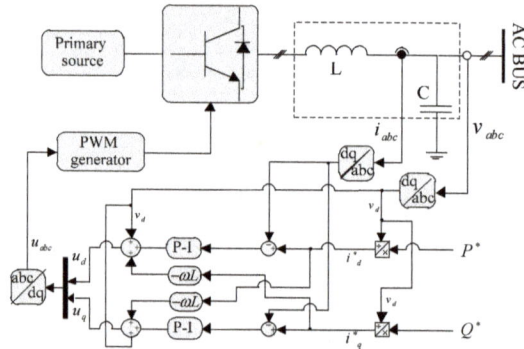

Figure 3. Structure of a grid-following converter control system.

4. Proposed Decentralized Approach

4.1. Mathematical Background

The consensus problem is a prevalent problem in distributed control. In the following two subsections, the ACA is explained.

4.1.1. Distributed Averaging

Let $g = (N, E)$ be a graph with N nodes and E edges. In node set $N = \{1, 2, \ldots, n\}$, consider each edge $\{i, j\} \in E$ is an unordered pair of distinct nodes. Let c_i^0 be a real number associated to node i at time $t = 0$. The average consensus problem calculates iteratively the average $(1/n)\sum_{i=1}^{n} c_i^0$ in a distributed way at every node (see Figure 4). The following iterative law, known as ACA, is proposed in the literature to solve this averaging problem [44]:

$$c_i^{k+1} = c_i^k + \sum_{j \in N_i} w_{ij}(c_j^k - c_i^k), \tag{1}$$

where $i = 1, 2, \ldots, n$; n is the number of nodes; c_i^k, c_i^{k+1} are the values of node i at iteration k and $k + 1$, respectively, and w_{ij} is the weight coefficient that enables communication between neighboring nodes i and j. If nodes i and j are connected together, $0 < w_{ij} < 1$, otherwise, $w_{ij} = 0$. N_i is the index of nodes connected to node i.

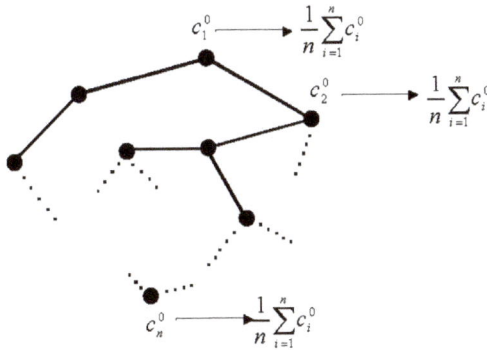

Figure 4. Principle of distributed averaging.

By considering $\mathbf{C}^k = [c_1^k, \ldots, c_i^k, \ldots, c_n^k]^T$, Equation (1) can be expressed in matrix form as follows:

$$\mathbf{C}_i^{k+1} = \mathbf{C}_i^k + \mathbf{A}\mathbf{C}_i^k = (\mathbf{I} + \mathbf{A})\mathbf{C}_i^k \rightarrow \mathbf{C}_i^{k+1} = \mathbf{D}\mathbf{C}_i^k, \tag{2}$$

where $\mathbf{I}$ is the identity matrix, and:

$$\mathbf{D} = \begin{bmatrix} 1 - \sum\limits_{j \in N_1} w_{1j} & \cdots & w_{1i} & \cdots & w_{1n} \\ \cdots & \cdots & \cdots & \cdots & \cdots \\ w_{i1} & \cdots & 1 - \sum\limits_{j \in N_i} w_{ij} & \cdots & w_{in} \\ \cdots & \cdots & \cdots & \cdots & \cdots \\ w_{n1} & \cdots & w_{ni} & \cdots & 1 - \sum\limits_{j \in N_n} w_{nj} \end{bmatrix}_{n \times n}. \tag{3}$$

The square matrix $\mathbf{D}$ is said to be *doubly stochastic* if its elements are non-negative and sums of each row and each column are equal to ones, i.e., with $\mathbf{1}_{1 \times n} = [1, 1, \ldots, 1]$, $\mathbf{1} \times \mathbf{D} = \mathbf{1}$ and $\mathbf{1} \times \mathbf{D}^T = \mathbf{1}$ [45]. Based on the Gerschgorin's Disks theorem, the eigenvalues of $\mathbf{D}$ are lower than or equal to one. According to the *Perron Frobenius Lemma* [46], one can write:

$$\lim_{k \to \infty} \mathbf{D}^k = \frac{\mathbf{1}^T * \mathbf{1}}{n}, \tag{4}$$

where n is the dimension of matrix $\mathbf{D}$. Combination of (2) and (4) leads to:

$$\lim_{k \to \infty} C_i^k = \frac{1^T * 1}{n} C_i^0. \tag{5}$$

From Equation (5), one can see that the system reaches to consensus when k approaches infinity. The speed of convergence depends on the design of $\mathbf{D}$. In practice, the exact equilibrium is not required, and the number of required steps for converging is approximately equal to:

$$k = \frac{-1}{\log_e\left(\frac{1}{\lambda_2}\right)}, \tag{6}$$

where e is the error tolerance and λ_2 is the second biggest eigenvalue of $\mathbf{D}$ [44]. Equation (6) shows that λ_2 determines the number of required steps to converge or equivalently the speed of the algorithm. To achieve the maximum speed and the optimal solution, the weight coefficients in matrix $\mathbf{D}$ must be determined in such a way to minimize λ_2.

4.1.2. Coefficient Setting

The employed method for setting the weight coefficients depends on the type of application, i.e., offline or online applications. If the system is exposed to changes of configuration, the optimization problem must be solved again at every change. Because of the multiple variables and constraints in this optimization problem and the required time to achieve the information of the new system configuration, the optimization is time consuming and is, therefore, suitable for offline applications. For online application, there is a requirement for a proper algorithm to adjust the weight coefficients near their optimum values. Normally, in online applications, the weight coefficients are determined by using a simple rule named *Uniform* method [44]. This method proposes the fixed coefficients that are calculated as follows:

$$w_{ij} = \begin{cases} 1/n, & j \in N_i \\ 1 - \sum_{j \in N_i} 1/n, & i = j \\ 0, & \text{otherwise.} \end{cases} \tag{7}$$

In above equations, n is the number of nodes. To achieve a higher convergence speed, another method named *Metropolis* is introduced in [47] that makes λ_2 near to its minimum value using an adaptive weight updating law. The updating rule is:

$$w_{ij} = \begin{cases} 1/(Max(n_i, n_j) + 1), & j \in N_i \\ 1 - \sum_{j \in N_i} 1/(Max(n_i, n_j) + 1), & i = j \\ 0, & \text{otherwise,} \end{cases} \tag{8}$$

where n_i and n_j are the number of nodes in the neighborhood of the node i and j, respectively. It is easy to show that these two methods guarantee the two required conditions for applying the *Perron Frobenius Lemma* to $\mathbf{D}$ (i.e., the sums of each column and row of the $\mathbf{D}$ are ones, and all its eigenvalues are equal or lower than one). To make a comparison between the speeds of the ACA with various coefficient setting rules, let consider the graph depicted in Figure 5. Now, let us define the initial values assigned to each node as follows: $c_1^0 = 100$, $c_2^0 = 100$, $c_3^0 = -50$, $c_4^0 = -100$, $c_5^0 = -50$.

By using Equation (2), the equilibrium point for unlimited iterations will be:

$$c_1^\infty = c_2^\infty = c_3^\infty = c_4^\infty = c_5^\infty = \frac{1}{5}\sum_{i=1}^{5} c_i^0 = 0. \tag{9}$$

It means that after using the consensus algorithm (Equation (1)), the number '0' exists in each node. Figure 6 shows the value in each node for 60 iterations. Considering an error tolerance equal to

0.01, the nodes have reached the consensus with 31 and 25 iterations for the uniform and metropolis methods, respectively.

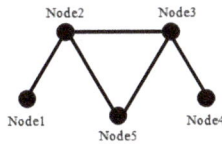

Figure 5. A typical studied graph.

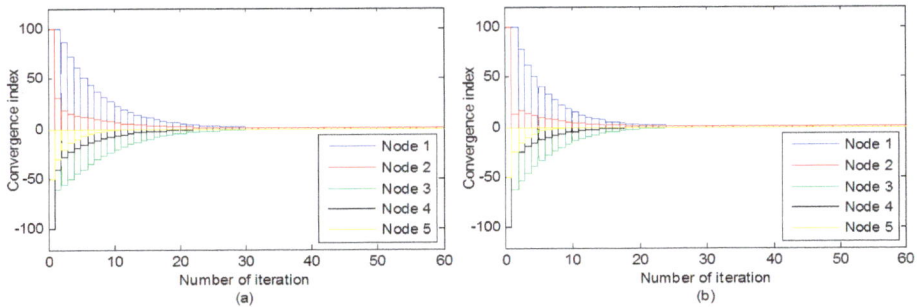

Figure 6. Comparison of converging speed of different methods: (**a**) uniform; (**b**) metropolis.

4.2. General Assumptions

The goal of this paper is to develop a decentralized multi-agent-based approach to restore the MG loads and generations with a proper sequence of actions after a total blackout. The local controllers (MCs and LCs) and the MG communication infrastructure are so important for the successful implementation of the decentralized restoration scheme.

Based on the abovementioned distributed averaging algorithm, it is assumed that each node of the graph can be considered as an agent. The edges of the graph can be considered as the communication links among the agents. It is assumed that each local controller is assigned to a specific agent. Thus, there are two kinds of agents: (i) MC agents and (ii) LC agents. Figure 7 shows the conceptual decentralized multi-agent based model for MG restoration. In this model, each MC agent has some local information such as the amount of the generation capacity of the corresponding microsource, connection situation of the corresponding microsource (connected or disconnected), its availability and preparedness for the restoration, and pre-defined priority for the restoration.

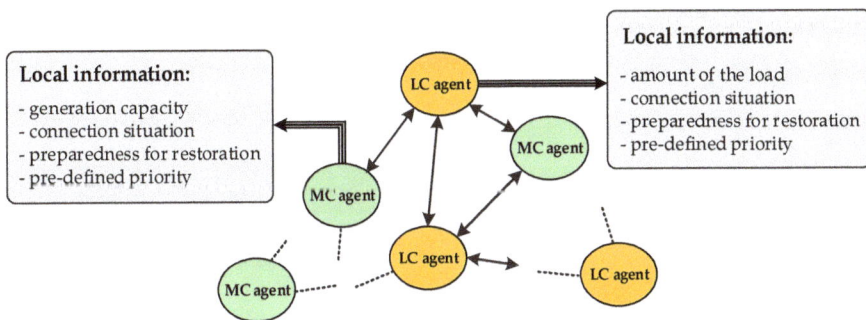

Figure 7. Conceptual decentralized multi-agent based model for MG restoration.

The LC agents have similar local information of their corresponding loads. The agents don't have any direct access to the global information of the system. An agent is only able to communicate with its neighbors. By using the ACA based communication law, the agents are able to share the local information, to access the global information, and accordingly to take a common decision for restoring the loads or generation units.

4.3. Information Sharing Process

The local initial information of each agent is placed within an initial matrix. The initial matrices just have the local information of the agents. Agent i is initialized with a $n \times 4$ matrix $\mathbf{M_i}$ where n is the number of agents. In $\mathbf{M_i}$, up to four non-zero elements may exist. These three non-zero elements are $M_i(i,1)$, $M_i(i,2)$, $M_i(i,3)$, and $M_i(i,4)$. $M_i(i,1)$ can be either 0 or i to show whether the generation unit or the load assigned to agent i is connected or not. Each agent can realize which generation units or loads are disconnected by checking the position of the zeros. $M_i(i,2)$ can be 0 or i to show that the disconnected load or generation unit is ready for restoration or not. In the case of MC agents, $M_i(i,3)$ shows the amount of the power that the generation unit can produce while in the case of LC agents, it represents the amount of the load power that agent i will consume. $M_i(i,4)$ can be 0, 1, 2, 3 that shows a pre-defined priority of the agents for restoration action. If $M_i(i,4) = 0$, the agent i has no pre-defined priority.

For example, let consider the initial matrices $\mathbf{M_i}$, $\mathbf{M_j}$, $\mathbf{M_n}$ for agent i, j, n, respectively, as follows:

$$\mathbf{M_i} = \begin{bmatrix} 0 & 0 & 0 & 0 \\ \vdots & \vdots & \vdots & \vdots \\ 0 & i & P_{G_i} & 0 \\ \vdots & \vdots & \vdots & \vdots \\ 0 & 0 & 0 & 0 \end{bmatrix}_{n \times 4} , \mathbf{M_j} = \begin{bmatrix} 0 & 0 & 0 & 0 \\ \vdots & \vdots & \vdots & \vdots \\ 0 & j & -P_{L_j} & 1 \\ \vdots & \vdots & \vdots & \vdots \\ 0 & 0 & 0 & 0 \end{bmatrix}_{n \times 4} , \mathbf{M_n} = \begin{bmatrix} 0 & 0 & 0 & 0 \\ \vdots & \vdots & \vdots & \vdots \\ \vdots & \vdots & \vdots & \vdots \\ \vdots & \vdots & \vdots & \vdots \\ n & 0 & 0 & 0 \end{bmatrix}_{n \times 4} . \tag{10}$$

The above initial matrices present the following information. MC-agent i is disconnected; it is ready for restoration; if it is connected, it can produce power equal to P_{G_i}; and it has no pre-defined priority. Similarly, the LC-agent j is disconnected; it is ready for restoration; if it is connected, it consumes P_{L_j}; and it has the highest priority to be connected. The agent n is connected, and there is no need for the restoration. Each agent has a similar principle to make the initial information matrix. By using the ACA (Equation (1)), all initial matrices will converge to the same matrix that is available for each agent. A typical final converged matrix can be:

$$\mathbf{M_{conv.}} = \begin{bmatrix} \vdots & \vdots & \vdots & \vdots \\ \dfrac{0}{n} & \dfrac{i}{n} & \dfrac{P_{G_i}}{n} & \dfrac{0}{n} \\ \vdots & \vdots & \vdots & \vdots \\ \dfrac{0}{n} & \dfrac{j}{n} & \dfrac{-P_{L_j}}{n} & \dfrac{1}{n} \\ \vdots & \vdots & \vdots & \vdots \\ \dfrac{n}{n} & \dfrac{0}{n} & \dfrac{0}{n} & \dfrac{0}{n} \end{bmatrix}_{n \times 4} . \tag{11}$$

Each element of the final converged matrix is equal to the average summation of the corresponding elements existed in the initial matrices. The actual amount of each element can be obtained by multiplying the element by n. According to the discovered global information ($\mathbf{M_{conv.}}$), the amount of disconnected loads and generations are available and after reaching the consensus, all agents take a common decision for restoring the loads or generation units. The following subsection is devoted to implementing the proposed method in which the function of agents and decision making process are described.

4.4. Implementation of the Proposed Multi-Agent Based Approach for MG Restoration

The function modules of the agent *i* is illustrated in Figure 8. Each agent has four main modules: (1) initialization; (2) information update; (3) information keep; and (4) exchange and decision making. The required steps for operation of the agents can be designated as follows:

Step 1: Initialization: in this step, the local initial information matrix of each agent ($\mathbf{M}_i^0$) is formed.

Step 2: Information sharing: in this step, each agent receives the information of its neighboring agents through a communication link and updates its information by using the ACA. After reaching the consensus, the common decision will be made.

Step 3: Decision making: when the agents reach the consensus in the sharing information process, a proper decision will be made. Decision making is one of the crucial parts of the agents' function blocks. This block must be designed to meet the initial restoration steps such as setting up the generation units with black start capability and energizing the restoration path. Moreover, in the next steps, this block must determine a proper sequence for connection of disconnected loads and generation units by providing the maximum amount of the restored loads in the shortest possible time.

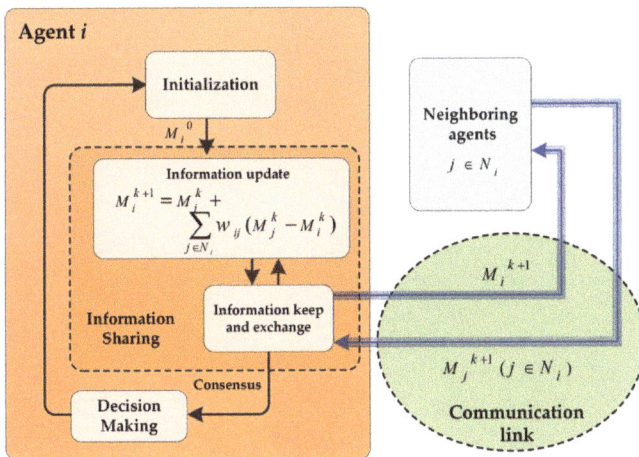

Figure 8. Function of agent *i* for MG restoration.

Decision Making Process

MG restoration process begins by setting up a generation unit with black start capability. In an inverter-based MG, the inverter with grid-forming control mode has the black start capability. It is assumed that in the MG, one inverter is in grid-forming control mode and the other ones operate in grid following mode. The output real power of the grid-following inverters is a constant value. However, the output real power of the grid-forming inverter varies based on the MG frequency (droop control). During the restoration process, the connection of a load or a generation unit changes the output power of the grid-forming inverter. The capacity and the instantaneous power of grid-forming inverter play a key role in determining a proper sequence of connecting the loads and generation units with a grid-following inverter.

To help better understand this, consider the flowchart of the decision making process shown in Figure 9. When the algorithm is run for the first time, the microsource with grid forming inverter is chosen and connected to energize the MG feeder. Then, a pre-defined interruption is required for damping of the frequency fluctuations. The interruption time depends on the inertia and damping factor of the MG. In the next run of the algorithm, after reaching the consensus, the information related to the output power of the grid-forming inverter P_{form} that is a small value (only for energizing the MG feeder) as well as the disconnected load units and available power of the generation units with grid-following inverter are available for all agents. The capacity of the grid-forming inverter P_{form}^{cap} is also specified. In this step, based on the discovered information, the largest possible amount of the load is chosen to be connected. This amount of the connected load makes the grid-forming inverter to work in the capacity limits, so in the next step, a generation unit with the grid-following inverter is chosen to releases the capacity of the grid-forming inverter by providing a fixed amount of power. This procedure will continue until all of the generation units and the maximum possible amount of the loads are connected.

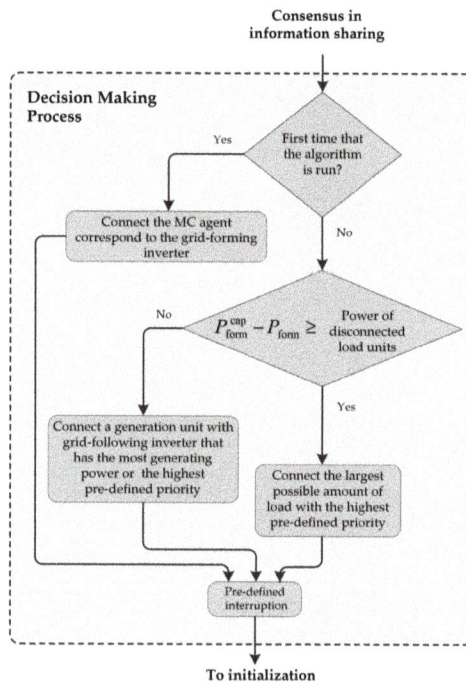

Figure 9. Flowchart of the decision making process.

5. Simulation Results and Discussion

In order to evaluate the dynamic behavior of an MG during restoration procedure, a benchmark LV MG network presented in Figure 10 is implemented in the simulation platform. The electrical data for this LV test system can be found in [48]. It is supposed that the MG is subjected to a total blackout. The studied MG includes nine local controllers, and each one is assigned to a specific agent. Figure 11 shows the two topologies for connection of the agents.

From (6), it can be observed that the speed of ACA is independent of the initial information matrix and it depends on how the agents are connected and how the weight coefficients are determined.

To verify the convergence speed of the ACA for the topologies of Figure 11, let define the initial values assigned to each agent as follows:

$$c_1^0 = 1, c_2^0 = -1, c_3^0 = 1, c_4^0 = -1, c_5^0 = 1, c_6^0 = -1, c_7^0 = 1, c_8^0 = -1, c_9^0 = 0. \tag{12}$$

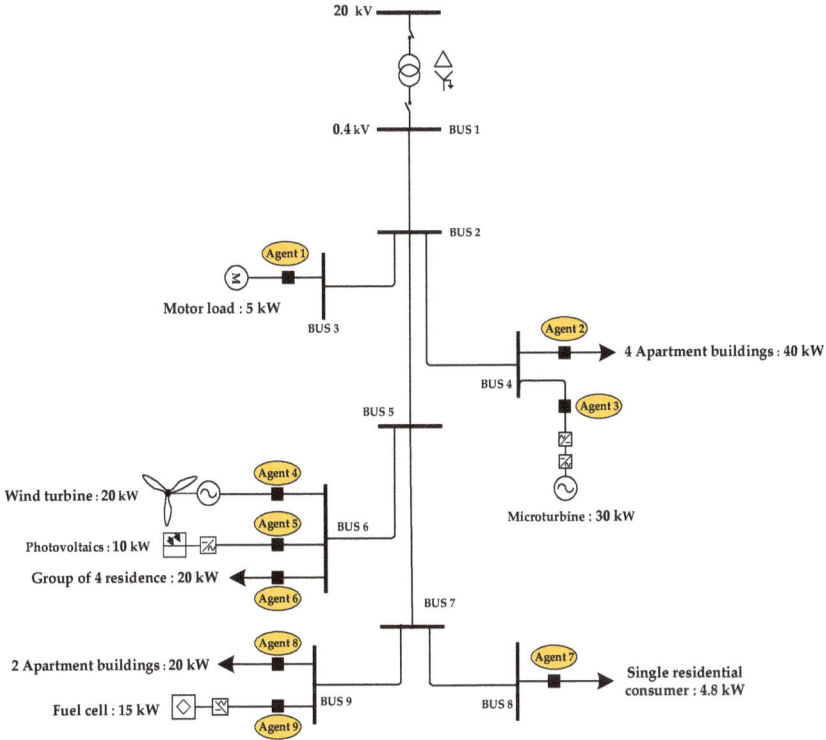

Figure 10. Benchmark LV MG network.

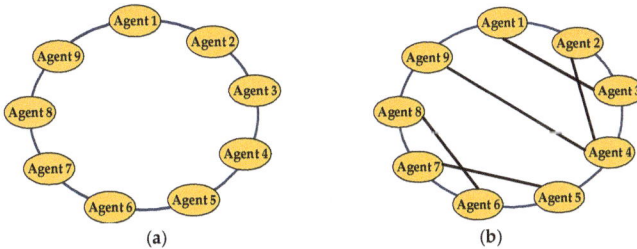

Figure 11. Different topologies for connection of the local controller agents: (**a**) topology a; (**b**) topology b.

By using Equation (2), the equilibrium point for unlimited iterations will be:

$$c_1^\infty = c_2^\infty = c_3^\infty = c_4^\infty = c_5^\infty = c_6^\infty = c_7^\infty = c_8^\infty = c_9^\infty = \frac{1}{9}\sum_{i=1}^{9} c_i^0 = 0. \tag{13}$$

It means that after using the consensus algorithm (Equation (1)), the number '0' exists in each agent. The metropolis method is used for determining the weight coefficient. Figure 12 shows the value in each agent for 40 iterations. Considering an error tolerance equal to 0.01, the agents have reached the consensus with 19 and 17 iterations for the topology (a) and (b), respectively.

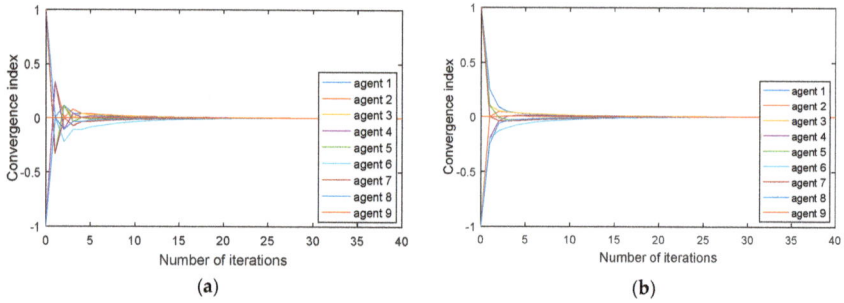

Figure 12. The convergence speed of the different topologies of Figure 11: (**a**) topology a; (**b**) topology b.

The time delay to reach the consensus can be estimated by:

$$T = \frac{N_{\text{iteration}} \times N_M \times N_b}{C},$$ (14)

where $N_{\text{iteration}}$ is the number of required iterations to reach the consensus, N_M is the size of the information matrix, N_b is the number of required bits to represent each element of the information matrix, and C is the communication link speed. For the topology (a) and (b), the system requires 19 and 17 iterations to converge, respectively. There are nine agents, $N_M = 9 \times 4$, and if 16 bits are used for representing each element of the information matrix, for a network with 5 Mbit/s, time delay for reaching the consensus for the topology (a), (b) are 0.002189 s and 0.001958 s, respectively. This time delay is very small compared with the pre-defined interruption time used in the decision making process. In the simulations, the interruption time is considered equal to 4 s. Therefore, the time delay for reaching the consensus can be neglected.

In order to provide the black start capability for the studied MG, the microturbine is equipped with the battery storage in the DC link, and its inverter operates in the grid-forming control mode. The inverters of the fuel cell and PVs systems operate in the grid following control mode. The wind turbine is connected directly to the grid through an induction generator. It should be noted that the secondary control is carried out locally by using a PI controller at microturbine control system aiming to restore the frequency and voltage to the nominal value after any restoration action. The restoration process is started with forming the initial information matrices of the agents. The initial information matrices are as follows:

$$M_1 = \begin{bmatrix} 0 & 1 & -5 & 2 \\ 0 & 0 & 0 & 0 \\ 0 & 0 & 0 & 0 \\ 0 & 0 & 0 & 0 \\ 0 & 0 & 0 & 0 \\ 0 & 0 & 0 & 0 \\ 0 & 0 & 0 & 0 \\ 0 & 0 & 0 & 0 \\ 0 & 0 & 0 & 0 \end{bmatrix}, M_2 = \begin{bmatrix} 0 & 0 & 0 & 0 \\ 0 & 2 & -40 & 1 \\ 0 & 0 & 0 & 0 \\ 0 & 0 & 0 & 0 \\ 0 & 0 & 0 & 0 \\ 0 & 0 & 0 & 0 \\ 0 & 0 & 0 & 0 \\ 0 & 0 & 0 & 0 \\ 0 & 0 & 0 & 0 \end{bmatrix}, M_3 = \begin{bmatrix} 0 & 0 & 0 & 0 \\ 0 & 0 & 0 & 0 \\ 0 & 3 & 30 & 0 \\ 0 & 0 & 0 & 0 \\ 0 & 0 & 0 & 0 \\ 0 & 0 & 0 & 0 \\ 0 & 0 & 0 & 0 \\ 0 & 0 & 0 & 0 \\ 0 & 0 & 0 & 0 \end{bmatrix}, M_4 = \begin{bmatrix} 0 & 0 & 0 & 0 \\ 0 & 0 & 0 & 0 \\ 0 & 0 & 0 & 0 \\ 0 & 4 & 20 & 0 \\ 0 & 0 & 0 & 0 \\ 0 & 0 & 0 & 0 \\ 0 & 0 & 0 & 0 \\ 0 & 0 & 0 & 0 \\ 0 & 0 & 0 & 0 \end{bmatrix}, M_5 = \begin{bmatrix} 0 & 0 & 0 & 0 \\ 0 & 0 & 0 & 0 \\ 0 & 0 & 0 & 0 \\ 0 & 0 & 0 & 0 \\ 0 & 5 & 10 & 0 \\ 0 & 0 & 0 & 0 \\ 0 & 0 & 0 & 0 \\ 0 & 0 & 0 & 0 \\ 0 & 0 & 0 & 0 \end{bmatrix}$$

$$M_6 = \begin{bmatrix} 0 & 0 & 0 & 0 \\ 0 & 0 & 0 & 0 \\ 0 & 0 & 0 & 0 \\ 0 & 0 & 0 & 0 \\ 0 & 0 & 0 & 0 \\ 0 & 6 & -20 & 0 \\ 0 & 0 & 0 & 0 \\ 0 & 0 & 0 & 0 \\ 0 & 0 & 0 & 0 \end{bmatrix}, M_7 = \begin{bmatrix} 0 & 0 & 0 & 0 \\ 0 & 0 & 0 & 0 \\ 0 & 0 & 0 & 0 \\ 0 & 0 & 0 & 0 \\ 0 & 0 & 0 & 0 \\ 0 & 0 & 0 & 0 \\ 0 & 7 & -4.8 & 0 \\ 0 & 0 & 0 & 0 \\ 0 & 0 & 0 & 0 \end{bmatrix}, M_8 = \begin{bmatrix} 0 & 0 & 0 & 0 \\ 0 & 0 & 0 & 0 \\ 0 & 0 & 0 & 0 \\ 0 & 0 & 0 & 0 \\ 0 & 0 & 0 & 0 \\ 0 & 0 & 0 & 0 \\ 0 & 0 & 0 & 0 \\ 0 & 8 & -20 & 0 \\ 0 & 0 & 0 & 0 \end{bmatrix}, M_9 = \begin{bmatrix} 0 & 0 & 0 & 0 \\ 0 & 0 & 0 & 0 \\ 0 & 0 & 0 & 0 \\ 0 & 0 & 0 & 0 \\ 0 & 0 & 0 & 0 \\ 0 & 0 & 0 & 0 \\ 0 & 0 & 0 & 0 \\ 0 & 0 & 0 & 0 \\ 0 & 9 & 15 & 0 \end{bmatrix}$$

By using ACA for the topology (a), after 0.002189 s, the agents share the initial information and reach the consensus. The final converged matrix that is available for all agents is as follows:

$$M_{conv.} = \begin{bmatrix} 0 & 0 & 0 & 0 & 0 & 0 & 0 & 0 & 0 \\ \dfrac{1}{9} & \dfrac{2}{9} & \dfrac{3}{9} & \dfrac{4}{9} & \dfrac{5}{9} & \dfrac{6}{9} & \dfrac{7}{9} & \dfrac{8}{9} & \dfrac{9}{9} \\ \dfrac{-5}{9} & \dfrac{-40}{9} & \dfrac{30}{9} & \dfrac{20}{9} & \dfrac{10}{9} & \dfrac{-20}{9} & \dfrac{-4.8}{9} & \dfrac{-20}{9} & \dfrac{15}{9} \\ \dfrac{2}{9} & \dfrac{1}{9} & 0 & 0 & 0 & 0 & 0 & 0 & 0 \end{bmatrix}^T$$

By checking the first and the second columns of $M_{conv.}$ (or second row of $M_{conv.}^T$), it can be found that all of the loads and generations are disconnected, and they are ready to be restored. Column 3 shows that the MC-agents have no predefined priority and they will be chosen based on their production capacity. Among the LC-agents, the LC-agent 2 (correspond to the apartment building disconnected from bus 4) has the highest priority to be connected. LC-agent 1 (correspond to the motor load) has the next priority, and the other LC-agents have no pre-defined priority for restoration. The first common decision of the agents in this step is to connect the MC-agent 3 (correspond to the microturbine) for energizing the MG feeder.

After connecting the microturbine and passing the interruption time (4 s), the initial information matrices are again formed. All of the initial matrices are same as the previous step except the initial matrix corresponds to the MC-agent 3. In this step, the final converged matrix is as follows:

$$M_{conv.} = \begin{bmatrix} 0 & 0 & 1 & 0 & 0 & 0 & 0 & 0 & 0 \\ \dfrac{1}{9} & \dfrac{2}{9} & 0 & \dfrac{4}{9} & \dfrac{5}{9} & \dfrac{6}{9} & \dfrac{7}{9} & \dfrac{8}{9} & \dfrac{9}{9} \\ \dfrac{-5}{9} & \dfrac{-40}{9} & \dfrac{30}{9} & \dfrac{20}{9} & \dfrac{10}{9} & \dfrac{-20}{9} & \dfrac{-4.8}{9} & \dfrac{-20}{9} & \dfrac{15}{9} \\ \dfrac{2}{9} & \dfrac{1}{9} & 0 & 0 & 0 & 0 & 0 & 0 & 0 \end{bmatrix}^T$$

From the above matrix, it can be seen that the microturbine has 30 kW capacity for supplying the loads. By considering the priority of the loads, the LC-agent 2 connects three apartment buildings and the microturbine will reach its capacity limit. After passing the interruption time, the initial information matrices are again formed, and the final converged matrix is expressed as follows:

$$M_{conv.} = \begin{bmatrix} 0 & 0 & 1 & 0 & 0 & 0 & 0 & 0 & 0 \\ \dfrac{1}{9} & \dfrac{2}{9} & 0 & \dfrac{4}{9} & \dfrac{5}{9} & \dfrac{6}{9} & \dfrac{7}{9} & \dfrac{8}{9} & \dfrac{9}{9} \\ \dfrac{-5}{9} & \dfrac{-10}{9} & 0 & \dfrac{20}{9} & \dfrac{10}{9} & \dfrac{-20}{9} & \dfrac{-4.8}{9} & \dfrac{-20}{9} & \dfrac{15}{9} \\ \dfrac{2}{9} & \dfrac{1}{9} & 0 & 0 & 0 & 0 & 0 & 0 & 0 \end{bmatrix}^{T}$$

The above matrix shows that the microturbine has reached its capacity limit and there is a need for connection of another MC-agent to release the capacity of the microturbine. In this step, among the MC-agents, the MC-agent 4 (correspond to the wind turbine) that has a higher production capacity is chosen to be connected. In the next step, the final converged matrix is as follows:

$$M_{conv.} = \begin{bmatrix} 0 & 0 & 1 & 1 & 0 & 0 & 0 & 0 & 0 \\ \dfrac{1}{9} & \dfrac{2}{9} & 0 & 0 & \dfrac{5}{9} & \dfrac{6}{9} & \dfrac{7}{9} & \dfrac{8}{9} & \dfrac{9}{9} \\ \dfrac{-5}{9} & \dfrac{-10}{9} & \dfrac{20}{9} & 0 & \dfrac{10}{9} & \dfrac{-20}{9} & \dfrac{-4.8}{9} & \dfrac{-20}{9} & \dfrac{15}{9} \\ \dfrac{2}{9} & \dfrac{1}{9} & 0 & 0 & 0 & 0 & 0 & 0 & 0 \end{bmatrix}^{T}$$

In this step, the microturbine has 20 kW free capacity. The remaining load of LC-agent 2 along with the LC-agent 1 and one of the residence groups correspond to the LC-agent 6 are connected. This process will continue until all of the MC-agents are connected, and the microturbine operates in its capacity limits. The remaining disconnected loads will be supplied when the MG is connected to the upstream network. Table 1 shows the sequence of restoration actions carried out based on the proposed decentralized multi-agent based scheme.

Table 1. Sequence of restoration actions.

Steps	Actions	time
Step 1	Connection of microturbine	$t = 1$ s
Step 2	Connection of three apartment buildings at bus 4	$t = 5$ s
Step 3	Connection of wind turbine	$t = 9$ s
Step 4	Connection of one apartment building at bus 4 Connection of motor load Connection of one of the residence groups at bus 6	$t = 13$ s
Step 5	Connection of fuel cell	$t = 17$ s
Step 6	Connection of three remaining residences at bus 6	$t = 21$ s
Step 7	Connection of PVs	$t = 25$ s
Step 8	Connection of one apartment building at bus 9	$t = 29$ s

After step 8, the final converged matrix is expressed as follows:

$$M_{conv.} = \begin{bmatrix} 1 & 1 & 1 & 1 & 1 & 1 & 0 & 0 & 1 \\ 0 & 0 & 0 & 0 & 0 & 0 & \dfrac{7}{9} & \dfrac{8}{9} & 0 \\ 0 & 0 & 0 & 0 & 0 & 0 & \dfrac{-4.8}{9} & \dfrac{-10}{9} & 0 \\ 0 & 0 & 0 & 0 & 0 & 0 & 0 & 0 & 0 \end{bmatrix}^{T}$$

By checking the first column of $\mathbf{M}_{conv.}$, it can be found that all of the generation units and loads are connected, except the load at bus 8 and a part of the load at bus 9. The microturbine also has reached its capacity limits, so the remaining loads will be supplied when the upstream network is available. At this point, the work of the proposed decentralized multi-agent based restoration scheme has been completed.

It should be noted that during the restoration procedure, if the produced (consumed) power of the generation units (loads) changes or they are not ready to be restored, the restoration decisions may be changed by providing a proper local initial information matrix. That's why it is emphasized that the proposed method uses online information of the system to determine the sequence of the restoration actions.

Dynamic simulations are carried out in the Matlab-Simulink environment. Figure 13 shows the microturbine real power during the restoration process. Each time that the loads are connected, the microturbine reaches to its capacity limit, and with the connection of generation units, its capacity is released. Figure 14 shows the frequency of the MG during the restoration process. The real power of the microsources is shown in Figure 15. The results show that for successful implementation of the proposed scheme, the generation unit with grid-forming inverter plays a key role and the proper time interval among the restoration actions is required.

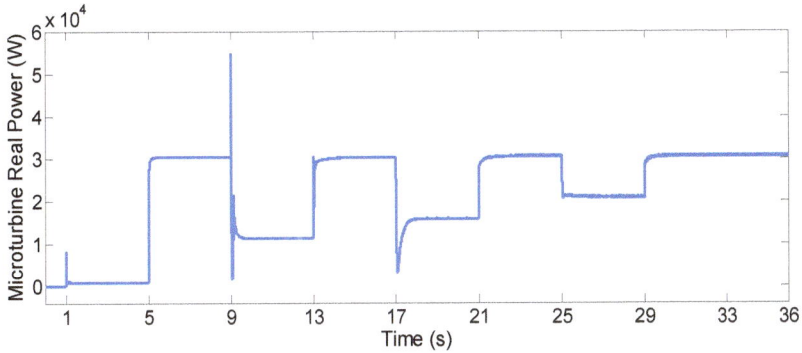

Figure 13. Microturbine real power during the restoration process.

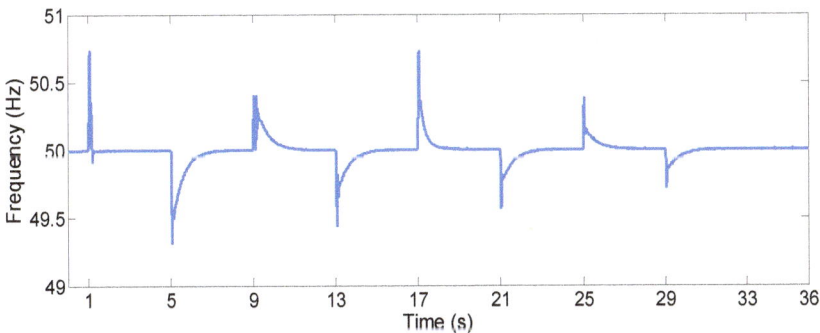

Figure 14. Frequency of the MG during the restoration process.

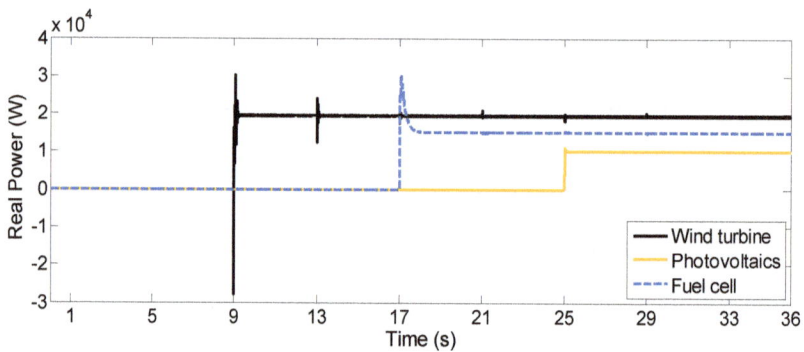

Figure 15. Real power of the microsources.

6. Conclusions

This paper proposed a decentralized multi-agent-based approach for MG restoration. In the proposed scheme, the MG local controllers were assigned to specific agents. The communication rule for sharing the local information of the agents and getting access to the global information was based on ACA. A proper restoration decisions strategy based on the discovered global information was developed. Compared to the centralized restoration schemes, the proposed method had the capability of surviving the single-point failure. Moreover, the online information of the generation/consumption of the MG was used to determine the proper sequence of restoration actions. The effectiveness of the proposed strategy is verified using a benchmark LV MG network.

Acknowledgments: João P. S. Catalão acknowledges the support by FEDER funds through COMPETE 2020 and by Portuguese funds through FCT, under Projects SAICT-PAC/0004/2015—POCI-01-0145-FEDER-016434, POCI-01-0145-FEDER-006961, UID/EEA/50014/2013, UID/CEC/50021/2013, and UID/EMS/00151/2013, and also funding from the EU 7th Framework Programme FP7/2007–2013 under GA No. 309048.

Author Contributions: All authors have worked on this manuscript together and all authors have read and approved the final manuscript.

Conflicts of Interest: The authors declare no conflict of interest.

References

1. Allen, E.H.; Stuart, R.B.; Wiedman, T.E. No Light in August: Power System Restoration Following the 2003 North American Blackout. *IEEE Power Energy Mag.* **2014**, *12*, 24–33. [CrossRef]
2. Xue, Y.; Xiao, S. Generalized congestion of power systems: Insights from the massive blackouts in India. *J. Mod. Power Syst. Clean Energy* **2013**, *1*, 91–100. [CrossRef]
3. Andrews, C.J.; Arsanjani, F.; Lanier, M.W.; Miller, J.M.; Volkmann, T.A.; Wrubel, J. Special considerations in power system restoration. *IEEE Trans. Power Syst.* **1992**, *7*, 1419–1427. [CrossRef]
4. Adibi, M.M. A Framework for Power System Restoration Following a Major Power Failure. In *Power System Restoration: Methodologies amp, Implementation Strategies*; Wiley-IEEE Press: Hoboken, NJ, USA, 2000; pp. 96–101. ISBN 978-0-470-54560-7.
5. Adibi, M.M.; Fink, L.H. Power system restoration planning. *IEEE Trans. Power Syst.* **1994**, *9*, 22–28. [CrossRef]
6. Adibi, M.M.; Kafka, R.J. Power system restoration issues. *IEEE Comput. Appl. Power* **1991**, *4*, 19–24. [CrossRef]
7. Adibi, M.M.; Kafka, L.R.J.; Milanicz, D.P. Expert system requirements for power system restoration. *IEEE Trans. Power Syst.* **1994**, *9*, 1592–1600. [CrossRef]
8. Moreira, C.L.; Resende, F.O.; Lopes, J.A.P. Using Low Voltage MicroGrids for Service Restoration. *IEEE Trans. Power Syst.* **2007**, *22*, 395–403. [CrossRef]
9. Lasseter, R.H. Smart Distribution: Coupled Microgrids. *Proc. IEEE* **2011**, *99*, 1074–1082. [CrossRef]
10. Kleinberg, M.R.; Miu, K.; Chiang, H.D. Improving Service Restoration of Power Distribution Systems Through Load Curtailment of In-Service Customers. *IEEE Trans. Power Syst.* **2011**, *26*, 1110–1117. [CrossRef]

11. Liang, C.; Khodayar, M.; Shahidehpour, M. Only Connect: Microgrids for Distribution System Restoration. *IEEE Power Energy Mag.* **2014**, *12*, 70–81. [CrossRef]

12. Ancona, J.J. A framework for power system restoration following a major power failure. *IEEE Trans. Power Syst.* **1995**, *10*, 1480–1485. [CrossRef]

13. Van Hentenryck, P.; Coffrin, C. Transmission system repair and restoration. *Math. Program.* **2015**, *151*, 347–373. [CrossRef]

14. He, X.; Liao, Z.; Guo, W.; Wen, F.; Liang, J.; Fu, J. A multi-objective model for transmission network restoration based on multi-agent and tabu search. In Proceedings of the Third International Conference on Electric Utility Deregulation and Restructuring and Power Technologies, Nanjing, China, 6–9 April 2008; pp. 2392–2397.

15. Iwasaki, K.; Aoki, H. Service restoration problem in distribution power system using improved ga. *Electr. Eng. Jpn.* **2009**, *166*, 10–19. [CrossRef]

16. Pham, T.T.H.; Besanger, Y.; Hadjsaid, N. New Challenges in Power System Restoration with Large Scale of Dispersed Generation Insertion. *IEEE Trans. Power Syst.* **2009**, *24*, 398–406. [CrossRef]

17. Gu, X. Others Reconfiguration of network skeleton based on discrete particle-swarm optimization for black-start restoration. In Proceedings of the Power Engineering Society General Meeting, Montreal, QC, Canada, 18–22 June 2006.

18. Liu, Q.; Shi, L.; Zhou, M.; Li, G.; Ni, Y. A new solution to generators start-up sequence during power system restoration. In Proceedings of the Third International Conference on Electric Utility Deregulation and Restructuring and Power Technologies, Nanjing, China, 6–9 April 2008; pp. 2845–2849.

19. Adibi, S.W.M. An Approach to Standing Phase Angle Reduction A report by the Power System Restoration Working Group. *Power Syst. Restor. Methodol. Implement. Strateg.* **2000**, 142–150.

20. Mota, A.A.; Mota, L.T.M.; Morelato, A. Restoration building blocks identification using a heuristic search approach. In Proceedings of the Power Engineering Society General Meeting, Montreal, QC, Canada, 18–22 June 2006.

21. Kostic, T.; Cherkaoui, R.; Germond, A.; Pruvot, P. Decision aid function for restoration of transmission power systems: conceptual design and real time considerations. *IEEE Trans. Power Syst.* **1998**, *13*, 923–929. [CrossRef]

22. Resende, F.O.; Gil, N.J.; Lopes, J.A.P. Service restoration on distribution systems using Multi-MicroGrids. *Eur. Trans. Electr. Power.* **2011**, *21*, 1327–1342. [CrossRef]

23. Choo, Y.C.; Lai, K.X.; Kashem, M.A.; Negnevitsky, M. MicroGrid (MV Network) restoration using distributed resources after major emergencies. In Proceedings of the Australasian Universities Power Engineering Conference (AUPEC), Hobart, Tasmania, Australia, 25–28 September 2005; Volume 1, pp. 99–104.

24. Wooldridge, M. *An Introduction to Multiagent Systems*; John Wiley & Sons: New York, NY, USA, 2009.

25. Chazelle, B. An Algorithmic Approach to Collective Behavior. *J. Stat. Phys.* **2015**, *158*, 514–548. [CrossRef]

26. Hatano, Y.; Mesbahi, M. Agreement over random networks. *IEEE Trans. Autom. Control* **2005**, *50*, 1867–1872. [CrossRef]

27. Wu, J.; Meng, Z.; Yang, T.; Shi, G.; Johansson, K.H. Sampled-Data Consensus over Random Networks. *IEEE Trans. Signal Process.* **2015**, *64*, 4479–4492. [CrossRef]

28. Xue, R.; Cai, G.; Xue, R.; Cai, G. Formation Flight Control of Multi-UAV System with Communication Constraints. *J. Aerosp. Technol. Manag.* **2016**, *8*, 203–210. [CrossRef]

29. Zhang, H.; Gurfil, P. Cooperative control of multiple satellites via consensus. In Proceedings of the 24th Mediterranean Conference on Control and Automation (MED), Athens, Greece, 21–24 June 2016; pp. 1102–1107.

30. Montijano, E.; Thunberg, J.; Hu, X.; Sagüés, C. Epipolar Visual Servoing for Multirobot Distributed Consensus. *IEEE Trans. Robot.* **2013**, *29*, 1212–1225. [CrossRef]

31. Grandi, R.; Falconi, R.; Melchiorri, C. Coordination and control of autonomous mobile robot groups using a hybrid technique based on Particle Swarm Optimization and Consensus. In Proceedings of the International Conference on Robotics and Biomimetics (ROBIO), Shenzhen, China, 12–14 December 2013; pp. 1514–1519.

32. Rokrok, E.; Golshan, M.E.H. Adaptive voltage droop scheme for voltage source converters in an islanded multibus microgrid. *IET Gener. Transm. Distrib.* **2010**, *4*, 562–578. [CrossRef]

33. Katiraei, F.; Iravani, R.; Hatziargyriou, N.; Dimeas, A. Microgrids management. *IEEE Power Energy Mag.* **2008**, *6*, 54–65. [CrossRef]

34. Bidram, A.; Davoudi, A. Hierarchical Structure of Microgrids Control System. *IEEE Trans. Smart Grid.* **2012**, *3*, 1963–1976. [CrossRef]

35. Lopes, J.P.; Moreira, C.L.; Madureira, A.G. Defining control strategies for microgrids islanded operation. *IEEE Trans. Power Syst.* **2006**, *21*, 916–924. [CrossRef]

36. Nagpal, M.; Moshref, A.; Morison, G.K.; Kundur, P. Experience with testing and modeling of gas turbines. In Proceedings of the Power Engineering Society Winter Meeting, Columbus, OH, USA, 28 January–1 February 2001; pp. 652–656.

37. Kim, I.-S. Robust maximum power point tracker using sliding mode controller for the three-phase grid-connected photovoltaic system. *Sol. Energy.* **2007**, *81*, 405–414. [CrossRef]

38. González-Longatt, F.; Amaya, O.; Cooz, M.; Duran, L. Dynamic behavior of constant speed wt based on induction generator directly connected to grid. In Proceedings of the 6th World Wind Energy Conference and Exhibition (WWEC), Buenos Aires, Argentina, 2–4 October 2007.

39. Zhu, Y.; Tomsovic, K. Development of models for analyzing the load-following performance of microturbines and fuel cells. *Electr. Power Syst. Res.* **2002**, *62*, 1–11. [CrossRef]

40. Padulles, J.; Ault, G.W.; McDonald, J.R. An integrated SOFC plant dynamic model for power systems simulation. *J. Power Sources* **2000**, *86*, 495–500. [CrossRef]

41. Palizban, O.; Kauhaniemi, K. Hierarchical control structure in microgrids with distributed generation: Island and grid-connected mode. *Renew. Sustain. Energy Rev.* **2015**, *44*, 797–813. [CrossRef]

42. Wang, Y.; Dong, X.; Wang, B.; Liu, J.; Guo, A.X. Black start studies for micro-grids with distributed generators. In Proceedings of the 12th IET International Conference on Developments in Power System Protection (DPSP 2014), Copenhagen, Denmark, 31 March–3 April 2014.

43. Rocabert, J.; Luna, A.; Blaabjerg, F.; Rodríguez, P. Control of Power Converters in AC Microgrids. *IEEE Trans. Power Electron.* **2012**, *27*, 4734–4749. [CrossRef]

44. Xiao, L.; Boyd, S. Fast linear iterations for distributed averaging. *Syst. Control Lett.* **2004**, *53*, 65–78. [CrossRef]

45. Marshall, A.W.; Olkin, I.; Arnold, B.C. Doubly Stochastic Matrices. Inequalities: Theory of Majorization and Its Applications. In *Springer Series in Statistics*; Springer: New York, NY, USA, 2010; pp. 29–77. ISBN 978-0-387-40087-7.

46. Horn, R.A.; Johnson, C.R. *Matrix Analysis*, 2nd ed.; Cambridge University Press: Cambridge, UK, 2012.

47. Xiao, L.; Boyd, S.; Kim, S.-J. Distributed average consensus with least-mean-square deviation. *J. Parallel Distrib. Comput.* **2007**, *67*, 33–46. [CrossRef]

48. Papathanassiou, S.; Hatziargyriou, N.; Strunz, K. Others A benchmark low voltage microgrid network. In Proceedings of the CIGRE Symposium: Power Systems with Dispersed Generation, Athens, Greece, 16–20 April 2005; pp. 1–8.

energies

MDPI

Article

Demand-Side Energy Management Based on Nonconvex Optimization in Smart Grid

Kai Ma [1,*], Yege Bai [1], Jie Yang [1,2], Yangqing Yu [1] and Qiuxia Yang [1]

[1] School of Electrical Engineering, Yanshan University, Qinhuangdao 066004, China;
     BaiYG9614@163.com (Y.B.); jyangysu@ysu.edu.cn or jyangysu@126.com (J.Y.); yyq_0802@163.com (Y.Y.);
     yangqiuxia@ysu.edu.cn (Q.Y.)
[2] Key Laboratory of System Control and Information Processing, Ministry of Education,
     Shanghai Jiao Tong University, Shanghai 200240, China
*   Correspondence: kma@ysu.edu.cn; Tel.: +86-335-838-7556

Received: 7 September 2017; Accepted: 28 September 2017; Published: 4 October 2017

Abstract: Demand-side energy management is used for regulating the consumers' energy usage in smart grid. With the guidance of the grid's price policy, the consumers can change their energy consumption in response. The objective of this study is jointly optimizing the load status and electric supply, in order to make a tradeoff between the electric cost and the thermal comfort. The problem is formulated into a nonconvex optimization model. The multiplier method is used to solve the constrained optimization, and the objective function is transformed to the augmented Lagrangian function without constraints. Hence, the Powell direction acceleration method with advance and retreat is applied to solve the unconstrained optimization. Numerical results show that the proposed algorithm can achieve the balance between the electric supply and demand, and the optimization variables converge to the optimum.

Keywords: demand-side energy management; multiplier method; Powell direction acceleration method; advance and retreat method; thermal comfort

1. Introduction

The power system includes generators, transformers, transmission, and distribution lines that deliver electricity power to terminal users. Smart grid enables real-time control and monitoring to provide distributed generation and storage. It can make grid operating reliably, economically and efficiently [1,2]. In smart grid, the energy providers can monitor the operating states of the loads in real time and control power supply directly. Demand-side energy management has been a hot topic in recent years [3,4]. Reasonable energy management can effectively promote the development of clean energy, save resources and reduce generation costs. In the process of the energy management, the consumers are encouraged to adjust the electricity purchase, optimize the load curve and improve the electricity efficiency [5–7]. Demand-side energy management is a mechanism which requires the consumers' response to pricing strategy [8–10]. The real-time price is an effective strategy to achieve demand-side response [11–13].

In [14], an energy management service for the smart building has been proposed to measure and predict the patterns of both energy generation and power load. Taking into account overall costs, climatic comfort level and timeliness, a mixed integer linear programming model and a heuristic algorithm were proposed to make consumers change the consumption profile during certain time interval [15]. In [16], an automatic rule creation based on the knowledge extraction of a smart building was proposed to optimize the consumers' electricity usage. In [17], the Lagrangian dual algorithm was employed to solve the nonconvex problem, and it came up with efficient demand response scheduling schemes. In [18], a complex telecommunication infrastructure was designed to manage the data exchange among the energy management system, generators, loads, and field sensors/actuators. In [19–21], the cost minimization of interactive consumers was studied based on the noncooperative game theory. The interaction between the consumers and energy provider was modeled with Stackelberg game theory [22–24]. Recently, convex optimization has been used for decreasing the consumers' total cost. In [25], distributed primal-dual algorithms were used to adjust the energy consumption and the price. And the primal-dual algorithm was used to analyze the volatility of electricity markets when considering the uncertainty in the consumer's value function [26]. In [27], an optimal and automatic residential energy consumption scheduling framework was proposed to provide the real-time price schedule to the consumers. In [28], the model of price response was established for the consumers with stochastic charging behaviors. In [29], a fully distributed control algorithm was proposed based on the saddle point dynamics and consensus protocols. In [30], the relationship between the operating states and energy consumption of the loads under forecast error was considered in an energy management problem. In the above studies, the cost functions of the consumers are assumed to be known in advance. However, the cost cannot be directly modeled when considering the comfort of the consumers and the operating state of the loads, such as the thermal comfort and the temperature settings of the heating, ventilation, and air conditioning (HVAC) systems.

In this study, we model energy management as a constrained optimization problem with non-convex objective function. And the Fanger thermal comfort cost which is unknown is included. The objective is to minimize the discomfort costs of the consumers and the generation costs of the providers. Meanwhile, it should keep balance between the consumers' total power consumption and the total generation. Each consumer's load operating state should be limited in upper and lower limits. Hence we propose an iterative algorithm to solve the optimization problem and study the influence of the tradeoff factor and the air conditioning's energy efficient ratio on the energy management scheme.

The rest of the paper is organized as follows. The energy management problem is formulated in Section 2. The algorithm is proposed in Section 3. Section 4 applies the algorithm to the energy management of HVAC systems. The simulation results and analysis are given in Section 5, and conclusions are summarized in Section 6.

2. Problem Formulation

In the process of the demand-side management, we consider an power system consisting of m consumers that are served by an utility company, as shown in Figure 1. The utility company announces the retail price through forecasting the consumers' power consumption. According to the announced price, the consumers can schedule the loads' operations to reduce the costs.

We suppose that an power grid with m loads and n buses. The operating states of consumer i's load ($i \in M = \{1, \cdots, m\}$) is x_i, and the generation on bus i ($i \in N = \{1, \cdots, n\}$) is q_i. The function $c_i(x_i)$ denotes the consumer i's discomfort cost caused by the load changes, and $w_i(q_i)$ denotes the generating cost. And the function $f_i(x_i)$ denotes the relationship between the energy consumption and the operating state. We suppose the lower limit and upper limit of the operating state of consumer i's load is x_i^{min} and x_i^{max}. The energy management can be formulated as the following optimization problem:

$$\max \quad -\tau \sum_{i=1}^{m} c_i(x_i) - (1-\tau) \sum_{i=1}^{n} w_i(q_i)$$

$$\text{s.t.} \quad \sum_{i=1}^{m} f_i(x_i) = \sum_{i=1}^{n} q_i$$

$$x_i^{min} \leq x_i \leq x_i^{max}, i = 1, 2, \cdots, m$$

where $\tau \in [0,1]$ is the parameter to achieve the tradeoff between the consumers' discomfort costs and the generating costs. The energy management problem is to minimize the costs of consumers and providers subject to the energy balance constraints and the operating state limits.

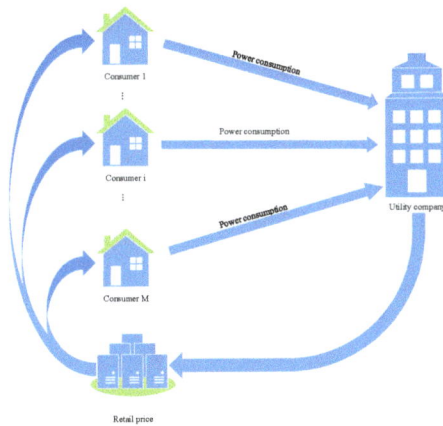

Figure 1. Demand-side management system.

3. Iterative Algorithms

In this section, an iterative algorithm is proposed to solve the above optimization problem. The algorithm, which includes multiplier method, Powell direction acceleration method, advance and retreat method and golden section method, is described in Figure 2.

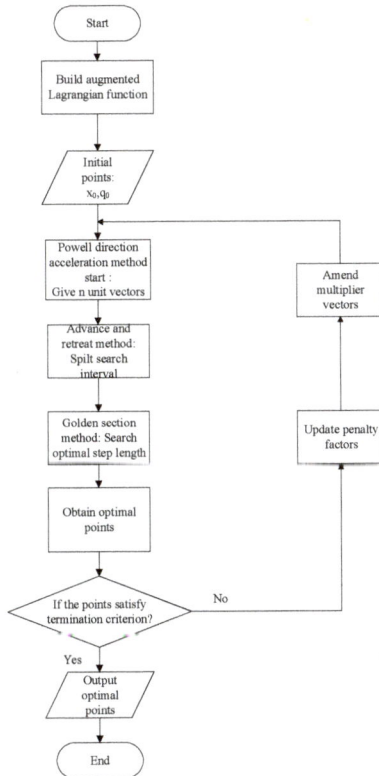

Figure 2. The flowchart of the iterative algorithm.

This iterative algorithm can solve the unknown and nonconvex optimization problem, and the specific algorithms are introduced as the following 4 parts.

Part 1: Multiplier Method

As a general constrained optimization problem, the constraints can be transformed to the objective. For the multiplier method, the constrained augmented Lagrange function can be established as:

$$M(x,q,\mu,\nu,\lambda,\sigma) = \tau \sum_{i=1}^{m} c_i(x_i) + (1-\tau)\sum_{i=1}^{n}\{[\max(0,\mu_i - \sigma(x_i - x_i^{\min}))]^2 - \mu_i^2\} + \frac{1}{2\sigma}\sum_{i=1}^{m}\{[\max(0,$$

$$\nu_i - \sigma(x_i^{\max} - x_i))]^2 - \nu_i^2\} - \lambda[\sum_{i=1}^{m} f_i(x_i) - \sum_{i=1}^{n} q_i] + \frac{\sigma}{2}[\sum_{i=1}^{m} f_i(x_i) - \sum_{i=1}^{n} q_i]^2$$

where λ, μ_i and ν_i are Lagrange multipliers, especially λ is denoted as the retail price. The multipliers are updated by

$$\lambda_{k+1} = \lambda_k - \sigma(\sum_{i=1}^{m} f_i(x_k) - \sum_{i=1}^{n} q_{ki}) \tag{1}$$

$$(\mu_{k+1})_i = \max[0,(\lambda_k)_i - \sigma(x_k - x_k^{\min})_i], i = 1,\cdots,m \tag{2}$$

$$(\nu_{k+1})_i = \max[0,(\lambda_k)_i - \sigma(x_k^{\max} - x_k)_i], i = 1,\cdots,m \tag{3}$$

The termination criterions are $\varphi_{k1} \leq \varepsilon$ and $\varphi_{k2} \leq \varepsilon$, where $\varepsilon > 0$ is the termination error. And φ_{k1} and φ_{k2} are given by

$$\varphi_{k1} = \{[\sum_{i=1}^{m} f_i(x_k) - \sum_{i=1}^{n} q_{ki}] + \sum_{i=1}^{m}[\min((x_k - x_k^{\min})_i, \frac{(\mu_k)_i}{\sigma})]^2\}^{0.5} \tag{4}$$

$$\varphi_{k2} = \{[\sum_{i=1}^{m} f_i(x_k) - \sum_{i=1}^{n} q_{ki}] + \sum_{i=1}^{m}[\min((x_k^{\max} - x_k)_i, \frac{(\nu_k)_i}{\sigma})]^2\}^{0.5} \tag{5}$$

The multiplier method includes 4 steps, as shown in Algorithm 1.

Algorithm 1 The multiplier algorithm.

Initialization:

The set of the initial points: x_0 and q_0;

The set of the initial multiplier vectors: λ_0, μ_0, and ν_0;

The set of the initial penalty factor: σ_1;

Amplification coefficient $c > 0$ and constant $\theta \in (0,1)$. $k = 1$.

Iteration:

The optimal solutions: x_k and q_k.

1: The initial points are x_{k-1} and q_{k-1}, then solve the unconstrained optimization problem:

$$\min M(x,q,\mu,\nu,\lambda,\sigma)$$

we can obtain the optimal points x_k and q_k.

2: Calculate φ_{k1} and φ_{k2} according to Equations (4) and (5). If $\varphi_{k1} < \varepsilon$ and $\varphi_{k2} < \varepsilon$, the optimal solutions are x_k and q_k, and the iteration terminates; else goto 3.

3: When $\frac{\varphi_{k1}}{\varphi_{k1-1}} \leq \theta$ and $\frac{\varphi_{k2}}{\varphi_{k2-1}} \leq \theta$, goto 4; else set $\sigma_{k+1} = c\sigma_k$ and goto 4.

4: Update multiplier vectors according to Equations (1)–(3), set $k = k+1$ and goto 1.

Part 2: Powell Direction Acceleration Method

In this paper, the explicit comfort function is hard to formulate, and it's impossible to take the derivative of an unknown objective function. Therefore, we consider a data-driven algorithm to solve the unconstrained optimization problem directly. The Powell direction acceleration method is one of the most effective data-driven methods. The basic idea of Powell method is to build the conjugated search direction in the next iteration by calculations from the previous iterations.

In the original Powell method, the new search direction will take place of the first component in the old direction vector. However, these new vectors could be linear dependent, and the optimum cannot be obtained. Hence we use the modified Powell method. The modified Powell method can judge whether the new search direction could be applied in the next iteration. If it cannot be applied, judge which direction in the original searching has the lowest objective value. Then let the new search direction replace the old one. In this way, the conjugated direction can be obtained.

In the ith iteration, set $f_1 = f(x_n^{(i)})$, $f_2 = f(x_n^{(i)})$, $f_3 = f(2x_n^{(i)} - x_0^{(i)})$, and $\Delta_m^{(i)} = \max\{f_{k-1}^{(i)} - f_k^{(i)}$, $k = 1, 2, \cdots, n\}$. Let $p_m^{(i)}$ be the search direction: $p^{(i)} = x_n^{(i)} - x_0^{(i)}$. If $f_3 < f_1$ and $(f_1 - 2f_2 + f_3)(f_1 - f_2 - \Delta_m^{(i)})^2 < 0.5\Delta_m^{(i)}(f_1 - f_3)^2$, replace $p_m^{(i)}$ with $p^{(i)}$. Else keep the original directions. The specific algorithm is given in Algorithm 2.

Algorithm 2 The Powell direction acceleration algorithm.

Initialization:

The set of the initial points: $X_0 = (x_0, q_0)^T$;

The control error is given as $\varepsilon > 0$;

$e_1, e_2, \cdots, e_n$ are unit vectors on the coordinate axis, and $k = 1$.

Iteration:

The optimal points: $X^* = X_n$.

1: Calculate $M_0 = M(X_0, \mu_k, \nu_k, \lambda, \sigma_k)$, let $p_i = e_i$, $i = 1, 2, \cdots, n$.

2: One-dimensional search:

$$M(X_{k-1} + \alpha_{k-1} p_k, \mu_k, \nu_k, \lambda_k, \sigma_k) = \min M(X_{k-1} + \alpha p_k, \mu_k, \nu_k, \lambda_k, \sigma_k)$$

Let $X_k = X_{k-1} + \alpha_{k-1} p_k$, $M_k = M(X_k, \mu_k, \nu_k, \lambda_k, \sigma_k)$.

3: If $k = n$, goto 4; If $k < n$, make $k = k+1$ and goto 2.

4: If $\|X_n - X_0\| \leq \varepsilon$, $X^* = X_n$, stop; Else goto 5.

5: Set $\Delta = \max(M_k - M_{k-1}) = M_m - M_{m+1}$, $M^* = M(2X_n - X_0, \mu_k, \nu_k, \lambda_k, \sigma_k)$.

6: If $M^* \geq M_0$ or $(M_0 - 2M_n + M^*)(M_0 - M_n - \Delta)^2 > 0.5(M_0 - M^*)^2 \Delta$, the search directions do

not change. Let $M_0 = M(X_n, \mu_k, \nu_k, \lambda_k, \sigma_k)$, $X_0 = X_n$, $k = 1$, goto 2; Else goto 7.

7: Set $p_k = p_k$, $k = 1, 2, \cdots, m$; $p_k = p_{k+1}$, $k = m+1, \cdots, n-1$, and $p_n = (X_n - X_0)/\|X_n - X_0\|$.

8: One-dimensional search:

$$M(X_n + \bar{\alpha} p_n) = \min M(X_n + \alpha p_n, \mu_k, \nu_k, \lambda_k, \sigma_k)$$

Set $X_0 = X_n + \bar{\alpha} p_n$, $M_0 = M(X_0, \mu_k, \nu_k, \lambda_k, \sigma_k)$, $k = 1$. goto 2.

Part 3: Advance and Retreat Method

Since the objective function is a multimodal and non-convex function, we should segment an unimodal interval before one-dimensional searching based on the specific advance and retreat algorithm, as shown in Algorithm 3.

Algorithm 3 The advance and retreat algorithm.

Initialization:

The set of the initial points: $X_0 = (x_0, q_0)^T$;

The initial step length is $\Delta x (> 0)$, and $t_0 = 0$.

Iteration:

The search interval.

1: Calculate $M_0 = M(X_0)$.

2: $X_1 = X_0 + \Delta x \cdot p_k$. Calculate $M_1 = M(X_1)$. $t_1 = t_0 + \Delta x$. If $M_1 \leq M_0$, goto 3. Else goto 6.

3: Let $t_2 = t_1 + \Delta x$, $X_2 = X_0 + t_2 \cdot p_k$. Calculate $M_2 = M(X_2)$.

4: If $M_1 \leq M_2$, $[t_0, t_2]$ is the search interval; Else goto 5.

5: $t_0 = t_1$, $t_1 = t_2$, $M_1 = M_2$, $\Delta X = 2\Delta X$, $t_2 = t_1 + \Delta x$, $X_2 = X_0 + t_2 \cdot p_k$. Calculate $M_2 = M(X_2)$,

then goto 4.

6: $\Delta x = -\Delta x$, $t = t_0$, $t_0 = t_1$, $t_1 = t$, $M = M_0$, $M_1 = M$, $t_2 = t_1 + \Delta x$, $X_2 = X_0 + t_2 \cdot p_k$.

Calculate $M_2 = M(X_2)$.

7: If $M_1 \leq M_2$, $[t_0, t_2]$ is the search interval; Else goto 8.

8: $t_0 = t_1$, $t_1 = t_2$, $M_1 = M_2$, $\Delta x = 2\Delta x$, $X_2 = X_0 + t_2 \cdot p_k$. Calculate $M_2 = M(X_2)$, goto 7.

Part 4: Golden Section Method

After segmented the interval, the optimal step length is calculated by Golden Section method, as shown in Algorithm 4.

Algorithm 4 The golden section algorithm.

Initialization:

The search interval: $[a, b]$; $\varepsilon > 0$.

Iteration:

The optimal stepsize: $\frac{a+b}{2}$.

1: Let $a_2 = a + 0.618(b - a)$, $X_2 = X_0 + a_2 \cdot p_k$, $M_2 = M(X_2)$.

2: Let $a_1 = a + 0.382(b - a)$, $X_1 = X_0 + a_1 \cdot p_k$, $M_1 = M(X_1)$.

3: If $|\frac{b-a}{b}| > \varepsilon$ and $|\frac{M_2 - M_1}{M_2}| > \varepsilon$, goto 4. Else the optimal result is $\frac{a+b}{2}$.

4: If $M_1 < M_2$, then $b = a_2$, $a_2 = a_1$, $M_2 = M_1$, $a_1 = a + 0.382(b - a)$, $X_1 = X_0 + a_1 \cdot p_k$.

Calculate $M_1 = M(X_1)$, goto 3; Else goto 5.

5: $a = a_1$, $a_1 = a_2$, $M_1 = M_2$, $a_2 = a + 0.618(b - a)$, $X_2 = X_0 + a_2 \cdot p_k$. Calculate $M_2 = M(X_2)$,

goto 3.

Remark 1. *The convergence of the algorithm has been proved in [31]. In the optimization problem with multi-dimensional variable, a global optimal point in each dimension can be obtained during the iterations. However, we cannot guarantee that the optimal points of all variables can be searched simultaneously in the same iteration, and the solution should be a sub-optimal solution in the calculation.*

4. Application to Energy Management of HVAC Systems

In this section, we apply the iterative algorithms to the energy management of HVAC systems. The discomfort of consumers are characterized by the Fanger thermal comfort model. In the research of professor P. O. Fanger from Denmark, the predicted mean vote (PMV) and the predicted percentage of dissatisfied (PPD) were proposed to describe the human body's comfort and satisfaction of the thermal environment, respectively. The Fanger thermal comfort model considers the thermal resistance of clothing, degree of human activities, the air temperature, the air velocity, the mean radiant temperature, and the moisture in the atmosphere. The PMV denotes the human body's hot and cold sensation, including seven grades: hot, warm, little warm, moderate, little cool, cool, cold. The corresponding values are: $+3, +2, +1, 0, -1, -2, -3$. In practice, different people could have different feelings in the same thermal environment. To describe this relationship, the PPD target was proposed in [32–34].

The mathematical expression of PMV is denoted as:

$$PMV = [0.303\exp(-0.036M) + 0.028]\{M - W - 3.05 \times 10^{-3}[5.733 - 6.99(M - W)] - Pa -$$
$$0.42[(M - W) - 58.15] - 1.7 \times 10^{-5}M(5867 - Pa) - 0.0014M(34 - t_a) - 3.96 \times 10^{-8}f_{cl} \times \quad (6)$$
$$[(t_{cl} + 273)^4 - (t_r + 273)^4] - f_{cl}h_c(t_{cl} - t_a)\},$$

where

$$f_{cl} = \begin{cases} 1.00 + 1.290I_{cl} & I_{cl} \leq 0.078, \\ 1.05 + 0.645I_{cl} & I_{cl} > 0.078 \end{cases}$$

and

$$h_c = \begin{cases} 2.38 \times (t_{cl} - t_a)^{0.25} & 2.38(t_{cl} - t_a)^{0.25} > 12.1\sqrt{V_{ar}} \\ 12.1 \times \sqrt{V_{ar}} & 2.38(t_{cl} - t_a)^{0.25} < 12.1\sqrt{V_{ar}} \end{cases}$$

where $t_{cl} = 35.7 - 0.028(M - W) - I_{cl}\{3.96 \times 10^{-8}f_{cl}[(t_{cl} + 273)^4 - (t_r + 273)^4] + f_{cl}h_c(t_{cl} - t_a)\}$.

The PPD target represents a percentage of the human's dissatisfaction of the environment, and the mathematical expression is given as:

$$PPD = 100 - 95 \times \exp[-(0.03353 \times PMV^4 + 0.2179 \times PMV^2)] \quad (7)$$

The explanation of the parameters is shown in Table 1, and the relationship between PPD and PMV is shown in Figure 3.

Table 1. The specific explanation of the parameters.

Parameters	Explanation
M	Human body's energy metabolic rate (W/m²)
W	Human body's mechanical work (W/m²)
Pa	Vapour pressure around body (Pa)
t_a	Air temperature (°C)
f_{cl}	Area coefficient of clothing
t_{cl}	Ttemperature of clothes (°C)
t_r	Indoor's mean radiant temperature (°C)
h_c	Convective heat transfer coefficient (W/(m²·K))
I_{cl}	Heat resistance of clothes ((m²·K)/W)
V_{ar}	Air velocity (m/s)

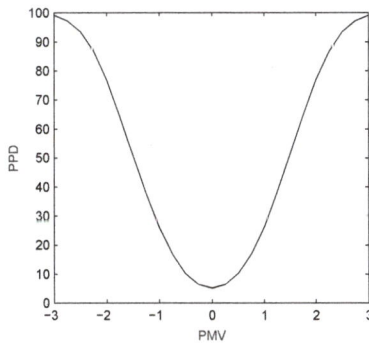

Figure 3. The relationship between PPD and PMV.

We can build the following function to describe the consumers' discomfort costs:

$$c(T_i) = \gamma_i \times \text{PPD} \tag{8}$$

where γ_i is a constant coefficient that transforms the PPD to the discomfort cost. The generating cost of the provider is given as [35]:

$$w(q) = \rho_1 q^2 + \rho_2 q + \rho_3 \tag{9}$$

where ρ_1, ρ_2, and ρ_3 are cost coefficients, which are determined by the power generation.

In the HVAC system, the relationship between the energy consumption and temperature is complicated. It could be influenced by many factors. For example, the cooling load includes the transmission load, the infiltration load, the solar load, and the internal load. The transmission load is the temperature transfer from outdoor to indoor through the components. The infiltration load is caused from the inflow of the air. The solar load is caused from the solar radiation. And the internal load is from the heat release of light, people and other electrical equipments [36], as shown in Figure 4.

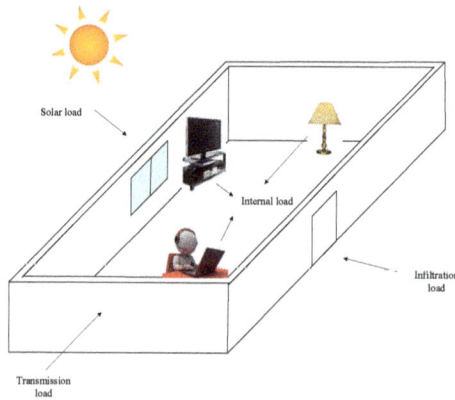

Figure 4. The cooling load system.

The transmission load is denoted as:

$$Q_i^{tl}(T_i) = \alpha S_i(T_o - T_i) \tag{10}$$

where $Q_i^{tl}(T_i)$ is the transmission load, T_o is outdoor temperature, α is the transfer constant in W/(m^2.°C), and S_i is the transmission area.

The infiltration load is calculated as:

$$Q_i^{il}(T_i) = \beta \zeta \phi_i(T_o - T_i) \tag{11}$$

where $Q_i^{il}(T_i)$ is the infiltration load, β is specific heat of air, ζ is the air density, and ϕ_i is the volumetric air velocity and satisfies:

$$\phi_i = A_i(I_0 + H_i I_1 |T_0 - T_i|) \tag{12}$$

where A_i is the effective infiltration area. I_0 and I_1 are determined by the wind speed and outdoor temperature. H_i is the hight of the building.

The solar load and internal load are independent of the actual temperature settings and can be denoted as Q^{sil}.

The total cooling load can be obtained:

$$Q_i^{cl}(T_i) = Q_i^{tl}(T_i) + Q_i^{il}(T_i) + Q^{sil} \tag{13}$$

In the HVAC system, the relationship between the cooling load and energy consumption is:

$$f_i(T_i) = \theta Q_i^{cl}(T_i) \tag{14}$$

where θ is the coefficient determined by the transformation from the cooling load to the energy consumption.

The relationship between temperature settings and energy consumption can be formulated as:

$$f_i(T_i) = b_1(T_0 - T_i)^2 + b_2(T_0 - T_i) + b_3 \tag{15}$$

where $b_1 = \theta \beta \zeta A_i H_i I_1$, $b_2 = \theta \alpha S_i + \psi \zeta A_i I_0$, and $b_3 = \theta Q^{sil}$.

Above all, the energy management model for the HVAC systems can be described as following optimization problem:

$$\max \quad -\tau \sum_{i=1}^{m} c_i(T_i) - (1-\tau) \sum_{i=1}^{n} w_i(q_i)$$

$$\text{s.t.} \quad \sum_{i=1}^{m} f_i(T_i) = \sum_{i=1}^{n} q_i$$

$$T_i^{min} \leq T_i \leq T_i^{max}, i = 1, 2, \cdots, m$$

where T_i is the indoor temperature. Each consumer's temperature setting is limited by $T_i^{min} \leq T_i \leq T_i^{max}$, where T_i^{min} and T_i^{max} are the minimal and maximal temperature settings, respectively.

5. Simulation Results

We consider two types of power systems that are installed with HVAC systems, e.g., the IEEE 9-bus system and IEEE 14-bus system shown in Figures 5 and 6, respectively. The equality constraints in the IEEE 9-bus system and IEEE 14-bus system are $\sum_{i=1}^{3} f_i(T_i) = \sum_{i=1}^{9} q_i$ and $\sum_{i=1}^{11} f_i(T_i) = \sum_{i=1}^{14} q_i$, respectively.

The parameter settings are shown in Table 2 [36], and the lower limit and the upper limit of the temperature setting for each consumer are 23 °C and 28 °C, respectively.

An important parameter of the HVAC system is energy efficiency ratio (EER). EER is the ratio of the actual cooling capacity to the actual input power during the cooling operation of the HVAC system, and the more efficient and power-saving HVAC has the higher EER. The EER is defined as $\frac{Q_i^{cl}(T_i)}{f_i(T_i)}$, which is the reciprocal of θ in Equation (14).

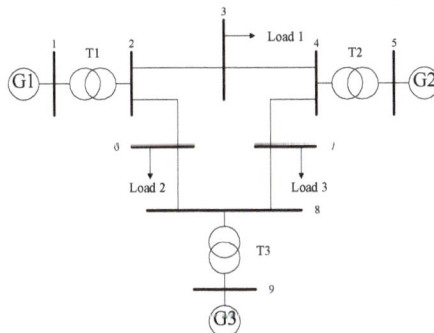

Figure 5. IEEE 9-bus system: 9 buses, 3 generators, and 3 loads ($n = 9$, $m = 3$).

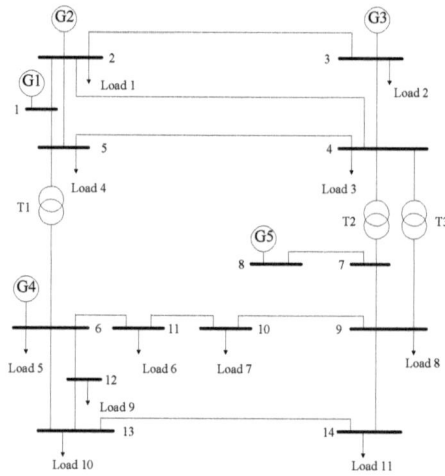

Figure 6. IEEE 14-bus system: 14 buses, 5 generators, and 11 loads ($n = 14$, $m = 11$).

Table 2. Parameter Settings.

Parameters	Values
Outdoor temperature (°C)	$T_o = 30$
Transmission area (m²)	$S_i \in [30, 60]$
Heat transfer constant (W/m²)	$\alpha = 15$
Specific heat of air (J/kg·°C)	$\beta = 1.006$
Air density (kg/m³)	$\zeta = 1.1839$
Wind speed coefficient	$I_0 = 0.343$
Outdoor heat coefficient	$I_1 = 1.12$
Effective infiltration area (m²)	$A_i \in [15, 45]$
Building height (m)	$H_i \in [8, 15]$
Solar and internal load (W)	$Q_i^{sil} \in [300, 4500]$

Taking the IEEE 9-bus system as an example, we discuss the impact of the tradeoff factor τ on the discomfort costs and power supply costs as well as the total costs. The results are given in Figure 7, from which, we can observe that the discomfort costs decrease with τ, and the generation costs increase with τ. When $\tau = 0.6$, we can obtain the minimum total costs. The parameter τ can achieve the tradeoff between consumers' discomfort costs and providers' generation costs. We can get minimum total costs through changing τ. The data of costs are shown in Table 3.

Table 3. The cost data.

τ	Discomfort Cost ($)	Generation Cost ($)	Total Costs ($)
0.1	8.1279	11.5264	19.6543
0.2	7.0753	11.6035	18.6788
0.3	6.7075	11.7344	18.4419
0.4	4.7092	12.5458	17.2550
0.5	5.0396	12.4054	17.4450
0.6	4.3398	12.8159	17.1557
0.7	3.3431	14.3050	17.6481
0.8	3.0488	15.4414	18.4902
0.9	3.0605	15.3121	18.3726

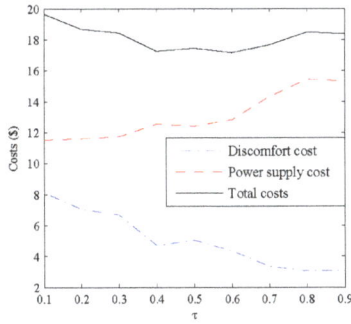

Figure 7. The impact of tradeoff factor.

Next, we assume $\tau = 0.6$ and evaluate the temperature settings, the power supply, and the retail price. The convergence of the temperature settings, the power supply, and the retail price are shown in Figures 8–10, respectively.

According to Figures 8–10, we can observe that all the optimization variables tend to be stable with the iterations and finally converge to the optimum.

It is observed from Tables 4 and 5 that the temperature settings satisfy the requirements for upper limits and lower limits. And the total Power consumption is equal to the power supply. Moreover, the retail price λ is 0.2147 \$/kWh, and the multipliers μ and ν are both zero. It means that the penalty terms are inactive at the optimum.

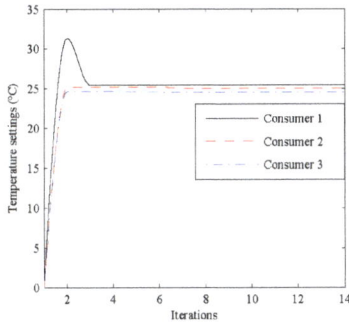

Figure 8. The convergence of the temperature settings.

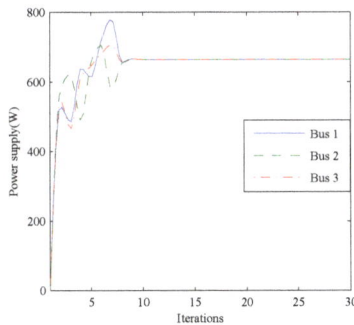

Figure 9. The convergence of the power supply.

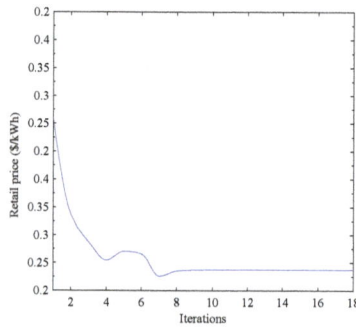

Figure 10. The convergence of the retail price.

Table 4. The temperature settings and energy consumption in IEEE 9-bus system.

Consumer i	Temperature (°C)	Power Consumption (kW)
1	25.4534	0.6986
2	25.0573	1.6701
3	24.5559	3.6015
1–3	/	5.9702

Table 5. The power supply on each bus in IEEE 9-bus system.

Buses i	Power Supply (W)
1	663.3679
2	663.3646
3	663.3654
4	663.3676
5	663.3669
6	663.3692
7	663.3686
8	663.3693
9	663.3683
1–9	59702

Next, we apply the energy management algorithm to the IEEE 14-bus system. It is observed from Figures 11–13 that the temperature settings, the power supply and the retail price can converge to the optimum in the IEEE 14-bus system. Comparing with the convergence results in the IEEE 9-bus system, more iterations are needed. Furthermore, the power supply on each bus is more than IEEE 9-bus system, as shown in Tables 6 and 7.

Table 6. The temperature settings and energy consumption in IEEE 14-bus system.

Consumer i	Temperature (°C)	Power Consumption (kW)
1	25.7925	0.3526
2	25.5883	0.6228
3	25.3969	0.9958
4	25.7143	0.7118
5	25.7578	0.7356
6	25.7708	0.7896
7	25.5143	1.2350
8	25.7187	1.0332
9	25.0050	2.2812
10	25.5317	1.5131
11	25.6491	1.4117
1–11	/	11.6825

Table 7. The power supply on each bus in IEEE 14-bus system.

Buses *i*	Power Supply (W)
1	834.4625
2	834.4609
3	834.4616
4	834.4629
5	834.4693
6	834.4612
7	834.4625
8	834.4650
9	834.4621
10	834.4595
11	834.4648
12	834.4673
13	834.4591
14	834.4637
1–14	11682

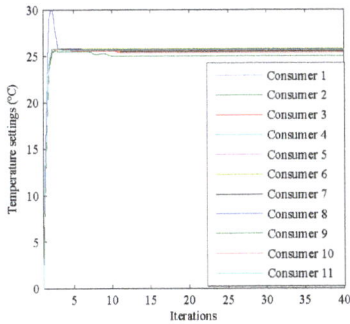

Figure 11. The convergence of the temperature settings.

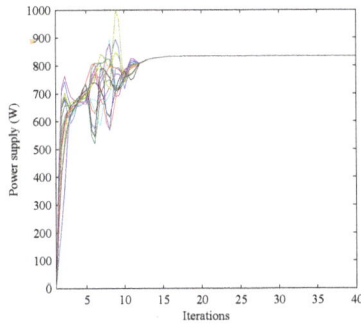

Figure 12. The convergence of the power supply.

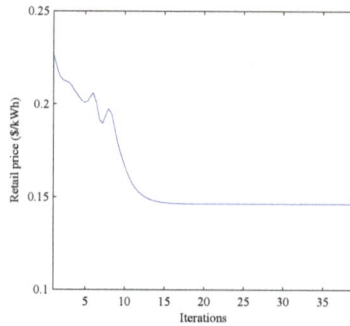

Figure 13. The convergence of the retail price.

Next, we will discuss the effect of EER on the energy management system. We take three different energy efficiency grades (EEGs) of the HVAC systems: EEG 1, EEG 2, and EEG 3. The corresponding EERs are 3.5, 3.3, and 3.1, respectively. From Figure 14, we can observe that the higher EEG can cause lower retail price. Figure 15 shows that the lower EEG is effective in saving power consumption and the cost. It shows that the energy management algorithm motivates the consumers to use more energy-efficient HVAC.

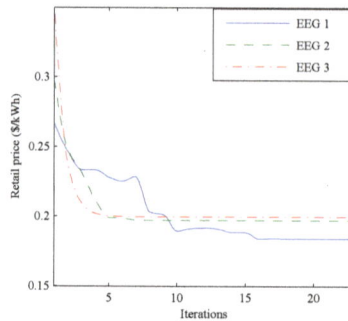

Figure 14. The retail prices under different EEGs.

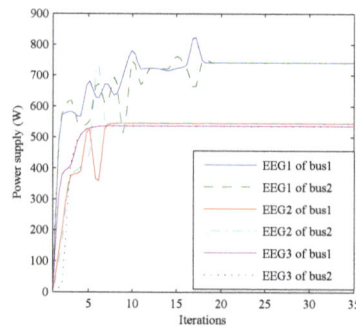

Figure 15. The power supply under different EEGs.

6. Conclusions

This work studies a demand-side energy management problem based on the nonconvex optimization algorithm. The objective is to minimize the discomfort costs and the generation costs by changing the operating states of the loads and the power supply. Specially, the discomfort costs are formulated based on the Fanger thermal comfort. The nonconvex algorithm includes the multiplier method, the Powell method, the advance and retreat method, and the golden section method. One of the major advantages of this algorithm is that it can be applied in solving the unknown objective function caused by the thermal comfort model. In the simulation, we analyze the influence of the tradeoff factor τ and the EER on the energy management. It is observed that the minimum costs can be achieved by changing the value of τ, and different EERs can cause different retail prices and power consumption using the proposed energy management algorithm. The simulation results also demonstrate the convergence of the iterative algorithm and the balance between the power supply and power consumption.

Acknowledgments: This research was supported in part by National Natural Science Foundation of China under Grants 61573303 and 61503324, in part by Natural Science Foundation of Hebei Province under Grant F2016203438, E2017203284, and E2016203092, in part by Project Funded by China Postdoctoral Science Foundation under Grant 2015M570233 and 2016M601282, in part by Project Funded by Hebei Education Department under Grant BJ2016052, in part by Technology Foundation for Selected Overseas Chinese Scholar under Grant C2015003052, and in part by a Project Funded by Key Laboratory of System Control and Information Processing of Ministry of Education under Grant Scip201604.

Author Contributions: Kai Ma contributed the idea and wrote the paper; Yege Bai performed the experiments; Jie Yang designed the experiments; Yangqing Yu analyzed the data; Qiuxia Yang contributed the analysis tools.

Conflicts of Interest: The authors declare no conflict of interest.

Abbreviations

The following abbreviations are used in this manuscript:

HVAC	Heating, Ventilation, and Air Conditioning
EER	Energy Efficient Ratio
EEG	Energy Efficient Grade
PHR	Powell-Hestenes-Rockafellar
PMV	Predicted Mean Vote
PPD	Predicted Percentage of Dissatisfied

References

1. Zaballos, A.; Vallejo, A.; Selga, J.M. Heterogeneous communication architecture for the smart grid. *IEEE Netw.* **2011**, *25*, 30–37.
2. Yu, Y.X.; Luan, W.P. Smart grid and its implementations. *Proc. CSEE* **2009**, *29*, 1–8.
3. Elaiw, A.M.; Xia, X.; Shehata, A.M. Hybrid DE-SQP and hybrid PSO-SQP methods for solving dynamic economic emission dispatch problem with valve-point effects. *Electr. Power Syst. Res.* **2013**, *103*, 192–200.
4. Verschae, R.; Kato, T.; Matsuyama, T. Energy Management in Prosumer Communities: A Coordinated Approach. *Energies* **2016**, *9*, 562.
5. Divshali, P.H.; Choi, B. Electrical Market Management Considering Power System Constraints in Smart Distribution Grids. *Energies* **2016**, *9*, 405.
6. Gao, B.; Liu, X.; Zhang, W.; Tang, Y. Autonomous Household Energy Management Based on a Double Cooperative Game Approach in the Smart Grid. *Energies* **2015**, *8*, 7326–7343.
7. Miceli, R. Energy Management and Smart Grids. *Energies* **2013**, *6*, 2262–2290.
8. Albadi, M.H.; El-Saadany, E.F. A summary of demand response in electricity markets. *Electr. Power Syst. Res.* **2008**, *78*, 1989–1996.
9. Zhao, H.T.; Zhu, Z.Z.; Yu, E.K. Study on demand response markets and programs in electricity markets. *Power Syst. Technol.* **2010**, *34*, 146–153.
10. Ma, K.; Liu, X.; Yang, J.; Liu, Z.; Yuan, Y. Optimal Power Allocation for a Relaying-Based Cognitive Radio Network in a Smart Grid. *Energies* **2017**, *10*, 909.

11. Mathieu, J.L.; Koch, S.; Callaway, D.S. State Estimation and Control of Electric Loads to Manage Real-Time Energy Imbalance. *IEEE Trans. Power Syst.* **2013**, *28*, 430–440.
12. Zhou, M.H.; Min, X.U. Researches on spot price based on optimal power flow and its algorithm. *Relay* **2006**, *34*, 63–67.
13. Sanduleac, M.; Lipari, G.; Monti, A.; Voulkidis, A.; Zanetto, G.; Corsi, A.; Toma, L.; Fiorentino, G.; Federenciuc, D. Next Generation Real-Time Smart Meters for ICT Based Assessment of Grid Data Inconsistencies. *Energies* **2017**, *10*, 857.
14. Rottondi, C.; Duchon, M.; Koss, D.; Palamarciuc, A.; Piti, A.; Verticale, G.; Schätz, B. An Energy Management Service for the Smart Office. *Energies* **2015**, *8*, 11667–11684.
15. Agnetis, A.; de Pascale, G.; Detti, P.; Vicino, A. Load Scheduling for Household Energy Consumption Optimization. *IEEE Trans. Smart Grid* **2013**, *4*, 2364–2373.
16. Gupta, P.K.; Gibtner, A.K.; Duchon, M.; Koss, D.; Schätz, B. Using knowledge discovery for autonomous decision making in smart grid nodes. In Proceedings of the IEEE International Conference on Industrial Technology, Seville, Spain, 17–19 March 2015; pp. 3134–3139.
17. Gatsis, N.; Giannakis, G.B. Residential demand response with interruptible tasks: Duality and algorithms. In Proceedings of the 2011 50th IEEE Conference on Decision and Control and European Control Conference, Orlando, FL, USA, 12–15 Dcember 2011; pp. 1–6.
18. Barbato, A.; Bolchini, C.; Geronazzo, A.; Quintarelli, E.; Palamarciuc, A.; Piti, A.; Rottondi, C.; Verticale, G. Energy Optimization and Management of Demand Response Interactions in a Smart Campus. *Energies* **2016**, *9*, 398.
19. Ma, K.; Hu, G.; Spanos, C.J. Distributed Energy Consumption Control via Real-Time Pricing Feedback in Smart Grid. *IEEE Trans. Control Syst. Technol.* **2014**, *22*, 1907–1914.
20. Mohsenian-Rad, A.H.; Wong, V.W.S.; Jatskevich, J.; Schober, R.; Leon-Garcia, A. Autonomous Demand-Side Management Based on Game-Theoretic Energy Consumption Scheduling for the Future Smart Grid. *IEEE Trans. Smart Grid* **2010**, *1*, 320–331.
21. Deng, R.; Yang, Z.; Chen, J.; Asr, N.R.; Chow, M.Y. Residential Energy Consumption Scheduling: A Coupled-Constraint Game Approach. *IEEE Trans. Smart Grid* **2014**, *5*, 1340–1350.
22. Chai, B.; Chen, J.; Yang, Z.; Zhang, Y. Demand Response Management with Multiple Utility Companies: A Two-Level Game Approach. *IEEE Trans. Smart Grid* **2014**, *5*, 722–731.
23. Chen, J.; Yang, B.; Guan, X. Optimal demand response scheduling with Stackelberg game approach under load uncertainty for smart grid. In Proceedings of the IEEE Third International Conference on Smart Grid Communications, Tainan, Taiwan, 5–8 November 2012; pp. 546–551.
24. Tushar, W.; Chai, B.; Yuen, C.; Smith, D.B. Three-Party Energy Management With Distributed Energy Resources in Smart Grid. *IEEE Trans. Ind. Electron.* **2014**, *62*, 2487–2498.
25. Samadi, P.; Mohsenian-Rad, A.H.; Schober, R.; Wong, V.W.S.; Jatskevich, J. Optimal Real-Time Pricing Algorithm Based on Utility Maximization for Smart Grid. In Proceedings of the First IEEE International Conference on Smart Grid Communications, Gaithersburg, MD, USA, 4–6 October 2010; pp. 415–420.
26. Roozbehani, M.; Dahleh, M.A.; Mitter, S.K. Volatility of Power Grids Under Real-Time Pricing. *IEEE Trans. Power Syst.* **2011**, *27*, 1926–1940.
27. Mohsenian-Rad, A.H.; Leon-Garcia, A. Optimal Residential Load Control With Price Prediction in Real-Time Electricity Pricing Environments. *IEEE Trans. Smart Grid* **2010**, *1*, 120–133.
28. Soltani, N.Y.; Kim, S.J.; Giannakis, G.B. Real-Time Load Elasticity Tracking and Pricing for Electric Vehicle Charging. *IEEE Trans. Smart Grid* **2015**, *6*, 1303–1313.
29. Bai, L.; Ye, M.; Sun, C.; Hu, G. Distributed control for economic dispatch via saddle point dynamics and consensus algorithms. In Proceedings of the IEEE Conference on Decision and Control, Las Vegas, NV, USA, 12–14 Dcember 2016; pp. 6934–6939.
30. Ma, K.; Hu, G.; Spanos, C.J. Energy Management Considering Load Operations and Forecast Errors with Application to HVAC Systems. *IEEE Trans. Smart Grid* **2016**, *PP*, 1–10.
31. Sun, W.; Yuan, Y.X. *Optimization Theory and Methods—Nonlinear Programming*; Springer: New York, NY, USA, 2006; Volume 1.
32. Van, H.J. Forty years of Fanger's model of thermal comfort: Comfort for all? *Indoor Air* **2008**, *18*, 182–201.
33. De Donato, S.R.; Graziani, M.; Mainetti, S. Evaluation of the predictive value of Fanger's PMV index study in a population of school children. Predicted mean vote. *Med. Lav.* **1996**, *87*, 51.

34. Gilani, I.U.H.; Khan, M.H.; Ali, M. Revisiting Fanger's thermal comfort model using mean blood pressure as a bio-marker: An experimental investigation. *Appl. Therm. Eng.* **2016**, *109*, 35–43.

35. Wood, A.J.; Wollenberg, B. 96/02779—Power generation operation and control, 2nd edition. *IEEE Power Energy Mag.* **1996**, *37*, 90–93.

36. Wang, N.; Zhang, J.; Xia, X. Energy consumption of air conditioners at different temperature set points. *Energy Build.* **2013**, *65*, 412–418.

energies

MDPI

Article

Location of Faults in Power Transmission Lines Using the ARIMA Method

Danilo Pinto Moreira de Souza [1], **Eliane da Silva Christo** [1,]*** and **Aryfrance Rocha Almeida** [2]

[1] Computational Modeling in Science and Technology (MCCT), Fluminense Federal University (UFF), Volta Redonda 21941-916, Brazil; danilopms@id.uff.br
[2] Technology Center, Federal University of Piauí (UFPI), Teresina 60455-760, Brazil; aryfrance@ufpi.edu.br
* Correspondence: elianechristo@id.uff.br; Tel.: +55-24-2107-3510

Received: 11 August 2017; Accepted: 6 October 2017; Published: 13 October 2017

Abstract: One of the major problems in transmission lines is the occurrence of failures that affect the quality of the electric power supplied, as the exact localization of the fault must be known for correction. In order to streamline the work of maintenance teams and standardize services, this paper proposes a method of locating faults in power transmission lines by analyzing the voltage oscillographic signals extracted at the line monitoring terminals. The developed method relates time series models obtained specifically for each failure pattern. The parameters of the autoregressive integrated moving average (*ARIMA*) model are estimated in order to adjust the voltage curves and calculate the distance from the initial fault localization to the terminals. Simulations of the failures are performed through the ATPDraw® (5.5) software and the analyses were completed using the RStudio® (1.0.143) software. The results obtained with respect to the failures, which did not involve earth return, were satisfactory when compared with widely used techniques in the literature, particularly when the fault distance became larger in relation to the beginning of the transmission line.

Keywords: transmission line; fault localization; time series; *ARIMA*; discrete wavelet transformer

1. Introduction

The behavior of the electricity sector is directly related to economic factors such as Gross Domestic Product (*GDP*). In this manner, the demand for electricity can be seen as a "thermometer" of the market. As such, growth of the economy as well as increases in purchasing power and quality of life must be accompanied by improvements in the power system, with the objective being compliance with current and future situations. The transportation of electric energy is carried out by means of transmission lines (*TLs*) which, because they span long distances and are present in great quantity, make the electric power system (*EPS*) more susceptible to perturbations which are caused mainly by natural phenomena, in particular atmospheric discharges. In the *EPS*, faults may occur in various components, among which *TLs* may be the most susceptible elements, especially considering their physical dimensions, functional complexity and the environment they are in, thus presenting greater difficulties in terms of maintenance and monitoring [1].

Keeping in mind the importance of having an electrical system where continuity, compliance, flexibility, and maintainability are observed and guaranteed, we have sought to improve and innovate with respect to techniques used in the protection and supervision equipment of the *EPS*, while also providing for the expansion of the electric sector and maintenance of system operation quality [2]. The development and improvement of algorithms that allow the analysis and diagnosis of failures in power systems can have an important economic impact, both for power utilities and consumers, as they enable the continuity and reliability of the electric sector. Intelligent, autonomous, online systems have been developed and applied to a significant degree to deal with this type of problem, since they enable fast and accurate diagnosis without the need for human intervention.

The transient voltage and current components are based on the charge of the capacitances of the faultless phases and the discharges of the fault phases. Transients can be detected in almost all occurrences of failures that require the functioning of the circuit breaker. The characteristics of transient phenomenon can be used in relay protection systems and in the location of faults. This technique is satisfactory when compared to techniques already used, such as the theory of traveling wave [3,4], and techniques that use the calculation of fault impedance [5–7]. The comparison is accomplished through the application of these two methods to several transient signals of various situations of simulated faults in a computational environment. The discrete wavelet transform (*DWT*) is used to decouple sinusoidal signals from the network and transient signals from faults. The technique is widely used for this purpose and has been demonstrated in the literature, with proven efficacy. For the decoupling of Fourier series, signals can be used very simply, although better results are obtained with more sophisticated methods such as *DWT* [8,9]. Unlike the Fourier analysis, which provides a global representation of the signal, the wavelet transform provides a local representation (in time and frequency) of a particular signal. This "location" in time allows disturbances in signals to be detected as soon as they begin [10].

The objective of this study is, through time series techniques, to model fault voltage data and thereby locate faults in transmission lines. The coefficients of autoregressive integrated moving average (*ARIMA*) models have different values depending on where the faults occur on the lines. When traveling waves are used, the main problem is to find the second reverse traveler wave from different disturbance signals [11]. Thus, with a database with different simulated situations, it is possible to adjust curves according to models and calculate fault distances for various situations.

2. Simulated Transmission Line

An *EPS* consists of power plants (hydroelectric, thermoelectric, thermonuclear, alternative sources, and small power plants), *TLs* (composed of towers, cables, and lifting and lowering substations), and end transmission lines (consisting of transformers, poles and cables and consumption measures). This complex system can involve hundreds or even thousands of kilometers, as is the case of Brazil, for example. Table 1 shows the statistical data in percentages for fault occurrences in the *EPS* components. Approximately 50% of absences occur in overhead lines [12]. In the Brazilian electricity system, the transmission lines represent 68% of the absences in the network [13].

Table 1. Percentage of absences for equipment in the electric power system (*EPS*).

Type of Equipment	Percentage Total
Aerial lines	50
Underground cables	9
Transformers and reactors	10
Power generators	7
Circuit breakers	12
Control equipment and transformers for instruments	12

The numbers shown in the Table 1 show the importance of more closely monitoring the *TLs* of an *EPS*, emphasizing the importance of this study. Therefore, the *JMARTI* [14] model is used to implement a transmission line in *ATPDraw*® (5.5) software. The simulated line has an extension of 200 km, nominal voltage of 500 kV, and fundamental frequency of 60 Hz, with distributed parameters dependent on the frequency and perfectly transposed in their totality [15].

Figure 1a shows a diagram where the position of the short circuit with respect to the generating terminal can be varied along the line. The monitoring of the voltage signal is performed at the same terminal. The distance *d* is the variable of interest in this study.

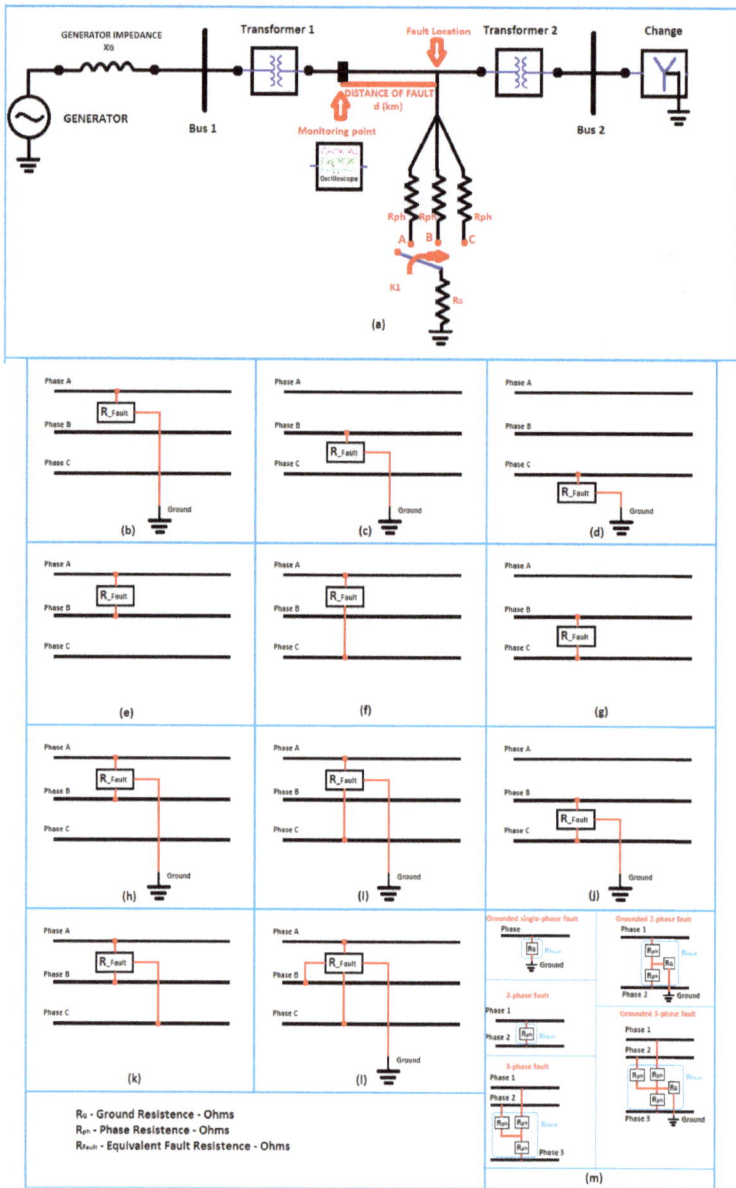

Figure 1. Simulated line topology in the *ATP*® software. Fault situations for a three-phase system.

All fault settings for a three-phase system are also shown. Situations of single-phase faults: Figure 1b–d. Situations of biphasic faults: Figure 1e–g. Situations of grounded biphasic faults: Figure 1h–j. Situation of three-phase fault: Figure 1k. Situations of grounded three-phase fault: Figure 1l. Settings of phase resistors R_{ph} and ground resistance R_G: Figure 1m.

Table 2 shows all situations of simulated faults with variations of the fault type (elements involved are phases *A*, *B*, *C* and ground *G*), signal sampling, line fault position, fault resistance, and incident angle of the disturbance.

Table 2. Situations of simulated faults.

Fault	Data Sampling (kHz)	Location *d* (km)	Failure Resistance (Ω)	Angle of Incidence θ (°)	Fault Situations
$A - G$	200	10; 45; 84; 155	20; 50; 80; 120; 150; 180; 200; 240	0; 45; 90; 135; 180; 225; 270; 315	256
$B - G$	200	10; 45; 84; 155	20; 50; 80; 120; 150; 180; 200; 240	0; 45; 90; 135; 180; 225; 270; 315	256
$C - G$	200	10; 45; 84; 155	20; 50; 80; 120; 150; 180; 200; 240	0; 45; 90; 135; 180; 225; 270; 315	256
$A - B$	200	10; 45; 84; 155	20; 50; 80; 120; 150; 180; 200; 240	0; 45; 90; 135; 180; 225; 270; 315	256
$A - C$	200	10; 45; 84; 155	20; 50; 80; 120; 150; 180; 200; 240	0; 45; 90; 135; 180; 225; 270; 315	256
$B - C$	200	10; 45; 84; 155	20; 50; 80; 120; 150; 180; 200; 240	0; 45; 90; 135; 180; 225; 270; 315	256
$A - B - G$	200	10; 45; 84; 155	20; 50; 80; 120; 150; 180; 200; 240	0; 45; 90; 135; 180; 225; 270; 315	256
$A - C - G$	200	10; 45; 84; 155	20; 50; 80; 120; 150; 180; 200; 240	0; 45; 90; 135; 180; 225; 270; 315	256
$B - C - G$	200	10; 45; 84; 155	20; 50; 80; 120; 150; 180; 200; 240	0; 45; 90; 135; 180; 225; 270; 315	256
$A - B - C$	200	10; 45; 84; 155	20; 50; 80; 120; 150; 180; 200; 240	0; 45; 90; 135; 180; 225; 270; 315	256

3. Theory of Traveling Waves

Disturbances occur in the transmission line of electric power, and are caused by a variety of electromagnetic phenomena such as atmospheric discharges. Sudden changes occur in the conditions of the electrical circuits that make up the transmission system, causing a redistribution of energy with the purpose of finding a new break-even point. Thus, traveling waves refer to the propagation of energy over a system. This energy is distributed by the system in its circuit elements, capacitors and inductors [16].

The propagation of traveling waves always occurs in the direction of all the terminals of the transmission line and causes the electrical transients perceived by the protection relays and other automation and control devices located in the operating centers of the system [16]. If any variation occurs on one terminal of a power transmission line, the other terminal will only feel the variation occurring when the wave travels the entire length of the line [17].

The remote terminal of the transmission line cannot influence the decisions about the system until the wave has traveled from the source of the local terminal to the remote terminal where, through its interaction with the transmission line, a response is produced that travels from back to the local source. In this way, electrical signals tend to propagate back and forth, like traveling waves, usually dissipating energy with losses in the material [17].

The traveling wave theory allows for definition of the reflection and refraction coefficients of the traveling wave in discontinuities as well as the wave propagation velocity and the transmission line surge impedance. It is noteworthy that during propagation along the line, traveling waves are attenuated mainly by resistive and leakage losses and may still suffer distortions in their waveform [18].

In order for the transient behavior of an electromagnetic wave on a transmission line to be adequately represented, it is necessary that the line parameters be evenly distributed over its length, since only this representation allows the theory of the traveling waves to be used to analyze the propagation of these electromagnetic phenomena in it [19].

It is important to note that transmission line models in which the parameters are constant are not adequate for simulation of the transmission line response over a large range of frequencies that are present in the signals during transient conditions [14]. Despite this, in practice, constant frequency distributed line models provide satisfactory results and are used in several transient studies in power systems, according to the *Alternative Transient Program Rule Book* [16].

The reflections and refractions of the waves that travel on the transmission lines are the result of discontinuities in the course of the wave. These discontinuities can not be caused by terminal impedances, short circuits, or circuit breakers.

In order to monitor the propagation time of the generated wave fronts, only the peaks of these waves are followed. This restriction to only a few points greatly facilitates the monitoring of this data. The lattice diagram shown in Figure 2 is a summary of the above because it focuses only on the propagation times between the point of origin of the fault and the terminals of the line.

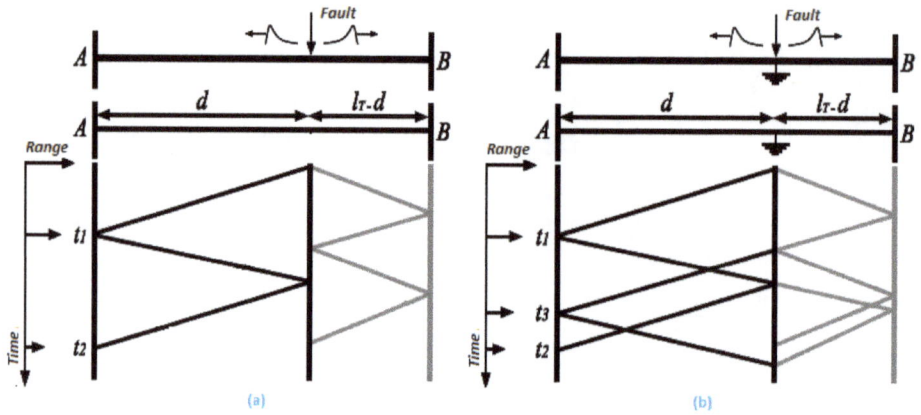

Figure 2. Peak propagation time. *A*: Monitoring terminal; *B* opposite terminal; *d*: fault distance; l_T: line length; t_1 and t_2: propagation times of wave fronts; t_3: refracted wave propagation time. (**a**) Ground fault; (**b**) Fault without ground [15].

The propagation time of a wave peak between its origin and the monitoring terminal depends on the line length to be traveled by the wave and the propagation speed of this wave. Its propagation speed is dependent on the inductance and capacitance of the line with $v = \frac{1}{\sqrt{LC}}$. However, this value is sufficiently near the rate of light propagation in the vacuum, and can be considered as $v = 3 \times 10^5 \frac{km}{s}$. In Figure 2a for example, the distance *d* can be calculated as shown in Equation (1). In Figure 2b one must take into account the refracted wave of the other part of the transmission line. Thus, the calculation is performed according to Equation (2).

$$d = \frac{(t_2 - t_1)v}{2} \tag{1}$$

$$d = l_T - \frac{(t_3 - t_1)v}{2} \tag{2}$$

where l_T is the line length in kilometers; and *d* is the length between the fault point and the terminal.

4. Wavelet Transform

The wavelet transform is a linear operation that decomposes a signal into different scales with different levels of resolution. The wavelet transform of the signal $f(t)$ *CWT* is defined by [10]:

$$CWT_f^{\Psi}(a, b) = \int_{-\infty}^{\infty} f(t) * \Psi_{a,b}(t)dt \tag{3}$$

where $\Psi_{a,b}(t)$ is a daughter wavelet, defined as [20]:

$$\Psi_{a,b}(t) = \frac{1}{\sqrt{a}}\psi\left(\frac{t - b}{a}\right); \ a \in \mathbf{R}^+ e\, b \in \mathbf{R} \tag{4}$$

In turn, $\Psi(t)$ is the chosen mother wavelet, *a* is the scaling factor, and *b* is the shift factor.

For computational use of *CWT* it is necessary to have discrete parameters of *a* and *b*. In the discrete case, the scaling and shift factors are represented as Equations (5) and (6) [21]:

$$a = a_0^m \tag{5}$$

$$b = nb_0 a_0^m \tag{6}$$

$$m, n \in \mathbf{Z}; \; a_0 \geq 1; \; b_0 \neq 0$$

Although the result of Equation (3) is a finite set of coefficients, it is still a continuous representation of the wavelet transform. When the function to be analyzed is given by discrete values $f(k)$, then we must use the discrete wavelet transform *DWT* defined by [10]:

$$DWT_f^{\Psi} m, n = a_0^{-\frac{m}{2}} \sum_k x[k] \, \Psi \left[a_0^{-m} k - nb_0 \right] \tag{7}$$

$$a_0 > 1; \; b_0 \neq 0$$

5. Time Series Models

A time series is any set of observations ordered in time, where each value has attached to itself an indicator of the time in which this value occurred or was observed [22]. According to [23], a time series is constructed when one an interest in:

- Investigating the generating mechanism of the time series;
- Making forecasts of future values of the series;
- Describing only the behavior of the series;
- Searching for relevant periodicities in the data.

A stochastic model that can be extremely useful in representing certain practically occurring series is the autoregressive model. In this model, the current value of the process is expressed as a finite linear aggregate of previous process values and the random shock a_t. Let us say the values of a process with moments by equally spaced times are $(t, t-1, t-2, \cdots)$ by $(Z_t, Z_{t-1}, Z_{t-2}, \cdots)$. Also, let $\tilde{Z}_t = Z_t - \mu$ be the series of muf deviations. Then, Equation (9) is called an autoregressive $(AR(p))$ process of order p.

$$\tilde{Z}_t = \phi_1 Z_{t-1} + \phi_2 Z_{t-2} + \cdots + \phi_p Z_{t-p} + a_t \tag{8}$$
$$\tilde{Z} = \phi_1 \tilde{Z}_1 + \phi_2 \tilde{Z}_2 + \cdots + \phi_p \tilde{Z}_p + a$$

Through the autoregressive operator B given by $\phi(B) = 1 - \phi_1 B - \phi_2 B^2 - \cdots - \phi_p B^p$, the autorregressive model in Equation (9) may be written economically as

$$\phi(B)\tilde{Z}_t = a_t \tag{9}$$

Another kind of model, of great practical importance in the representation of observed time series is the finite moving average (*MA*) process, where $\tilde{z}_t$ is linearly dependent on the finite number q of previous a's. Thus, Equation (10) is called a moving average *MA* (q).

$$\tilde{Z}_t = a_t - \theta_1 a_{t-1} - \theta_2 a_{t-2} - \cdots - \theta_q a_{t-q} \tag{10}$$

In the same way, we have a moving average operator defined by $\theta(B) = 1 - \theta_1 B - \theta_2 B^2 - \cdots - \theta_q B^q$. The moving average model may be written as Equation (11).

$$\tilde{Z}_t - \theta(B)u_t \tag{11}$$

The union of the autoregressive model of order p with the moving average model of order q can sometimes benefit the assembly of the time series. This leads to the mixed autoregressive-moving average *ARMA* (p, q) model:

$$\tilde{Z}_t = \phi_1 \tilde{Z}_{t-1} + \cdots + \phi_p \tilde{Z}_{t-p} + a_t - \theta_1 a_{t-1} - \cdots - \theta_q a_{t-q} \tag{12}$$

Equation (12) can be written with the B operator.

$$\phi(B)\tilde{Z}_t = \theta(B)a_t \tag{13}$$

In some cases, it is necessary to make a distinction between the terms of the series to exclude trends, in accordance with Equation (14). The result is an *ARIMA* (p, d, q) model, where the term I expresses a differentiation of order d.

$$\phi(B)W_t^d = \theta(B)a_t \tag{14}$$

where W_t^d are differentiations on the terms of series Z shown in Equation (15).

$$W_t = \Delta \tilde{Z}_t = \tilde{Z}_t - \tilde{Z}_{t-1} \tag{15}$$
$$W_t^d = \Delta^d \tilde{Z}_t$$

When W_t presents deterministic seasonal behavior of period s, a model that can be used is shown in Equation (16).

$$\phi(B)\Phi(B^s)(\Delta_s)^D(\Delta)^d Z_t = \theta(B)\Theta(B^s)a_t \tag{16}$$

where $\Phi(B^s) = 1 - \Phi_1 B^s - \cdots - \Phi_P B^{sP}$ is the seasonal autoregressive operator of P order; $\Theta(B^s) = 1 - \Theta_1 B^s - \cdots - \Theta_Q B^{sQ}$ is the seasonal moving-averages operator of Q order; $\phi(B) = 1 - \phi_1 B - \cdots - \phi_p B^p$ is the autoregressive operator of p order; $\theta(B) = 1 - \theta_1 B - \cdots - \theta_q B^q$; is the moving-averages operator of q order; and $\Delta_s = (1 - B^s)$ is the seasonal difference operator. In $\Delta_s^D = (1 - B^s)^D$, D indicates the number of seasonal differences. The Equation (16) is denoted by seasonal autoregressive integrated moving average (*SARIMA*) $(p, d, q)(P, D, Q)_s$.

Model Evaluation Criteria

For the process *ARMA* (k, l), the Bayesian information criterion (*BIC*) is given by Equation (17) [24].

$$BIC(k, l) = ln\sigma_{k,l}^2 + (k+l)\frac{lnN}{N} \tag{17}$$

where $\sigma_{k,l}^2$ is a maximum likelihood estimate of the residual variance of the model with N observations. It seeks to minimize *BIC* through the adjustments of k and l.

For the estimation of the error, Equation (18) is used, where the absolute error module committed in the extermination of the fault location is divided by the total length of the line.

$$Error(\%) = 100 \left| \frac{R_t - C_t}{T_t} \right| \tag{18}$$

where R is the actual distance value of the fault, C is the value calculated for this distance, and T is the total length of the line. This calculation is performed for all calculated fault distances.

6. Results

The proposed method, illustrated by Figure 3, consists of using a discrete wavelet transform *DWT* in order to decouple the transient signal from the sinusoidal signal characteristic of the transmission line. These decoupled signals are used in *ARIMA* models to establish mathematical relationships between fault distances and calculated coefficients. The RStudio® (1.0.143) software is used for the computational implementation of *DWT* [25] and *ARIMA* models.

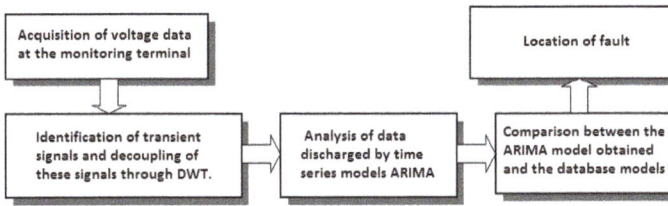

Figure 3. Schematic diagram for fault location with the proposed model. *ARIMA*: autoregressive integrated moving average; *DWT*: discrete wavelet transform.

Figure 4 illustrates a fault situation showing the behavior of the disturbance in the three phases that make up the system. Although the fault does not involve phase *C*, it is affected because there is a coupling between the three phases. However, the highest voltage values in the involved phases are evident. The figure further illustrates the decoupled disturbance signal of the characteristic sinusoidal signal of the transmission line.

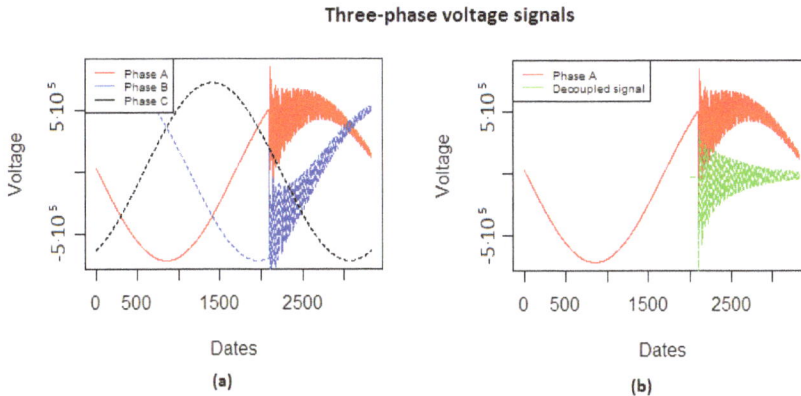

Figure 4. (a) Voltage signals with incident angle of 90° and fault resistance of 240 Ω at a distance of 10 km from the monitoring terminal; (b) Disconnected disturbance signal.

As mentioned, the objective of this work is to relate the distances of occurrences of faults with the curves of *ARIMA* models. The Table 3 shows some results obtained, showing the distance–coefficient relationship. All cases are two-phase faults with an incidence angle of 90°.

It can be seen from Table 3 that the coefficients of the obtained *SARIMA* models are equal for the same fault distances. For example, for faults occurring at 10 km, the obtained models are $SARIMA(2,0,2)(2,0,4)_{19}$ where $\phi_1 = 1.283$, $\phi_2 = -0.349$, $\theta_1 = -0.344$, $\theta_2 = 0.028$, $\Phi_{19} = -1.460$, $\Phi_{38} = -0.918$, $\Theta_{19} = -1.333$, $\Theta_{38} = -0.736$, $\Theta_{57} = 0.044$, and $\Theta_{76} = 0.073$ in all faults whose angle of incidence is 90°, regardless of the fault resistance values

In Table 4 it should be noted that the traveler wave method presents a smaller error than the other methods when it is at the beginning of the line. According to [26], when traveling waves are used, the main problem is to find the second reverse traveler wave from different disturbance signals. In this case, the proposed model presented a satisfactory result, because as the distance of the fault increases, the relative error becomes smaller than for the other methods. Fault resistances do not influence the behavior of the model, and therefore results are shown for only two resistance values. Figure 5 shows a example of the a fault situation.

Table 3. Some examples of models obtained.

Term	10 km		45 km		84 km		155 km	
	20 Ω	240 Ω	20 Ω	240 Ω	20 Ω	240 Ω	20 Ω	240 Ω
ϕ_1	1.2830		1.5420		1.58880		1.4780	
ϕ_2	−0.3496		−0.5945		−0.6358		−0.5263	
ϕ_3	0.0000		0.0000		0.0000		0.0000	
θ_1	−0.3440		−0.1750		−0.3322		−0.6510	
θ_2	0.0283		0.0000		0.0000		0.0000	
Φ_1	−1.4600		0.0000		0.0000		0.0000	
Φ_2	−0.9181		0.0000		0.0000		0.0000	
Θ_1	−1.3330		0.0000		0.0000		0.0000	
Θ_2	−0.7397		0.0000		0.0000		0.0000	
Θ_3	0.0437		0.0000		0.0000		0.0000	
Θ_4	0.0730		0.0000		0.0000		0.0000	
Constant	17.9800		14.48		10.6400		14.65	
Bayesian information criterion (*BIC*)	15,369.26		13,025.38		7984.45		4889.99	

Table 4. Comparison between models through the *Error* value (Equation (18)).

Model	5–10 km	20–45 km	70–84 km	155 km
Time series	0.13	0.37	0.13	0.86
Stockwell transformer [26]	0.07	0.10	0.65	–
Neural networks [15]	–	0.75	–	–
Independent components [11]	–	1.90	–	–

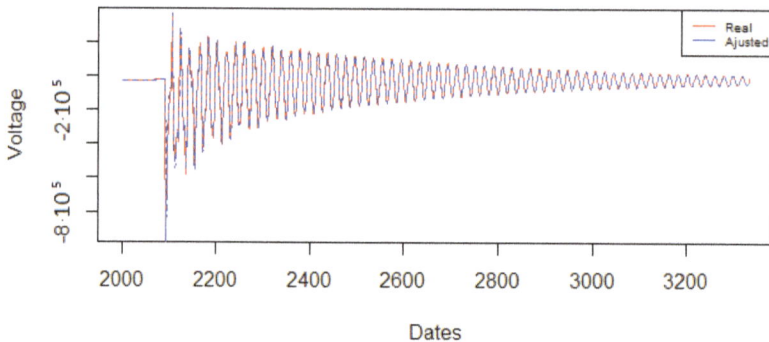

Figure 5. Missing data with distance of 10 km and with missing resistance of 240 Ω.

The auto.arima function of RStudio® (1.0.143) software uses a variation of the Hyndman and Khandakar algorithm presented by [27] that combines unit root tests and minimization of *BIC* to obtain a model *ARIMA*.

7. Discussion and Conclusions

The proposed method is effective in detecting simulated data changes from different fault locations on the line. However, it is also sensitive to variations in the angles of incidence of the onset of disturbances in sine-wave AC signals. This is particularly true when the fault distance becomes larger in relation to the beginning of the transmission line. With improvements in the algorithm it may be possible, in addition to identifying positioning of short circuits in the line, to also identify the angle of said angle of incidence in order to improve the sensitivity of relays used for this type of monitoring.

The algorithm is insensitive to changes in the value of fault resistance. This is in fact an important factor for the adopted methodology, since the fault resistance is highly random and variable, depending on environmental conditions and type and location of the fault. The fault resistance influences the transient signal damping behavior (vertical variations in the Cartesian plane of the signal), but the location of the source of the disturbance is related to the signal oscillation frequency (horizontal variations in the Cartesian plane of the signal).

Another important factor refers to the type of fault in relation to the number of elements involved, whether they are single-phase, two-phase, two-phase ground or three-phase. Another important factor refers to the type of fault in relation to the number of elements involved, that are either single-phase, two-phase, or three-phase. Failure data involving the ground component present greater volatility, since these types of transient waves, besides suffering successive reactions between the point of origin and the line terminals, also undergo refractions. Some of the signals reflected from the terminals exceed the point of origin of the disturbance to the opposite terminal, causing the data to become interlaced. For future works, we suggest evaluating models more sensitive to the heteroscedastic series. A possible alternative to eliminate the data coming from the opposite terminal would be the encapsulation of signal filters in the *DWT* used for the separation.

Acknowledgments: The authors thank the Federal University Fluminense—UFF and the Federal University of Piauí for their teachers. They are also grateful to the Coordination for the Improvement of Higher Education Personnel—CAPES for post-graduate financial assistance.

Author Contributions: Danilo Pinto Moreira de Souza carried out the computational implementation and the writing of the article; Eliane da Silva Christo contributed to the analysis of the time series, the revision of the text and the orientation of the construction of the work; and Aryfrance Rocha Almeida contributed to the analysis of signals using the wavelet transformer, provided the simulation data used in this work, and contributed to the orientation of the construction of the work.

Conflicts of Interest: The authors declare no conflict of interest.

Abbreviations

The following abbreviations are used in this manuscript:

AR	autoregressive
ARIMA	autoregressive integrated moving average
ARMA	autoregressive moving average
BIC	Bayesian information criterion
CWT	continuous wavelet transformer
DWT	discrete wavelet transformer
EPS	electric power system
GDP	Gross Domestic Product
MA	moving average
SARIMA	seasonal autoregressive integrated moving average
TL	transmission line

References

1. Abreu, S.S. Localização de Faltas em Linhas de Transmissão Aéreas pelo Método das Ondas Viajantes Utilizando Filtros Digitais e Transformada Wavelet. Master's Thesis, Programa de Pós-Graduação em Engenharia Elétrica-Universidade Federal de Minas Gerais (PPGEE-UFMG), Belo Horizonte, Brazil, 2005. (In Portuguese)

2. Formiga, D.A. *Estimação de Fasores para Proteção de Sistemas Elétricos Baseada em Mínimos Quadrados e Morfologia Matemática*; Universidade Federal do Rio Grande do Norte (UFRN): Natal, Brazil, 2012. (In Portuguese)

3. He, Z.; Liu, X.; Li, X.; Mai, R. A Novel Traveling-Wave Directional Relay Based on Apparent Surge Impedance. *IEEE Trans. Power Deliv.* **2015**, *30*, 1153–1161.
4. Rajendra, S.; McLaren, P.G. Travelling-Wave Techniques Applied to the Protection of Teed Circuits: Principle of Travelling Wave Techniques. *IEEE Trans. Power Appar. Syst.* **1985**, *104*, 3544–3550.
5. Hänninen, S. *Single Phase Earth Faults in High Impedance Grounded Networks: Characteristics, Indication and Location*; VTT Technical Research Centre of Finland: Espoo, Finland, 2001.
6. Personal Vázquez, E.; García, A.; Parejo, A.; Larios Marín, D.F.; Biscarri Triviño, F.; León de Mora, C. A Comparison of Impedance-Based Fault Location Methods for Power Underground Distribution Systems. *Energies* **2016**, *9*, 1022.
7. Bo, Z.Q.; Weller, G.; Redfern, M.A. Accurate fault location technique for distribution system using fault-generated high-frequency transient voltage signals. *IEE Proc. Gener. Transm. Distrib.* **1999**, *146*, 73–79.
8. Souza, D.P.M.; Christo, E.S.; Almeida, A.R. Séries de Fourier Aplicadas à Localizaçao de Faltas em Linhas de Transmissão de Energia Elétrica. In Proceedings of the Anais do III Simpósio de Matemática da Região Sul Fluminense (SIMA/UFF), Rio de Janeiro, Brazil, 23–29 July 2017. (In Portuguese)
9. Souza, D.P.M.; Christo, E.S.; Almeida, A.R. Localização de Faltas em Linhas de Transmissão por Séries Temporais. In Proceedings of the Anais do XXXVII Congresso Nacional de Matemática Aplicada e Computacional (CNMAC/SBMAC), São José dos Campos, Brazil, 19–22 September 2017. (In Portuguese)
10. Silveira, P.M.; Seara, R.; Zürn, H.H. Localização de faltas por ondas viajantes–uma nova abordagem baseada em decomposição Wavelet. In Proceedings of the Anais do XVI Seminário Nacional de Produção e Transmissão de Energia Elétrica—SNPTEE'2001, Campinas, Brazil, 21–26 October 2001. (In Portuguese)
11. Almeida, A.R.; Almeida, O.M.; Silva, J.P.; Alves, M.H.S.; Abreu, F.C.M. Localização de Faltas em Sistemas de Transmissão de Alta Tensão a partir de Registros Oscilográficos Usando Análise de Componentes Independentes. In Proceedings of the Simpósio Brasileiro de Sistemas Elétricos—SBSE, Foz do Iguaçu, Brasil, 22–25 Apirl 2014. (In Portuguese)
12. Paithankar, Y.G.; Bhide, S.R. *Fundamentals of Power System Protection*; PHI Learning Pvt. Ltd.: New Delhi, India, 2010.
13. Mamede Filho, J.; Mamede, D.R. *Proteção de Sistemas Elétricos de Potência*; Grupo Gen-LTC: São Paulo, Brasil, 2000; pp. 272–281. (In Portuguese)
14. Marti, J.R. Accurate Modeling of Frequency Dependent Transmission Lines in Electromagnetic Transient Simulations. *IEEE Trans. Power Appar. Syst.* **1996** *1*, 1475–1550.
15. Souza, S.C.A.; Braga, A.P.S.; Leão, R.P.S.; Almeida, O.M.A.; Almeida, A.R.; Abreu, F.C.M. *Uso de Redes Neurais Artificiais e Transformada de Stockwell na Localização de Faltas em Linhas de Transmissão*; Universidade Federal do Ceará (UFC): Fortaleza, Brazil, 2015. (In Portuguese)
16. Souza, T.B.P. Análise de Ondas Viajantes em Linhas de Transmissão para Localização de Faltas: Abordagem via Transformada Wavelet. Master's Thesis, Federal University of Pará (UFPA): Belém, Brazil, 2007. (In Portuguese)
17. Hedman, D.E. *Teoria das Linhas de Transmissão-II*; Universidade Federal de Santa Maria: Santa Maria, Brazil, 1978. (In Portuguese)
18. Naidu, S.R. *Transitórios Eletromagnéticos em Sistemas de Potência*. Universidade Federal da Paraíba: João Pessoa, Brazil, 1985. (In Portuguese)
19. Greenwood, A. *Electrical Transients in Power Systems*; Wiley-Interscience: New York, NY, USA, 1991.
20. Robertson, D.C.; Camps, O.I.; Mayer, J.S.; Gish, W.B. Wavelets and electromagnetic power system transients. *IEEE Trans. Power Deliv.* **1996**, *11*, 1050–1056.
21. Chang, H.; Nguyen V.L. Statistical Feature Extraction for Fault Locations in Nonintrusive Fault Detection of Low Voltage Distribution Systems. *Energies* **2017**, *10*, 611.
22. Valins, T.F. Relé Digital de Distância Baseado na Teoria de Ondas Viajantes e Transformada Wavelet. Master's Thesis, Escola de Engenharia São Carlos da Universidade de São Paulo, São Carlos, Brazil, 2005. (In Portuguese)
23. Morettin, P.A.; Toloi, C. Modelos para Séries Temporais. In *Análise de Séries Temporais*; Blucher: Reichenau, Brazil, 2006; pp. 3–70.
24. Choi, B. *ARMA Model Identification*; Springer Science and Business Media: New York, NY, USA, 2012; pp. 58–87.

25. Aldrich, E. *Wavelets: A Package of Functions for Computing Wavelet Filters, Wavelet Transforms and Multiresolution Analyses*; R Package Version 0.3-0; University of Washington: Seattle, WA, USA, 2013.
26. Souza, S.C.A.; Braga, A.P.S.; Almeida, A.R.; Abreu, F.C.M.; Almeida, O.M. *Uso da Transformada de Stockwell e ondas Viajantes na Localização de Faltas em Linhas de Transmissão*; Universidade Federal do Ceará (UFC): Fortaleza, Brazil, 2014. (In Portuguese)
27. Hyndman, R.J.; Khandakar, Y. Automatic time series for forecasting: The forecast package for R. *J. Stat. Softw.* **2008**, *27*, 5–22.

energies

MDPI

Article

Aggregation Potentials for Buildings—Business Models of Demand Response and Virtual Power Plants

Zheng Ma *, Joy Dalmacio Billanes and Bo Nørregaard Jørgensen

Center for Energy Informatics, Mærsk Mc-Kinney Møller Institute, University of Southern Denmark, Campusvej 55, 5230 Odense M, Denmark; joyb@mmmi.sdu.dk (J.D.B.); bnj@mmmi.sdu.dk (B.N.J.)
* Correspondence: zma@mmmi.sdu.dk; Tel.: +45-65-50-3579

Received: 13 August 2017; Accepted: 16 October 2017; Published: 20 October 2017

Abstract: Buildings as prosumers have an important role in the energy aggregation market due to their potential flexible energy consumption and distributed energy resources. However, energy flexibility provided by buildings can be very complex and depend on many factors. The immaturity of the current aggregation market with unclear incentives is still a challenge for buildings to participate in the aggregation market. However, few studies have investigated business models for building participation in the aggregation market. Therefore, this paper develops four business models for buildings to participate in the energy aggregation market: (1) buildings participate in the implicit Demand Response (DR) program via retailers; (2) buildings with small energy consumption participate in the explicit DR via aggregators; (3) buildings directly access the explicit DR program; (4) buildings access energy market via Virtual Power Plant (VPP) aggregators by providing Distributed Energy Resources (DER)s. This paper also determines that it is essential to understand building owners' needs, comforts, and behaviours to develop feasible market access strategies for different types of buildings. Meanwhile, the incentive programs, national regulations and energy market structures strongly influence buildings' participation in the aggregation market. Under the current Nordic market regulation, business model one is the most feasible one, and business model two faces more challenges due to regulation barriers and limited monetary incentives.

Keywords: demand response; virtual power plant; energy flexibility potential; aggregators; business model; building energy flexibility

1. Introduction

Energy stability and flexibility are essential for the entire power system [1]. Flexibility is the ability of electricity systems to maintain the balance between energy supply and demand [2]. Flexibility addresses generation-load imbalance, reduces peak load, power outage, electricity cost, and improves grid reliability [3].

Energy aggregation provides an efficient solution for providing flexibility in power systems. Two models have been discussed broadly that can provide aggregation potentials in the electricity system: Demand Response (DR) and Virtual Power Plants (VPPs). Various stakeholders in the electricity market can participate in the energy aggregation market with new roles or new presence. For instance, consumers convert to prosumers, and new market players such as service aggregators appear.

Buildings as prosumers have an important role in the energy aggregation market due to their potential flexible energy consumption and distributed energy resources [4]. However, energy flexibility provided by buildings can be very complex, and depends on many factors. Meanwhile, different types of buildings can provide different energy flexibilities [5]. Energy flexibility programs that buildings can participate in are defined by regulations and policies. The immaturity of the current aggregation market with unclear incentives is still a challenge for buildings (especially with small energy consumption) to participate in the aggregation market. Various energy flexibility programs also impede buildings' motivation. So far, only a few studies have investigated business models for buildings' participation in the aggregation market.

To fill this gap, this paper aims to (1) understand two existing business models in the energy aggregation market (DR and VPPs) including market players and their relationships; (2) evaluate building energy aggregation potentials; and (3) develop business models for different types of buildings to participate in the energy aggregation market.

This paper is organized as: Section 2 discusses two aggregation models (DR and VPPs), stakeholders and their relationships in DR and VPPs. Section 3 presents three types of buildings and their energy flexibility resources. Section 4 introduces methods applied in this paper including a business model canvas, an evaluation tool for business model analysis, Strengths, Weaknesses, Opportunities, and Threats (SWOT) analysis, and TOWS analysis (a derivative of SWOT analysis). Section 5 presents four business models for buildings to participate in the energy aggregation market. Section 6 use the Nordic electricity market as a case study to discuss potentials and challenges that affect buildings' participation in the four business models, and provide suggestions. Section 7 provides conclusion and further research.

2. Aggregation Models

Two aggregation models broadly discussed in practice that can provide flexibility in the electricity system are Demand Response (DR) and Virtual Power Plants (VPPs).

2.1. Demand Response (DR)

DR is defined by the European Commission as "voluntary changes by end-consumers of their usual electricity use patterns—in response to market signals" [6]. It is a shift of electricity usage in response to price signals or certain requests [7]. DR reduces peak load, electricity cost, and improves system reliability [8]. Electricity consumers can participate in energy-load balance through DR [9]. Controllable appliances in buildings that contribute to DR include commercial buildings like (heating, ventilation and air-conditioning) HVAC systems, home appliances (e.g., dishwashers, dryers, and freezers) [10], energy storage (e.g., batteries of electric vehicles, heat pumps, and refrigeration) [11], and industrial processes (e.g., roller press) [12].

DR Programs

There are two types of DR programs: explicit and implicit demand response. The two types of DR programs are activated at different times and serve different purposes in markets. Consumers can participate in both programs. Consumers typically receive a lower bill by participating in a dynamic pricing program (implicit DR), and receive a direct payment for participating in an explicit demand response program [13].

Explicit DR (also called incentive-based DR program) is divided into traditional-based (e.g., direct load control, interruptible pricing) and market-based (e.g., emergency demand response programs, capacity market programs, demand bidding programs, and ancillary services market programs) [14].

In explicit DR, demand competes directly with supply in wholesale, balancing, and ancillary services markets through services by aggregators or as single large consumers. Load requirements (size of energy consumption) need to comply to participate in DR programs [15]. Therefore, small consumers only can participate by contracting with DR service providers. DR service providers can either be third-party aggregators or customer retailers. Through incentive-based programs, consumers receive direct payments to change their electricity consumption upon request (e.g., to consume more or less) [15].

Explicit DR is more flexible in terms of helping DR service providers acquire DR resources [16]. Direct load control enables DR service providers to control appliances within a short notice [15]. Explicit DR provides a valuable and reliable operational tool for system operators to adjust load to resolve operational issues [13].

On the other hand, implicit DR (sometimes called price-based DR program) refers to the voluntary program in which consumers are exposed to time-varying electricity prices or time-varying network tariffs (such as a day/night tariff) [15]. Compared to explicit DR with direct load control, implicit DR provides less flexibility from the perspective of energy suppliers [16]. Price-based programs depend on the cost of electricity production at different times, and on consumers' own preferences and constraints [15]. In some Nordic countries, customers have opportunities to participate in priced-based programs (e.g., time-of-use (TOU), critical peak pricing, and real-time pricing) [15]. For instance, in real-time pricing, consumers reduce electricity usage at peak periods or shift their usage to off-peak periods [9]. These prices are always part of their supply contract [13].

Market players in DR markets can include producers, grid operators (Transmission System Operators (TSOs), Distribution System Operations (DSOs)), retailers, aggregators, Balance Response Parties (BRPs), policymakers, and consumers (building owners and occupants). New actors (e.g., aggregators) and new roles (e.g., retailers' aggregation service) appear in the energy market. The main relationships between actors in the DR market are shown Table 1.

Table 1. Actors in demand response.

Actors	Offers	To
Aggregator	Pay for BRPs' energy loss	BRP
	Market access DR incentives	Consumer
	Ancillary services Tariff	Transmission System Operator (TSO)
	Network balancing services Tariff	Distribution System Operation (DSO)
Supplier/retailer	Incentives and contract package for the implicit DR program	Consumers
Regulator	DR incentives DR regulations DR awareness	All actors
Consumer	Demand profile	Aggregator
	Direct control	Supplier/retailer
	Large consumers can directly provide energy flexibility to the DR market	Demand Response (DR) market

2.2. Virtual Power Plants

Virtual Power Plants (VPPs) aggregate DER units and offer them to the energy market [17]. The aggregated DERs maintain reliability of renewable energy resources [18] and address grid congestion [19]. VPPs can be managed by third-party aggregators, BRPs, or suppliers [20,21]. VPPs provide a variety of services to power plant operators, industries, public services, energy suppliers, and grid operators. VPPs create new business opportunities for aggregators and suppliers [21]. In Denmark, DONG Energy implements VPPs known as the Power Hub that integrates DR with large industrial companies to balance the power systems [22].

VPPs are aggregated DERs forming a Local Virtual Plant (LVPP). Then the aggregated LVPPs form a Regional Virtual Plant (RVPP) [22]. LVPPs provide various opportunities to stakeholders, such as energy trade, network services, and balancing services [21,22]. VPPs focus on the physical aspect of DERs and their impact on the electrical system [20]. Meanwhile, VPP units at different locations [19] are coordinated using networking infrastructure [23].

2.2.1. Components of VPPs

A VPP is comprised of generation units [24], energy storage, and Information Communication Technology (ICT) [21]. Generation technology in VPPs consists of DER portfolios (supply-side and demand response) [24]. Supply-side in DER portfolios are Distributed Generation (DG) units [24], such as Combined Heat and Power (CHP), biomass and biogas, small power plants, solar, and wind generation [21]. DR in DER portfolios consists of flexible loads and energy storage [24]. Flexible loads refer to loads or consumption patterns shifted in response to price signals (e.g., heating, cooling, and charging of electrical vehicles) [18]. VPPs require energy storage to store energy such as Hydraulic Pumped Energy Storage (HPES), Compressed Air Energy Storage (CAES), Flywheel Energy Storage (FES), Superconducting Magnetic Energy Storage (SMES), Battery Energy Storage System (BESS), and electric vehicles [18]. VPPs are coordinated through ICT systems that help to reduce transmission system losses, relieve congestions, and provide grid stability [25]. This ICT infrastructure includes Energy Management Systems (EMS), and Supervisory Control and Data Acquisition (SCADA) systems. VPPs can monitor energy flows of DERs, storage facilities, and controllable loads [10].

2.2.2. Systems-of-Interests in Virtual Power Plants

The operation of VPP systems serves trade, balancing, and network support according to the Systems-of-Interests (SoI):

(1) Virtual Power Plants for Trade

VPP systems provide energy trade opportunities to VPP owners. VVPs optimize and aggregate DERs' capacity (DG units and DR) and provide DERs with visibility and market access [18,21]. VPP owners submit bids and optimize DERs' revenue in the wholesale market [26].

The DER owners can receive more benefits by collectively participating in wholesale energy markets compared to participating individually. Moreover, volume threshold for power producers may prevent small DER owners to trade their energy individually [22]. Practically, both DER owners and participants in demand side response are represented by RVPP operators as a single entity in the wholesale market [22].

(2) Virtual Power Plants for Balancing

VPPs can participate in the energy balancing market by employing available DER units, storage devices, and controllable loads [22]. The balancing market is the regulating market in the Nord Pool market structure. BRPs might be particularly interested in this VPP operation, due to imbalance responsibilities.

VPPs can contribute short, medium, and long term balancing of energy flow by the operations of virtual synchronous generators and demand side management. The duration of primary control is presented in seconds, and VPPs can contribute with fast power response obtained from rotating (synchronous) generators, super capacitors, and fast batteries [22]. VPPs can also contribute to secondary control by (1) increasing the generation of reserve DER units (e.g., micro-CHP) for a period of minutes; and (2) decreasing the demand through employment of controllable loads for a few hours until top-down power supply is recovered [22].

(3) Virtual Power Plants for Network Services

Due to the increase of load or generation, network operators need to either expand network capacity or prevent overload or congestion [22]. VPPs can provide grid services to TSOs/DSOs to support load and congestion management and improve power quality [24]. VPPs can also provide services to DSOs' local system management [21]. In addition, VPPs provide system services (e.g., black start, voltage control) to TSOs [24].

2.2.3. The VPP Stakeholders

The main actors in VPPs are VPP aggregators. Third-party aggregators manage VPPs [20], aggregate DERs, storages and adjustable loads [15] and offer them to different market participants (e.g., TSOs, BRPs) [27]. There are large and small (e.g., DERs, prosumers) energy producers. DERs are small energy generators located in low-voltage grid expecting high return of investment [22]. Energy consumers can provide adjustable loads, DERs, or storages to VPP aggregators based on their energy flexibility resources. BRPs can also play a role of an aggregator. For example, NEAS Energy, an independent BRP, acts as an aggregator by aggregating various generation units (e.g., CHP, wind, hydro, solar) in Denmark [28]. The main relationship between actors in the VPP aggregation market is shown Table 2.

Table 2. Actors in Virtual Power Plants (VPPs).

Actor	Offers	To
VPP aggregator	Market access Ancillary services Balancing services Buy and sell electricity Network services	DER owners TSO BRP Wholesale Market DSO
DER owner	Produce electricity Direct control	VPP aggregator VPP aggregator
BRP	Settle the imbalance Accurate forecast of supply and demand Bilateral contracts [29]	Market VPP aggregator VPP aggregator
Policy maker	Energy rules	All actors

3. Buildings and Energy Flexibility

Buildings are responsible for a large percentage of the global energy consumption (e.g., about 45% energy consumption in Denmark is from buildings, shown in Figure 1) and are therefore good candidates for providing energy aggregation potentials to the grid. Buildings can participate in the energy aggregation market via different channels. Meanwhile, energy consumption may vary between residential, commercial, and industrial buildings due to differences in building features.

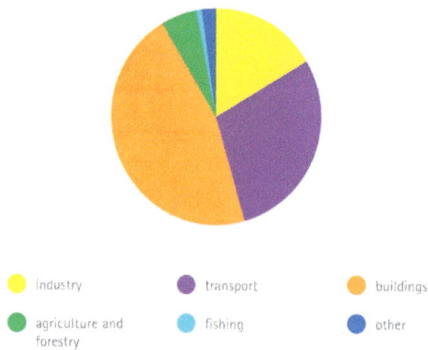

industry   transport   buildings

agriculture and   fishing   other
forestry

Figure 1. Danish energy consumption by sector in 2012 [30].

Residential buildings are defined as small electricity consumers due to their numerosity [31]. In Europe, residential buildings constitute 75% of the total number of buildings and 16% are high-rise buildings constructed within the period of 1960–1980 [32]. A majority of residential buildings consist of standard building technologies such as heating, hot water, cooling, ventilation, and lighting. However, residential building appliances' differ between each other. Common, potentially controllable appliances in residential buildings are dishwashers, washing machines, clothes dryers, freezers and refrigerators, heat pumps, and electric vehicles. The consumptions of appliances are different. For instance, statistics (shown in Figure 2) show the energy consumption by appliances in residential buildings in North America.

Residential buildings can provide energy flexibility. Flexibility potentials by home appliances vary. For instance, freezers and refrigerators provide less flexibility because more than 30 min interruption to freezer or refrigerator operation may cause spoilage [33]. Compared to other appliances, heating and ventilation provide more flexibility by shifting the temperature, especially during the day when households are empty [31].

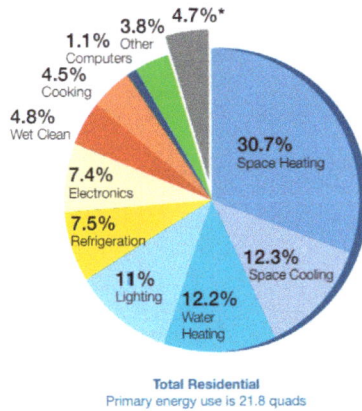

Figure 2. Residential primary energy end-use in the USA 2005 [34].

Commercial buildings include hospitals, hotels, stores, and offices. An example of the energy consumption by appliances in commercial building is shown in Figure 3. Some commercial buildings are more reluctant to participate to DR (e.g., reschedule their usage of power) due to the effect on their business routines and profits [35]. For instance, hotels and hospitals operate 24/7, and are reluctant to shift their usage of power due to consideration of their profits or occupants' comfort. Small or medium size commercial buildings (e.g., stores, offices) might participate in direct load control programs, while hospitals, hotels, and other large commercial buildings can participate in more interruptible programs.

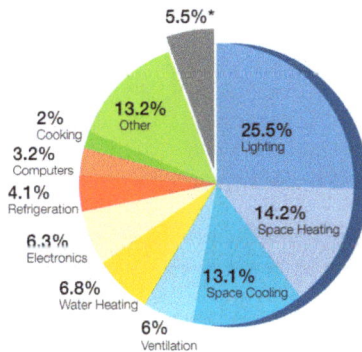

Figure 3. Commercial primary energy end-use in the USA 2005 [36].

Industries are usually large energy consumers. There are different types of industries engaging in different processes (e.g., steel, textile, food industries) and technologies [12]. Industrial buildings are often equipped with wind turbines or CHP to generate their own electricity. The energy flexibility potentials differ from one industry to another (e.g., Table 3 shows potentials for load flexibility of different processes in agriculture and industry in Denmark). For instance, refrigeration companies have particularly high load shift potentials with duration of several hours, and there are several approaches to obtain energy flexibility from refrigerators/freezers [37]. As with some commercial buildings that have large energy consumption rates, industrial buildings usually are reluctant to reschedule their usage of power considering their big profits [35].

Table 3. Potential for load flexibility of different processes in agriculture and industry [37].

Industry	Electricity Consumption, GWh/Year (2001)			Flexibility Potential, MW		
	Eastern Denmark	Western Denmark	Total	East	West	Total
Agriculture	405	2150	2555	13	69	82
Food and beverage	518	1738	2526	13	43	56
Textile	14	194	208	0	4	4
Wood industry	123	281	404	2	6	8
Paper and printing industry	228	527	755	5	11	16
Chemical industry	1116	1079	2195	17	16	33
Stone, clay, and glass industry	211	719	930	4	15	20
Iron and steel mills	528	117	645	26	6	32
Foundries	-	196	196	0	10	10
Iron and metal	447	1304	1751	20	59	79
Trade & Service	1507	2206	3173	54	79	134

4. Methods

This paper applies three analysis methods to review and evaluate aggregation potentials for buildings. The method of 'business model canvas' is adopted to describe potential scenarios that buildings can participate in energy aggregation markets with different values and channels. SWOT and TOWS analyses aim to evaluate feasibility and barriers of different business models in the current situation and in future trends of energy systems.

4.1. Business Model Canvas

Buildings can provide aggregation potentials to energy market via different channels with different values. Meanwhile, there are different involved market players, structures of revenue and cost, and dependencies among scenarios. Therefore, this paper adopts the 'business model' concept to discuss different scenarios for buildings to participate in the energy aggregation markets.

The term 'business model' has been massively applied and discussed in recent years. A business model is "a system of resources and activities which create a value that is useful to the customer and the sale of this value makes money for the company" [38]. It is part of a company's business strategy that describes how a company creates, delivers, and captures value within economic, social, cultural or other contexts [39].

So far, several business model generation methods have been developed. For example, Mullins and Komisar's five-pillar model [40] (including revenue model, gross margin model, operating model, working capital model, investment model). However, this model has barriers to be applied for complex analysis, due to little attention to the value offered to customers. The business model developed by Afuah [41] divides the model into six components (shown in Figure 4). However, this model also does not describe the value offered to customers.

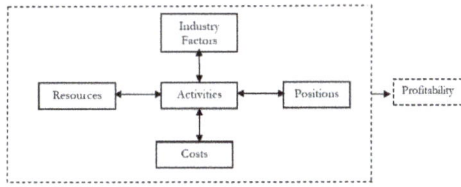

Figure 4. Business model by Afuah.

This paper adopts the 'business model canvas' by Osterwalder and Pigneur [42] which divides the business model into nine components: customer segments, customer relationships, distribution channels, value proposition, key resources, key activities, partners, cost structure and revenue streams (shown in Figure 5). Compared to other methods, the business model canvas is a more powerful visualized and flexible tool that is popularly adopted by both industry and academia.

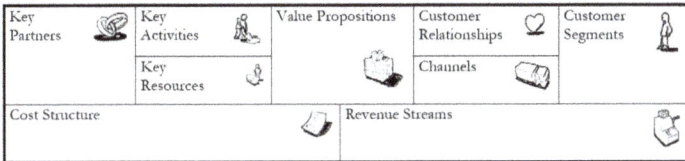

Figure 5. Business model canvas by Osterwalder and Pigneur.

4.2. Evaluation Tool for Business Model Analysis

To evaluate the value of potential business models, this paper modifies the evaluation tool developed by [43]. The tool is an organized and transparent system that facilities the work of the evaluators of potential business models [43]. This tool includes all factors that affect the business models and can be adapted to specific needs/strategies. This evaluation tool is originally intended for smart cities business models, and this paper modifies it for evaluating business models of buildings' participation in the aggregation market. The modified evaluation tool can be applied to general business model analysis, and the decision-making (value criteria) can be specific for different scenarios (shown in Table 4).

The value of the business model (VBM) in this paper is calculated as follows:

$$
\text{Value of business model} = \text{value proposition} \times \text{customer segment} \times (\text{partners} \\ + \text{resources} + \text{revenue streams} - \text{cost} + \text{customer relationship} + \text{channels} + \text{activities}) \tag{1}
$$

Table 4. TOWS analysis and strategy options.

Elements from Business Model Canvas	Value Criteria
Value Proposition	1: if provide significant more benefits to customers compared to existing solutions (product/service) 0.5: if provide around half more benefits to customers compared to existing solutions 0.1: if not provide visible benefits to customers compared to existing solutions
Customer segment	Value of customer segment = size × purchasing power *Size:* 1: if majority of the total potential customers can be targeted, otherwise the percentage of the total potential customers can be targeted *Purchasing power:* 1: high 0.5: medium 0.1: low

Table 4. *Cont.*

Elements from Business Model Canvas	Value Criteria
Partners	1: if the partner is the existing partner 0.5: if it is new but easy to reach 0: if it is new but difficult to reach Note: total value = ∏(value of individual partners), because the more partners you need to have, the more risk exists
Resources	1: if it is an existing resource 0.5: if it is new but easy to reach 0: if it is new but difficult to reach Note: total value = Σ(individual resource)/number of compulsory resources
Revenue streams	Depends mainly on customers' familiarity and companies' affordability 1: if it is familiar to customers and fits to companies' normal business 0.5: if it partly familiar to customers and companies need to make small changes 0: if it is totally new to customers and companies
Cost	1: if large spending for devices and personals 0.5: if within the range of affordable spending 0: if based on existing devices and personals
Customer relationship	Mainly depends on how simple and easy the approach is. 1: if it is for keeping existing customers 0.5: if it is for growing existing customers 0.1 if it for getting new customers Note: If it is easy to get new customers, you can move it to 0.5 or even 1. Total value = Σ(individual customer relationship)/number of compulsory customer relationships
Channels	1: if it is an existing channel 0.5: if it is new but easy to establish 0: if it is new but difficult to establish
Activities	1: if it is an existing activity or similar to the existing activities 0.5: if it is new but easy to conduct 0: if it is new but difficult to conduct Note: total value = Σ(individual activity)/number of activity, because the more activities you need to manage, the more difficult the task

4.3. SWOT Analysis and TOWS Analysis

The discussion of buildings' aggregation potentials needs to be integrated with the context of the specific electricity market. For instance, the demand response status among EU member countries is divided into three groups: (1) who have yet to seriously engage with DR reforms; (2) who are in the process of enabling DR through retailers only; (3) who enable both DR and independent aggregation [6]. Therefore, business models for buildings to participate in energy aggregation markets are strongly influenced by national electricity market structures.

This paper uses the Nordic electricity market as an example, and applies SWOT and TOWS analyses to evaluate feasibility and barriers of different business models. SWOT is an acronym for strengths, weaknesses, opportunities, and threats. It is a structured planning method that evaluates internal and external factors of an organization, project, or business venture. Strengths and weaknesses aim to examine organization's internal situation, opportunities and threats focus on external environment. SWOT analysis is broadly used in planning and decision-making.

However, SWOT analysis does not show relationships between internal and external factors. Therefore, this paper applies TOWS analysis to match internal factors and external factors to identify relevant strategic options. The strategy options in TOWS are described in Table 5.

Table 5. TOWS analysis and strategy options.

Strategy Options	Opportunities	Threats
Strengths	**S-O Strategies** Strategies that use strengths to take advantages of opportunities	**S-T Strategies** Strategies that use strengths to avoid threats
Weaknesses	**O-W Strategies** Strategies that take advantages from opportunities for mitigating weaknesses	**W-T Strategies** Strategies that mitigate weakness and avoid threat

5. Results

Buildings are commonly defined into three types (residential, commercial, and industrial). Due to the requirement of volume threshold for aggregation markets (e.g., the minimum bid to provide primary service in Demark is 0.3 MW), this paper divides buildings into two categories according to their energy consumptions: small and large energy consumers. The majority of residential buildings and some commercial buildings are small energy consumers. Comparatively, industrial buildings and some commercial buildings are large energy consumers.

There are four business models proposed for buildings to participate in the energy aggregation market in this paper (shown in Table 6).

Table 6. Four Business Models of Buildings' Participation in the Aggregation Market.

Aggregation Market	Types	Business Model	Direct Participants	Indirect Building Participants
Demand Response	Implicit DR (price based)	1—buildings participate in the implicit DR program via retailers	Retailers	All buildings
	Explicit DR	2—buildings (small energy consumers) participate in the explicit DR via aggregators	Independent aggregator	Buildings with small energy consumption
		3—buildings (large energy consumers) directly access the explicit DR program	Buildings with large energy consumption	-
Virtual Power Plants	Trading, balancing, network services	4—buildings access the energy market via VPP aggregators by providing DERs	VPP aggregators	DER owners (buildings which equip the DERs)

5.1. Business Model 1—Buildings Participate in the Implicit DR Program via Retailers

All buildings can participate in the implicit DR program. In this business model (shown in Table 7), buildings receive the DR program package as part of their electricity supply contract with their electricity retailer. Therefore, buildings can obtain a lower bill. For instance, buildings can reduce electricity usage at peak periods or shift their usage to off-peak periods.

Retailers can provide different DR program packages due to buildings' own preferences and constraints, and improve consumers' satisfaction rate. For instance, customers' satisfaction rate can be increased due to lower bills. Retailers might get new customers by providing an explicit DR package as a competitive offer. On the other hand, retailers need to provide consulting services to customers. Retailers usually do not have professional knowledge in the DR domain, and DR services are a new business model for retailers. Therefore, retailers need to hire experts and additional staff for the DR business.

5.2. Business Model 2—Buildings (Especially with Small Energy Consumption) Participate in the Explicit DR Program via Aggregators

In Business Model 2 (shown in Table 8), buildings, especially those with low energy consumption, can obtain direct payment by participating in explicit DR programs via aggregators.

Table 7. Business Model 1—buildings participate in the implicit DR program via retailers.

Partners	Activities	Value Proposition	Customer Relation	Customers
• Regulators • Billing companies • Electricity retailers	• Customer analysis to provide different DR offers; • Customer education to promote the offers; • Customer consulting due to customer constraints; • Billing system integration • Staffs/expert recruitment	Buildings receive a lower electricity bill	• Different DR offers due to buildings' own preferences and constraints for existing customers • Increase existing customers' satisfaction rate due to lower bills	All buildings
	Resources		**Channels**	
	• Price signal • Regulators' support		Part of the electricity supply contract	
	Cost Structure		**Revenue Streams**	
	Integration of DR offers into electricity supply contract (which might need DR experts and facility purchasing) Price signal sending to customers (facilities and staffs)		Optional choices for existing customers	

Table 8. Business Model 2—buildings (small energy consumption) participate in the explicit DR via aggregators.

Partners	Activities	Value Proposition	Customer Relation	Customers
• Regulators • BRPs • DSOs • TSOs • Control system providers • Energy suppliers (retailers)	• Access customers via energy suppliers or other channels • Provide consulting and analysis of customer demand pattern • Participate in DR market (wholesale, balancing or ancillary market) • Control customers' appliances • Payment to customers for energy flexibility	Buildings receive direct payment by participating in the explicit DR market via aggregators	• Payment system • Incentives by regulation, TSOs and DSOs • Consulting service (e.g., training, building energy behavior analysis) • Control system operations and maintenance	Buildings (who are small energy consumers)
	Resources		**Channels**	
	• Local control system • Customer data (demand pattern) • Market information		• energy consulting directly by aggregators • access customers via energy suppliers (retailers)	
	Cost Structure		**Revenue Streams**	
	• DR control system (customer side and aggregator side) • Payment to customers • Tarifts to DSOs and TSOs • Payment/compensation to BRPs • Market access fees to the DR markets		• Payment from the DR market (including reserve capacity payment from TSO) • Incentive from TSO/DSO and regulators	

Aggregators can maintain good relationships with customers through (1) an efficient and customer-friendly payment system and control system; (2) a training and consulting service, including DR knowledge and market information sharing. Meanwhile, the customized DR contracts should be based on customers' energy constraints and preferences; (3) the participation in the DR market needs customers to install direct load control systems. Therefore, aggregators can provide discount or free control systems, and maintenance services to customers; (4) Aggregators provide backup for individual loads as part of pooling activities that can increase the overall reliability, and reduce risk for individual consumers.

Aggregators generate revenue by providing DR services to the market (e.g., wholesale market, regulating market, and ancillary service). Aggregators might also receive incentives from regulators, TSOs and DSOs, depended on market regulations and structures.

5.3. Business Model 3—Buildings (with Large Energy Consumption) Directly Access Explicit DR Program

Buildings with large energy consumption are the energy flexibility providers who can directly participate and compete directly with producers in the DR market (wholesale market, regulating market, or ancillary service) (shown in Table 9).

To participate in wholesale and balancing markets, large energy consumers need to comply with market rules. Meanwhile, to participate in the reserve market as an ancillary service, buildings need to allow TSOs to directly control energy flexibility resources of buildings (e.g., building energy management systems).

Buildings can receive direct payment by providing flexibility via direct participation in explicit DR programs, and may get incentives from regulators, DSOs and TSOs.

5.4. Business Model 4—Buildings Access the Energy Market via VPP Aggregators by Providing DERs

In this business model, buildings (which have DERs) are able to obtain direct payment from VPP aggregators by providing energy flexibility. The volume threshold for power producers may prevent small DER owners to trade their energy individually, and VPP aggregators aggregate DERs and flexible loads as a single entity in the wholesale market that can help DER owners collectively to participate in the market with lower risk.

Buildings can have different types of DERs. Therefore, DER owners can participate in different aggregation markets (shown in Table 10). For instance, residential buildings usually only have PVs. Due to response requirements for different markets (e.g., primary service in Denmark requires a response in 15 seconds with a minimum of 0.3 MW), aggregation potentials that can be provided by DER owners depend mainly on the types of DERs.

VPP aggregators can provide customized market access strategies for different types of DER owners. Meanwhile, VPP aggregators should provide accurate forecast information of supply and demand, and user-friendly control systems, because this influences DER owners' daily business or energy usage patterns.

The main reason for DER owners to participate in the energy flexibility market is monetary benefits. Therefore, VPP aggregators need to provide an efficient and fair payment system that also affects DER owners' satisfaction and motivation.

Table 9. Business Model 3- buildings directly access the explicit DR program.

Partners	Activities	Value Proposition	Customer Relation	Customers
• Technology providers • TSOs • BRPs • Energy consulting • DSOs • Regulators	• Install energy control system; • Analysis and integration of DR into the existing building management system; • Directly participate in the DR markets.	Buildings receive direct payment by providing energy flexibility to the market	• Allow direct load control by TSOs in the reserve market as ancillary service; • Comply to market rules in wholesale and balancing markets	DR market (wholesale market, and ancillary service to TSOs)
	Resources		**Channels**	
	• Energy flexibility from appliances in the buildings; • Building energy control systems		• Direct participation in the wholesale and balancing market; • Bidding in the reserve market (there are rules for bidding and ancillary capacity, and control in the reserve market)	
Cost Structure			**Revenue Streams**	
• Employees' salary or expert consulting • Control system installation/upgrade • Market access fee (rules for participation in wholesale and balancing markets) • Fees to BRPs by contract • Tariffs to DSOs and TSOs • Cost due to energy behavioral changes (influence production or occupants' satisfaction in buildings)			• Payment by providing demands in wholesale and balancing markets • Reserve capacity payment from TSOs • Incentive from TSOs/DSOs, and regulators	

Table 10. Business Model 4- buildings access the energy market via VPP aggregators by providing DERs.

Partners	Activities	Value Proposition	Customer Relation	Customers
• Technology providers • TSOs • BRPs • Energy consulting • DSOs • Regulators	• Install control system • Customer service (analysis and package deals)	Buildings can access the market with direct payment and low risk	• Customized market access strategy • Payment system • Forecast information • Direct control system	Building with DERs (e.g., PV, micro-CHP)
	Resources		**Channels**	
	• Accurate forecast of supply and demand • DERs		• Direct contact; • Via DER technology/equipment providers; • Via electricity retailers	
Cost Structure			**Revenue Streams**	
• VPP control system • Employees' salary (including expert payment) • Market access fee (rules to participate in the wholesale and balancing markets) • Fees to BRPs by contract • Tariffs to DSOs and TSOs			• Trading via wholesale market • Balancing service offered to BRPs • Reserve capacity payment from TSOs • Network service offered to DSOs	
Note				

- This business model canvas only focuses on the 'independent VPP aggregators'. The scenario of the existing retailers/BRP's new role as 'VPP aggregators' is not considered.
- The correlation between building types and installed DERs is not considered. However, it is important to be aware that different building types can equip different DERs due to preferences and constraints of buildings. For example, a majority of residential buildings are only equipped with PV systems, but industrial buildings, such as greenhouses, might be equipped with CHPs. The differences influence market contribution from DER owners.

6. Case Study—The Aggregation Potential for Buildings in the Nordic Electricity Market

The Nordic electricity market of Denmark, Finland, Norway, and Sweden, comprises of a wholesale market and a retail market. Each Nordic country is an integral part of the free Nordic electricity market [44]. The wholesale market trade is via the Nord Pool Spot market. The Nord Pool market is owned by the TSOs in the Nordic countries. There are two electricity market places in the Nord Pool Spot market: Elspot (Day-ahead) and Elbas (Intra-day), and a regulating power market. The national TSOs in each country are responsible for the electricity retail markets. In Denmark, Energinet.dk is responsible, and consumers can freely choose their electricity retailers.

In the current Nordic electricity market, there is no independent aggregator, and demand response has been enabled within the ancillary services [13]. In some Nordic countries, customers have opportunities to participate in priced-based programs (e.g., time-of-use (TOU), critical peak pricing, and real-time pricing) [15].

There are nearly 15 million electricity customers in the Nordic electricity market, and the four business models show that there are opportunities and benefits for electricity consumers to participate in the energy aggregation market. However, there are also barriers and constraints under the current energy market structure.

Opportunities: there is a market need for buildings' energy flexibility, due to market (e.g., imbalance payment) and grid (grid capacity) demands. Meanwhile, technologies including control systems, forecast software, and DERs, are more advanced, cheaper, and user-friendly compared to before. Therefore, market players, such as aggregators and buildings, participate much more easily in the aggregation market. In many countries, regulators, TSOs, or DSOs have provided incentives for participation in the aggregation market.

Threats: there are still regulation barriers for market players to access the aggregation market. For instance, there is no DR market in some countries such as Denmark, and DR participation is limited to the small consumers and only large consumers can participate in the wholesale market. Meanwhile, monetary benefit is not significantly visible to encourage buildings to participate in the aggregation market, especially with the compromise of comfort and low return on investment.

Strengths: the majority of buildings have potentials to provide flexibility to the energy market, either by changing energy usage pattern (adjustable loads) or by giving direct control of their appliances or DERs to aggregators.

Weaknesses: Return on investment is the main concern for energy consumers. Small energy consumers, e.g., residential buildings, still lack investment incentives to purchase controllable appliances, control systems, and DERs. Energy consumers might also be conservative due to potential effects on their daily business or energy usage patterns. Meanwhile, limited capacity of energy flexibility provided by small energy consumers (e.g., residential buildings) might prevent their access to the aggregation market or not have visible monetary benefits.

6.1. Value of the Business Model (VBM) in the Nordic Electricity Market

For the aggregation market development, it is necessary to investigate which business model brings more value to the whole market. By applying the evaluation of business model analysis (1), the value for the four business models are shown in Table 11.

Under the current situation of the Nordic electricity market, Business Model 1—'buildings participate in the implicit DR program via retailers' significantly provides the highest value to the whole aggregation market. This business model not only covers all buildings, is highly supported by regulations, but also requests few changes in the existing market structure.

Comparatively, Business Model 2—'buildings participate in the explicit DR via aggregators' brings the lowest value to the whole aggregation market. The reason is (1) there are no independent aggregators in the Nordic electricity market as yet, and stakeholders are conservative to this business model (the value of 'partners' is only 0.425); (2) buildings with small energy consumption provide limited energy flexibility, and the benefit is not visible under the current Nordic market regulation (the values of 'value proposition' is 0.5); meanwhile, (3) buildings with small energy consumption usually do not install building control systems, and the cost to participate the market and install control systems is high (the values of 'cost' is −1.4).

Business Models 2 and 4 both present buildings' participation in the aggregation market via aggregators. The comparison results show that buildings that equip DERs have more potentials and incentives to participate the aggregation market.

Table 11. Value of four developed business models in the Nordic electricity market.

Business Model	1—buildings participate in the implicit DR program via retailers	2—buildings participate in the explicit DR via aggregators	3—buildings directly access to the explicit DR program	4—buildings access the energy market via VPP aggregators by providing DERs
Value Proposition	1	0.5	1	1
Customer Segment	1	0.21	0.21	0.19
Partners	1	0.025	1	0.125
Resources	1	0.83	0.75	0.75
Revenue Streams	0.5	1.1	1.2	1.7
Cost	−0.5	−1.4	−0.5	−1.4
Customer Relationship	0.75	0.425	0.75	0.875
Channels	1	1	1.5	1.5
Activities	1	0.6	0.67	1
Value of Business Model	3.75	≅0.17 (0.1659)	≅1.34 (1.3377)	≅0.77 (0.7695)

6.2. Recommendation for Encouraging Building Participation

With TOWS analysis (shown in Table 12), this paper presents the following suggestions to encourage buildings to participate in the aggregation market:

- Regulation needs to be adjusted to allow buildings easy access to the aggregation market;
- Incentives from regulators, TSOs/DSOs can encourage buildings to participate in the energy aggregation market;
- Clear monetary benefits (e.g., payment) needs to be defined;
- Financial support, e.g., loans, renting, cost reduction strategies and packages, for installation of control systems, DERs, and controllable appliances;
- Easy and user-friendly control systems with accurate forecast and analysis;
- Customized service (e.g., payment and control solutions) for different types of buildings;
- Selective market access for buildings which can have visible benefit from the aggregation market (e.g., large energy consumers or industrial buildings with large capacity of DERs);
- Utilization of ADR (automatic DR) in buildings with challenges of privacy, user acceptance, and security needs to be addressed.

Table 12. Combined SWOT and TOWS analyses of buildings' energy aggregation potentials.

	Opportunities	Threats
1. European Union (EU) climate and energy goals 2. Technology readiness 3. Market demands 4. Constraints of grid capacity 5. Cost reduction 6. Incentives		• Regulation barriers • Limited monetary benefits • Slow Return on investment (ROI)
Strenghts	**S-O Strategies**	**S-T Strategies**
• Flexible load • Installed DERs • Advanced appliances	• Cost reduction strategies and packages of control system and DER equipment; • Easy and user-friendly control systems	• Regulation changed to allow buildings easy access to the aggregation market; • Clear monetary benefits and incentives; • Analysis and service (including training) regarding consumer behavior.
Weaknesses	**O-W Strategies**	**W-T Strategies**
• No investment support • Constraints of daily business and energy usage pattern • Low capacity of energy flexibility	• Aggregation of small consumers by DR and VPP programs; • Incentives from regulators, TSOs/DSOs; • Software support for forecast and analysis.	• Financial support for equipment control system (e.g., loans, renting); • Selective market access for buildings which can have visible benefits from the aggregation market (e.g., large energy consumers, or industrial buildings with large capacity of DERs)

7. Conclusions

This paper develops and discusses four business models for buildings (e.g., residential, industrial, and commercial) to participate in the aggregation markets by providing flexible loads and DERs. With a case study of the Nordic electricity market, an evaluation of the four business models is conducted with the SWOT analysis and evaluation tool of business model analysis. The evaluation result shows that there are opportunities for buildings to participate in the aggregation market and constraints for different types of buildings. Under the current Nordic market regulation, the most feasible business model is: buildings participate in the implicit DR program via retailers. Meanwhile, the regulation barriers and limited monetary incentives impede buildings' participation. Therefore, the business model of 'buildings with small energy consumption participate in explicit DR via aggregators' possesses more challenges compared to other three business models.

This study contributes to the literature in several unique ways. First, this study demonstrates four business models with explicit description about how the flexibility potentials of buildings can be utilized in different aggregation scenarios.

Second, by investigating buildings' participation in the four scenarios, this study contributes to the literature regarding the correlation between buildings' flexibility and aggregation market access. This study finds that the flexibility resources and potentials are different for different types of buildings, and building owners have different needs and behaviors. Thus, it is essential to understand building owners' needs, comforts, and behaviors to develop feasible market access strategies for different types of buildings.

Third, the importance and implication of incentive programs, national regulations and energy market structures to the buildings' participation are identified. Incentive programs can enhance buildings' participation. In addition, the involvement of governments and regulators in the aggregation market can provide incentives, increase DR awareness and participation. However, the aggregation market is still immature, and regulations and polices of aggregation markets are various across countries. For instance, in Europe, the countries Belgium, France, Ireland, and the UK have created the regulative framework to enable both DR and independent aggregators, whereas other European countries have not yet engaged with DR reforms, e.g., Portugal and Spain. Therefore, the business models of aggregation potentials for buildings need to be based on national regulations and energy market structures.

Acknowledgments: This study was conducted as part of Energy in Buildings and Communities Programme (EBC) Annex 67 Energy Flexible Buildings.

Author Contributions: Zheng Ma and Bo Nørregaard Jørgensen conceived and designed the methodologies; Joy Dalmacio Billanes contributed literature analysis; Zheng Ma wrote the paper and performed the business model development and analysis.

Conflicts of Interest: The authors declare no conflict of interest.

References

1. European Commission. *The Future Role and Challenges of Energy Storage*; Europa.eu: Brussels, Belgium, 2017.
2. Papalexopoulos, A.; Hansen, C.; Frowd, R.; Tuohy, A.; Lannoye, E. Impact of the transmission grid on the operational system flexibility. In Proceedings of the 2016 Power Systems Computation Conference (PSCC), Genoa, Italy, 20–24 June 2016; pp. 1–10.
3. Wu, H.; Shahidehpour, M.; Alabdulwahab, A.; Abusorrah, A. Thermal generation flexibility with ramping costs and hourly demand response in stochastic security-constrained scheduling of variable energy sources. *IEEE Trans. Power Syst.* **2015**, *30*, 2955–2964. [CrossRef]
4. Ma, Z.; Badi, A.; Jørgensen, B.N. Market opportunities and barriers for smart buildings. In Proceedings of the 2016 IEEE Green Energy and Systems Conference (IGSEC), Long Beach, CA, USA, 6–7 November 2016; pp. 1–6.

5.	Ma, Z.; Billanes, J.D.; Kjærgaard, M.B.; Jørgensen, B.N. Energy flexibility in retail buildings: From a business ecosystem perspective. In Proceedings of the 2017 14th International Conference on the European Energy Market (EEM), Dresden, Germany, 6–9 June 2017; p. 6.

6.	Annala, S.; Honkapuro, S. Demand response in Australian and European electricity markets. In Proceedings of the 2016 13th International Conference on the European Energy Market (EEM), Porto, Portugal, 6–9 June 2016; pp. 1–5.

7.	Du, H.; Liu, S.; Kong, Q.; Zhao, W.; Zhao, D.; Yao, M.G. A microgrid energy management system with demand response. In Proceedings of the 2014 China International Conference on Electricity Distribution (CICED), Shenzhen, China, 23–26 September 2014; pp. 551–554.

8.	Minou, M.; Stamoulis, G.D.; Thanos, G.; Chandan, V. Incentives and targeting policies for automated demand response contracts. In Proceedings of the 2015 IEEE International Conference on Smart Grid Communications (SmartGridComm), Miami, FL, USA, 2–5 November 2015; pp. 557–562.

9.	Liu, Z.; Zeng, X.J.; Ma, Q. Integrating demand response into electricity market. In Proceedings of the 2016 IEEE Congress on Evolutionary Computation (CEC), Vancouver, BC, Canada, 24–29 July 2016; pp. 2021–2027.

10.	Son, J.; Hara, R.; Kita, H.; Tanaka, E. Energy management considering demand response resource in commercial building with chiller system and energy storage systems. In Proceedings of the 2014 International Conference on Power Engineering and Renewable Energy (ICPERE), Bali, Indonesia, 9–11 December 2014; pp. 96–101.

11.	De Vries, L.J.; Verzijlbergh, R. Organizing flexibility: How to adapt market design to the growing demand for flexibility. In Proceedings of the 2015 12th International Conference on the European Energy Market (EEM), Lisbon, Portugal, 19–22 May 2015; pp. 1–5.

12.	Ma, Z.; Friis, H.T.A.; Mostrup, C.G.; Jørgensen, B.N. Energy flexibility potential of industrial processes in the regulating power market. In Proceedings of the 6th International Conference on Smart Cities and Green ICT Systems, Porto, Portugal, 22–24 April 2017; pp. 109–115.

13.	Bertoldi, P.; Zancanella, P.; Boza-Kiss, B. *Demand Response Status in EU Member States*; Europa.eu: Brussels, Belgium, 2016.

14.	Sebastian, S.; Margaret, V. Application of demand response programs for residential loads to minimize energy cost. In Proceedings of the 2016 International Conference on Circuit, Power and Computing Technologies (ICCPCT), KK Dist, India, 18–19 March 2016; pp. 1–4.

15.	Lamprinos, I.; Hatziargyriou, N.D.; Kokos, I.; Dimeas, A.D. Making demand response a reality in Europe: Policy, regulations, and deployment status. *IEEE Commun. Mag.* **2016**, *54*, 108–113. [CrossRef]

16.	Yu, M.; Hong, S.H.; Beom, K.J. Incentive-based demand response approach for aggregated demand side participation. In Proceedings of the 2016 IEEE International Conference on Smart Grid Communications (SmartGridComm), Sydney, Australia, 6–9 November 2016; pp. 51–56.

17.	Gharesifard, B.; Başar, T.; Domínguez-García, A.D. Price-based coordinated aggregation of networked distributed energy resources. *IEEE Trans. Autom. Control* **2016**, *61*, 2936–2946. [CrossRef]

18.	Ghavidel, S.; Li, L.; Aghaei, J.; Yu, T.; Zhu, J. A review on the virtual power plant: Components and operation systems. In Proceedings of the 2016 IEEE International Conference on Power System Technology (POWERCON), Wollongong, Australia, 28 September–1 October 2016; pp. 1–6.

19.	Vandoorn, T.L.; Zwaenepoel, B.; de Kooning, J.D.M.; Meersman, B.; Vandevelde, L. Smart microgrids and virtual power plants in a hierarchical control structure. In Proceedings of the 2011 2nd IEEE PES International Conference and Exhibition on Innovative Smart Grid Technologies, Manchester, UK, 5–7 December 2011; pp. 1–7.

20.	Goutard, E.; Passelergue, J.C.; Sun, D. Flexibility marketplace to foster use of distributed energy resources. In Proceedings of the 22nd International Conference and Exhibition on Electricity Distribution (CIRED 2013), Stockholm, Sweden, 10–13 June 2013; pp. 1–4.

21.	Saboori, H.; Mohammadi, M.; Taghe, R. Virtual Power Plant (VPP), definition, concept, components and types. In Proceedings of the 2011 Asia-Pacific Power and Energy Engineering Conference, Wuhan, China, 22–28 March 2011; pp. 1–4.

22.	Bakari, K.E.; Kling, W.L. Development and operation of virtual power plant system. In Proceedings of the 2011 2nd IEEE PES International Conference and Exhibition on Innovative Smart Grid Technologies, Manchester, UK, 5–7 December 2011; pp. 1–5.

23. Hatziargyriou, N.D.; Tsikalakis, A.G.; Karfopoulos, E.; Tomtsi, T.K.; Karagiorgis, G.; Christodoulou, C.; Poullikkas, A. Evaluation of Virtual Power Plant (VPP) operation based on actual measurements. In Proceedings of the 7th Mediterranean Conference and Exhibition on Power Generation, Transmission, Distribution and Energy Conversion (MedPower 2010), Agia Napa, Cyprus, 7–10 November 2010; pp. 1–8.

24. Plancke, G.; de Vos, K.; Belmans, R.; Delnooz, A. Virtual power plants: Definition, applications and barriers to the implementation in the distribution system. In Proceedings of the 2015 12th International Conference on the European Energy Market (EEM), Lisbon, Portugal, 19–22 May 2015; pp. 1–5.

25. Zehir, M.A.; Bagriyanik, M. Smart energy aggregation network (SEAN): An advanced management system for using distributed energy resources in virtual power plant applications. In Proceedings of the 2015 3rd International Istanbul Smart Grid Congress and Fair (ICSG), Istanbul, Turkey, 29–30 April 2015; pp. 1–4.

26. Ravichandran, S.; Vijayalakshmi, A.; Swarup, K.S.; Rajamani, H.S.; Pillai, P. Short term energy forecasting techniques for virtual power plants. In Proceedings of the 2016 IEEE 6th International Conference on Power Systems (ICPS), New Delhi, India, 4–6 March 2016; pp. 1–6.

27. Kessels, K.; Claessens, B.; hulst, R.D.; Six, D. The value of residential flexibility to manage a BRP portfolio: A Belgian case study. In Proceedings of the 2016 IEEE International Energy Conference (ENERGYCON), Leuven, Belgium, 4–8 April 2016; pp. 1–6.

28. The European Association for the Promotion of Cogeneration (COGEN Europe). *The Role of Aggregators in Bringing District Heating and Electricity Networks Together: Integrated Supply Maximising the Value of Energy Assets*; Europa.eu: Brussels, Belgium, 2014; pp. 1–3. Available online: http://www.cogeneurope.eu/medialibrary/2014/12/03/4bb831e2/Case%20Study%20-%20Neas%20Energy%20-%20December%202014%20FINAL.pdf (accessed on 15 July 2017).

29. Puglia, L.; Patrinos, P.; Bernardini, D.; Bemporad, A. Reliability and efficiency for market parties in power systems. In Proceedings of the 2013 10th International Conference on the European Energy Market (EEM), Stockholm, Sweden, 28–30 May 2013; pp. 1–8.

30. Nordic Energy Research. Denmark: Danish Energy Used Mostly in Buildings. 2013. Available online: http://www.nordicenergy.org/figure/energy-consumption-by-sector/danish-energy-used-mostly-in-buildings/ (accessed on 12 January 2017).

31. Thavlov, A.; Bindner, H.W. An aggregation model for households connected in the low-voltage grid using a VPP interface. In Proceedings of the IEEE PES ISGT Europe 2013, Lyngby, Denmark, 6–9 October 2013; pp. 1–5.

32. Kimmo, I.; Sirvio, A. *Sustainable Buildings for the High North. Existing Buildings—Technologies and Challenges for Residential and Commercial Use*; ePooki: Oulu, Finland, 2015; Available online: http://www.oamk.fi/epooki/2015/high-north-project-existing-buildings/ (accessed on 15 March 2017).

33. Safdarian, A.; Fotuhi-Firuzabad, M.; Lehtonen, M. Benefits of Demand Response on Operation of Distribution Networks: A Case Study. *IEEE Syst. J.* **2016**, *10*, 189–197. [CrossRef]

34. Energy Information Administration (EIA). Residential Site Energy Consumption by End Use. Available online: http://buildingsdatabook.eren.doe.gov/ChapterIntro2.aspx (accessed on 15 June 2016).

35. Yang, Z.; Wang, L. Demand Response Management for multiple utility companies and multi-type users in smart grid. In Proceedings of the 2016 35th Chinese Control Conference (CCC), Chengdu, China, 27–29 July 2016; pp. 10051–10055.

36. Energy Information Administration (EIA). Commercial Site Energy Consumption by End Use. Available online: http://buildingsdatabook.eren.doe.gov/ChapterIntro3.aspx (accessed on 23 February 2017).

37. Ea Energianalyse for Energunet. Dk; Dansk Energi. *Kortlægning af Potentialet for Fleksibelt Elforbrug I Industri, Handel og Service*; Energinet.dk: Vejle, Denmark, 2011.

38. Štefan, S.; Richard, B. Analysis of business models. *J. Compet.* **2014**, *6*, 19–40.

39. Duening, T.N.; Hisrich, R.D.; Lechter, M.A. Chapter 1—Fundamentals of Business. In *Technology Entrepreneurship*; Academic Press. Boston, MA, USA, 2010; pp. 1–27.

40. Mullins, J.; Komisar, R. *Getting to Plan B: Breaking through to a Better Business Model*; Harvard Business Press: Boston, MA, USA, 2009.

41. Afuah, A. *Business Models: A Strategic Management Approach*; McGraw-Hill/Irwin: New York, NY, USA, 2003.

42. Osterwalder, A.; Pigneur, Y. *Business Model Generation*; John Wiley & Sons: Hoboken, NJ, USA, 2009.

43. Díaz-Díaz, R.; Muñoz, L.; Pérez-González, D. The business model evaluation tool for smart cities: Application to smart santander use cases. *Energies* **2017**, *10*, 262. [CrossRef]

44. Ma, Z.P.Z.; Jørgensen, B.N. The international electricity market infrastructure-insight from the nordic electricity market. In Proceedings of the 13th European Energy Market Conference (EEM 2016), Porto, Portugal, 6–9 June 2016; p. 5.

energies

MDPI

Article

Reschedule of Distributed Energy Resources by an Aggregator for Market Participation

Pedro Faria *, João Spínola and Zita Vale

Research Group on Intelligent Engineering and Computing for Advanced Innovation and Development (GECAD),
Polytechnic Institute of Porto (IPP), Rua DR. Antonio Bernardino de Almeida, 431, 4200-072 Porto, Portugal;
jafps@isep.ipp.pt (J.S.); zav@isep.ipp.pt (Z.V.)
* Correspondence: pnfar@isep.ipp.pt; Tel.: +351-228-340-511; Fax: +351-228-321-159

Received: 18 January 2018; Accepted: 20 March 2018; Published: 22 March 2018

Abstract: Demand response aggregators have been developed and implemented all through the world with more seen in Europe and the United States. The participation of aggregators in energy markets improves the access of small-size resources to these, which enables successful business cases for demand-side flexibility. The present paper proposes aggregator's assessment of the integration of distributed energy resources in energy markets, which provides an optimized reschedule. An aggregation and remuneration model is proposed by using the k-means and group tariff, respectively. The main objective is to identify the available options for the aggregator to define tariff groups for the implementation of demand response. After the first schedule, the distributed energy resources are aggregated into a given number of groups. For each of the new groups, a new tariff is computed and the resources are again scheduled according to the new group tariff. In this way, the impact of implementing the new tariffs is analyzed in order to support a more sustained decision to be taken by the aggregator. A 180-bus network in the case study accommodates 90 consumers, 116 distributed generators, and one supplier.

Keywords: aggregator; clustering; demand response; distributed generation

---

1. Introduction

The number of aggregators operating in energy markets has been on the rise since the end of the last decade [1]. Companies like Voltalis (Paris, France), REstore (Antwerp, Belgium), and EnerNOC (Boston, MA, USA) are currently major aggregators of flexibility and are the usual participants in energy markets [2]. These companies provide tools for energy services (e.g., optimization, monitoring, consultancy) to consumers, which reduces unnecessary or inefficient consumption. The aggregators, after an analysis of the consumer's load profile, conciliates the energy reductions of the consumers with its participation in the energy markets. In this way, a cooperative relation between the aggregators and the resources is achieved.

Demand Response (DR) and distributed generators are the flexibility resources with more interest and development in current power systems, which opens a path for others to raise as well, for instance, electric vehicle and storage units [3,4]. Demand response is divided in two types including price and incentive-based where the first corresponds to the response of consumers given a price signal (price variation) and, the latter, to the response of consumers given monetary incentives (tax relief, payment) [5–10]. These two types of demand response are used by different entities and to distinct consumers. Namely, the grid operators and aggregators use incentive-based while retailers tend to use more price-based strategies. Distributed generators have been significantly promoted in recent years through feed-in tariffs to make these resources more attractive to consumers [11–13]. Due to this initiative, the number of prosumers (consumers that own generation means) raised significantly in several countries (e.g., Portugal, Germany, UK). However, the high participation of these resources in energy markets has not yet been achieved mostly because of their small generation capacity and intermittent production. The complement that demand response and distributed generators give to each other provide the aggregator with sufficient tools to manage these resources and allow their indirect participation in energy markets [2,14–18].

1.1. Related Literature

Several energy markets are not adjusted to demand response participation due to the requirements needed to participate either in terms of minimum capacity or event duration. For example, in Finland, secondary reserve has 5 and 10 MW minimum capacity in automatic and manual actuation, respectively [19–21]. In the same country, it is possible to find more adjusted conditions in the primary reserve with a minimum capacity of 100 kW. Another example is the Californian independent system operator, CAISO (Folsom, CA, USA), with minimum requirement of 100 and 500 kW to consumer's participation [22]. In such market approaches, the aggregator or a single entity can deliver the requested reduction amount. In both cases, the market operator is not concerned about the way that consumers are aggregated and enumerated.

The need for an aggregator entity arises as a solution for the participation of small-size consumers when considering that it can create a virtual energy amount that enables enough energy to be negotiated in the market by the aggregator. This participation of the aggregator should ensure that the revenues obtained are sufficient to reward the participating resources while providing profit for the aggregator. In incentive-based programs, resources are remunerated bearing in mind their availability and utilization where it is considered a period that consumers make their loads available to be modified and, in the second, payment is made when load modification is actually done [23,24]. These are current approaches for the remuneration of consumers participating in demand response programs. However, for the research in the present paper, the questions are how much to pay to the consumers, how many distinct tariffs to implement, and which consumers should be in each tariff.

Aiming the aggregation and the remuneration of demand response resources, several works appear in the literature with most of them addressing only one of these topics and others addressing both in a single methodology, which are namely previous works from the authors of the present paper [25,26]. In fact, most of the recent and relevant literature in demand response, namely review papers, still insist in the demand response opportunities and flexibility options that are more and more evident with the increase of technology that supports demand response by providing examples of practical evidence of DR implementations and identifying the most relevant barriers without referring to possible innovative approaches for aggregation and remuneration [6,27]. Most of the identified barriers are related to market structures and incentives regarding the incentivizing consumers to participate in DR programs [28].

In Reference [29], a hierarchical DR architecture is proposed in order to control and coordinate various DR categories. In Reference [30], the author refers to the way that incentivizing DR with flat incentives implies with the revenues of retailers. In fact, in the beginning, DR incentives are needed but in a large implementation, DR must be remunerated by adequate and fair market mechanisms. In Reference [31], DR and generators are compared regarding the actual costs in real markets, which refers to the actual remuneration cost for DR. In Reference [32], the authors deal with the comfort in a building in order to determine the flexibility of consumption. A multi-agent approach is proposed for the bids and auctions establishment. The consumers are assumed to take part of the negotiation. None of the referred works proposed a model that implements the remuneration for DR participation, which consists of addressing the consumers' benefits and offering an advantageous remuneration for them. Moreover, the aggregation is done according to the open call to the previously enrolled consumers.

In Reference [25], the authors proposed a methodology in which aggregation and remuneration is done in an integrated approach in order to support the aggregator decisions. In Reference [26], a complementary approach is defined in order to analyze the profits of the aggregator, which supports the participation in the market by comparing the situations of using or not additional suppliers with DR use.

However, in the previous works, after defining the groups and the tariffs, the aggregation and the remuneration, it was assumed that the operation costs are still minimized. In the present paper, the proposed methodology contributes to making an evaluation of the new optimal scheduling of the resources using the new tariffs. New decision aspects are raised for the aggregator namely because some consumers are lower remunerated with the re-scheduling even if the aggregator operation costs are the same.

1.2. Proposed Aggregator

The present paper proposes a methodology to address market participation of an aggregator in energy markets by considering two types of distributed energy resources including demand response and distributed generation. Due to the small size of these resources, a Virtual Power Player (VPP) is considered the aggregator for DR and DG resources making it possible for them to participate in the electricity market. This aggregator defines the groups and the tariffs for each group to be scheduled in each different context or period of the day, which receives incomes from the market and forms the consumers to fulfill their load and paying to DG and DR resources by also obtaining some profits. The DG and DR are scheduled according to the available forecasts that are assumed to be adequately accurate. It is also a task assumed for the VPP to accommodate the deviations of the DG and the DR resources due to their unpredictability.

After an initial schedule of resources, these are aggregated and assigned a tariff for each of the groups formed with these tariffs considered forward when performing the second schedule of resources. This allows the aggregator to schedule resources in line with a group tariff that is applied to all resources in that group and decided whether to participate or not with bids in the energy markets. Aggregation is made through k-means clustering algorithm while the group tariff corresponds to the average price of the resources in the group.

The present section approached the most relevant concepts related to the developed methodology and the activities that it intends to represent. In Section 2, the proposed methodology is presented and explained in detail while Section 3 shows the mathematical formulation used. In Section 4, the case study used to verify the usefulness of the methodology is presented, and finally, in Section 5, the conclusions obtained from the methodology implementation are presented.

2. Proposed Methodology

The present section details the proposed methodology that can be divided into three stages per scheduling phase, which is illustrated in Figure 1. Phase one is presented in Section 2.1 while phase 2 is presented in Section 2.2.

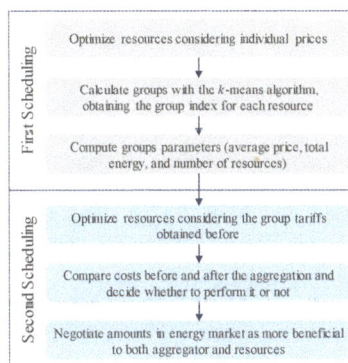

Figure 1. Proposed methodology scheme.

2.1. Phase One

According to Figure 1, phase one is divided into three stages. In the first stage, each resource has an individual price that represents the cost to the aggregator to schedule it in terms of demand response or distributed generation. The optimization model considers demand response programs such as curtailment. The mathematical formulation for the minimization of the aggregator's operation costs is detailed in Section 3. The results of the optimization include the amounts scheduled in generators and the energy to be reduced from the consumers in order for the load-generation relation to be balanced in each of the periods considered. Additionally, the resources for aggregation are obtained from this stage.

In the second stage, the aggregation stage, the resources that have contribution higher than zero in the aggregator's schedule (stage 1), are in the aggregation process. The other resources with zero contribution are not considered in either aggregation and remuneration stages. This ensures that non-participating resources do not affect the results of the aggregation and remuneration and, therefore, neither influence the prices for the participants. This second stage of the methodology uses k-means to obtain the group indexes for each distributed resource. The number of groups to be formed is a parameter that the aggregator can modify in consonance with its operation context. This aggregation analysis is made for each period. The distributed resource types have distinct aggregation processes. However, consumers and generators are not aggregated together. In this way, different data inputs are considered for aggregation in both cases. The clustering stage is very relevant for the groups definition. Normally, the consumers' tariffs are defined for all the consumers of the same type (domestic, commercial, industrial, etc.). In the proposed approach, the aggregator is able to request the simulation of several number of tariff groups, according to the number of DR programs. The clustering algorithm will provide it.

When aggregating consumers, the data input consists of:

- the energy scheduled in the curtailment program for each consumer,
- and the price of these reductions (input parameter of the objective function).

In the case of distributed generators, the data input in the aggregation process is:

- scheduled energy for each unit,
- and individual price (input parameter of the objective function).

The third stage, which is the final stage of phase one, corresponds to computing the group parameters. These can be defined as the relevant features that the aggregator needs to form a bid in the energy markets including average price and total energy scheduled. The number of resources is important information to the aggregator so that it knows the resources in each group.

In conclusion, the results of phase one are the consumers and generators in each group and the tariff for each group.

2.2. Phase Two

In the second phase, we have three stages, according to Figure 1. The first stage corresponds to the same procedure performed in the first stage of phase one. The difference is that now, in the first stage of phase two, the prices for the resources entering the scheduling are the ones resulting from stage three of phase one for each group, which involve the resulting tariffs. In this way, after scheduling phase two, the aggregator can compare each period based on the operation costs and conclude if it is or not beneficial to proceed with phase two or adopt the results of phase one. This allows for a more efficient operation of the aggregator by maintaining its capability of market participation.

In the second stage, the resources are aggregated for each period. The clustering algorithm considered is implemented and given as input for the energy schedule and price (these features are considered for both the distributed generators and consumers participating in the demand response program). The information about aggregation is also available for the aggregator in terms of power, tariffs, the number of resources in each group, the resources that participated in the scheduling, and the group assignment.

In the final stage of phase two, the aggregator goes to market considering the groups formed and respective group tariffs as available bids. In addition, the results provided give the aggregator the possibility to check which consumers have been positively or negatively affected by the re-scheduling made in Phase two. In this context, the aggregator must ensure that the bids guarantee fair payment of distributed energy resources and its services.

The provided overall results are relevant for the aggregator to consider in terms of the participation in energy markets even though the present paper doesn't consider the different specific market opportunities that can be available for the aggregator to participate. In fact, the consideration of market negotiation should be accompanied by several involved parties' agreements so that the rest of the grid wouldn't be impaired in its stability and energy quality.

The methodology proposed in the present paper provides a solution for the management of aggregator's activities including the optimal scheduling of resources with an aggregation and remuneration model to complement its participation in market. In fact, the optimization of the energy resources used is made in the present paper in order to support the proposed aggregation and remuneration methodology. It is not intended to be the focus of the paper since it can be found in other previous works [25,26]. Also, the clustering algorithm and its input features as used in the present paper have been previously used in Reference [26]. It is included in the present paper in order to support the overall proposed methodology framework. In this way, the present paper is innovative by presenting a rescheduling model for the decrease of aggregator's operation costs based on the aggregation and remuneration model applied to distributed resources. Section 3 details the mathematical formulation used to guarantee the resource's optimized scheduling.

3. Scheduling Formulation

The optimization problem is labeled as mixed-integer linear programming (MILP) since discrete and continuous variables are considered. The scheduling problem is relatively simple considering the program's definition and respective modelling. However, the problem's size implicates an analysis of the best option. The proposed methodology was implemented in TOMSYM™ optimization environment, which was developed in MATLAB™. The algorithm was run in a 64-bit computer system with 16 GB RAM and 2.1 GHz processor.

In Equation (1), the objective function is considered for optimization, which involves the demand response programs mentioned before as well as the distributed generators and external suppliers. This objective function is considered for both schedules (phase one and phase two) of resources. However, in the second, the $C^{DG}_{(p,t)}$, $C^{red}_{(c,t)}$, and $C^{cut}_{(c,t)}$, are updated for certain resources, which include the ones that participated in the aggregation and remuneration processes. The variables of the problem, as presented in the objective function (1) are: Energy schedule for external supplier s, in period t, Energy schedule for distributed generator p, in period t, and Energy schedule for load curtailment in consumer c, in period t. This means that the output of the optimization problem corresponds to the power amounts present in Equation (1).

$$Min\ OC = \sum_{s=1}^{S} P^{Sup}_{(s,t)}.C^{Sup}_{(s,t)} + \sum_{p=1}^{P} P^{DG}_{(p,t)}.C^{DG}_{(p,t)} + \sum_{c=1}^{C} P^{cut}_{(c,t)}.C^{cut}_{(c,t)} \qquad (1)$$

As mentioned before, the system's balance is insured by defining the constraint represented in Equation (2). This maintains the balance between load and generation, which considers the contributions of demand response and distributed generation.

$$\sum_{s=1}^{S} P^{Sup}_{(s,t)} + \sum_{p=1}^{P} P^{DG}_{(p,t)} = \sum_{c=1}^{C} \left[P^{Load}_{(c,t)} - P^{cut}_{(c,t)} \right] \qquad (2)$$

The aggregator, when establishing a contract with the resources, specifies an energy amount that both agree or in real-time monitors the resource's availability. In this case, the resource and the aggregator have previously agreed upon a given amount of flexibility for each period. In this way, the external supplier limits are represented by Equation (3), for the distributed generators by Equation (4) and for the curtailment program by Equations (5) and (6).

$$P_{(s,t)}^{minSup} \leq P_{(s,t)}^{Sup} \leq P_{(s,t)}^{maxSup} \tag{3}$$

$$P_{(p,t)}^{minDG} \leq P_{(p,t)}^{DG} \leq P_{(p,t)}^{maxDG} \tag{4}$$

$$P_{(c,t)}^{mincut} \leq P_{(c,t)}^{cut} \leq P_{(c,t)}^{maxcut} \tag{5}$$

$$P_{(c,t)}^{cut} = P_{(c,t)}^{maxcut} . \lambda_{(c,t)}^{cut}, \ \lambda_{(c,t)}^{cut} \epsilon \{0,1\} \tag{6}$$

This simple mathematical formulation guarantees the correct execution of the aggregator's activities and programs at play. This optimization minimizes the aggregator's cost considering the individual cost (in a first scheduling) and aggregate cost (in a second scheduling) of each resource.

Regarding the aggregation process, this is based on k-means clustering algorithm, which have as inputs the energy scheduled and individual price including the number of groups wanted by the aggregator. The algorithm is based on the minimization of distances between resources and centroids, which is shown in Equation (7). Centroids are points that represent the center of a given group, are initially randomly set, and, in the following iterations, can be computed given a certain rule (e.g., average position of the objects in the group). The distances are then computed for each resource in relation to the centroids (number of centroids equals the number of groups desired) where the nearest are placed in that group. This is an iterative process where resources can change group between iterations.

$$J(T,M) = \sum_{i=1}^{K} \sum_{j=1}^{N} \gamma_{(i,j)} \parallel x_{(j)} - m_{(j)} \parallel^2 \tag{7}$$

where T represents a partition matrix (matrix with the group index for each object), M the cluster prototype or centroid matrix, x an object of a given set (this set corresponds to the resources considered), and m the centroid at the given iteration. The binary variable Y assumes the value one when the object j belongs to the cluster i represented by the centroid and zero when otherwise. The k-means algorithm insures that every object considered is assigned to a group and that all groups have at least one object (none are empty). The k-means algorithm is already developed in MATLAB as a function of k-means. This function returns several outputs including group index for each object, centroid matrix over iterations, distances from objects to each centroid, and the sum of distances for each centroid. In this context, only the first output is needed for the proposed methodology.

The remuneration of resources considers an arithmetic average tariff by the group based on the resource's prices in that group. In this way, there is an average price that some consumers will be encouraged to participate while others may be unsatisfied due to low payment bearing in mind their initial individual price. However, price equality is assured for all resources belonging to the same group. Moreover, in a previous work [25], a maximum-based tariff was proposed in which the highest price in the group was represented as the group tariff where consumers were either satisfied to be remunerated at their price request or encouraged to participate since the tariff was higher than the initial price point.

4. Case Study

The proposed case study is composed of a distribution network with 180 bus [33]. In terms of resources, the network has 116 distributed generators, 90 consumers, and a single external supplier. All consumers can participate in the curtailment program and each one has a distinct price, as shown by Figure 2. It is important to note that although the labels in Figure 2 and its axis rises until 88 due to visualization optimization, the chart includes 90 bars as expected with one for each consumer. Realistic case studies are an important part of the proposed methodology effectiveness and adaption, which insure that it outputs valid solutions for both resources and the aggregator.

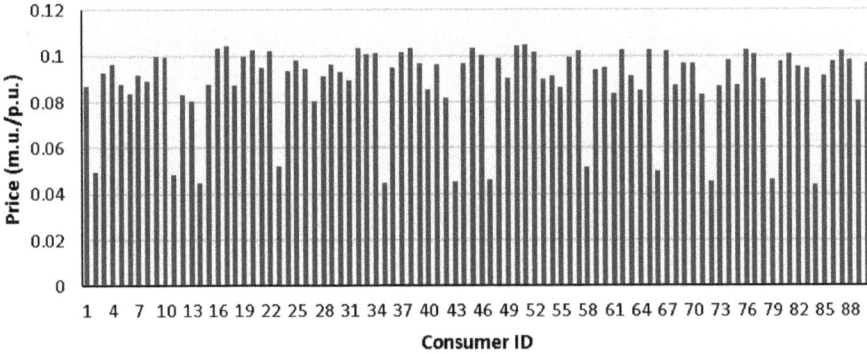

Figure 2. Curtailment prices for consumers.

In generation, the prices are the same for resources of the same type. For instance, all photovoltaic units have an equal price for scheduling. In Table 1, the features of generation resources are presented. In the second column of this table, the installed capacity of each type of generation resource is shown. Moreover, some types of resources have distinct levels of installed power such as in the case of wind and photovoltaic units. In the third column, the number of resources of each resource type are shown and considered for the level of installed power.

Table 1. Generation resources features.

Generation Resource	Inst. Power (kW)	# Units	Price (m.u./p.u.)
Photovoltaic	200	11	0.1560
	150	9	
	25	13	
	20	24	
	15	3	
Small hydro	3010	1	0.1014
Biomass	450	1	0.1231
Co-generation	2100	1	0.0796
Wind	300	2	0.0964
	200	2	
	100	38	
	20	11	
Total	6590	116	-
External supplier	10,000	1	Dynamic

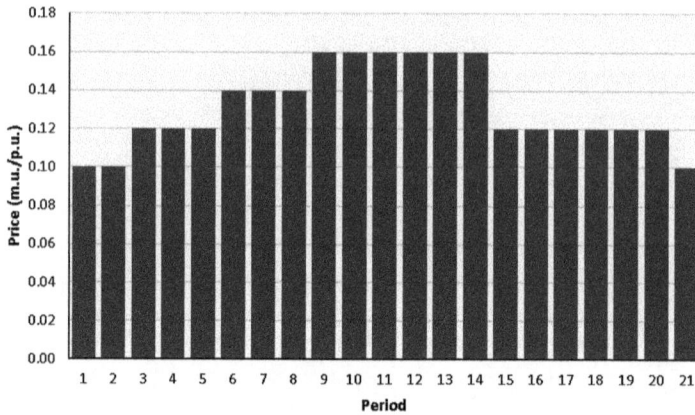

Figure 3. Dynamic tariff considered for the external supplier.

The last column of the table presents the linear prices of each type of resource in exchange for their contribution for scheduling. Besides the distributed generators, Table 1 also presents the features of the external supplier considered in the same terms of the previous mentioned generators. However, it is important to notice that a dynamic tariff is considered for the external supplier and it is shown in Figure 3.

The consumers' curtailment capacity is the same throughout all periods and takes the values as shown in Figure 4. The consumers are classified into five different types including domestic, large commerce, large industrial, medium commerce, and small commerce.

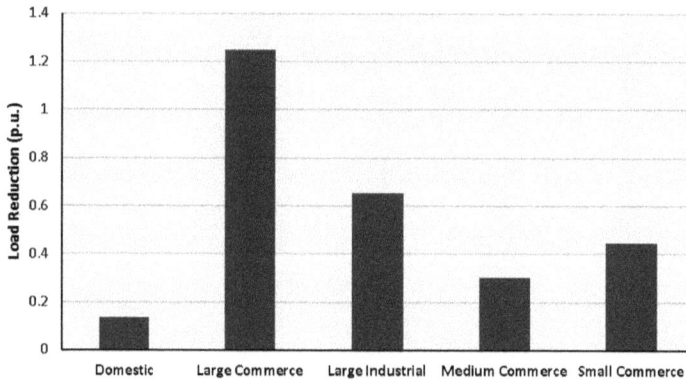

Figure 4. Curtailment capacity in all periods by type of consumer.

This flexibility is used to manage consumption according to the energy prices of the other resources and the available generation at a given period. This type of approach to demand response (load curtailment programs) is often used in power systems by system operators to balance generation and consumption in times where the security and reliability of the network are at risk. In this way, the aggregator can provide relevant services to the system operator and enabling an indirect participation of distributed energy resources (generators and consumers) in the operation of the system.

This section presented the case study evaluated in the present paper and its results are shown in the results section. The proposed case study is adjusted to the operation of an aggregator and represents a realistic approach to the real activities that an aggregator develops.

5. Results

The scheduling of resources is analyzed considering distinct number of groups to be formed in the aggregation process such as the operation costs obtained in the rescheduling of resources, which are distinct considering the number of groups formed after the first scheduling.

Table 2 presents the results obtained in terms of operation costs and number of resources that changed price from the first to second scheduling. The operation costs obtained in the first scheduling is always the same since the initial conditions are maintained equal. The reschedule depends on the number of groups formed after the first schedule. In Table 2, the evaluated periods are chosen based on the power variation that occurred in the reschedule. This includes periods where there are differences between the first and the second scheduling for the available resources. Moreover, it shows the number of resources where the prices were changed between the first and second scheduling due to the aggregation and remuneration processes implemented after the first scheduling. For instance, in period 11 with a total number of groups equal to 3, a total of 80 distributed generators and 75 consumers changed their price. The value in parentheses represents the number of resources that changed the energy schedule while the value between brackets reflects the number of resources where the energy price was raised and lowered, respectively (number of raised prices, number of lowered prices). For instance, given the previous example, only one distributed generator changed the energy schedule and no consumers changed in the second scheduling. Furthermore, of the 80 distributed generators that changed price, 53 had a raise and 27 had a decrease on the price in the second scheduling. Similarly, out of the 75 consumers, 27 had a raise and 48 had a decrease on the price.

Table 2. Summary of results from rescheduling.

		Total Number of Groups			
		3	4	5	6
1st Schedule (m.u.)			43.7693		
Reschedule (m.u.)		43.6797	43.7286	43.8317	43.8075
Evaluated Periods		[10,11,20,21]	[10,20,21]	[20]	[20]
Changes in Distributed Generators	10	75 (1) [53,22]	74 (1) [53,21]	-	-
	11	80 (1) [53,27]	-	-	-
	20	55 (1) [53,2]	53 (1) [53,1]	49 (1) [48,1]	49 (1) [48,1]
	21	55 (1) [53,2]	44 (1) [42,2]	-	-
Changes in Demand Response	10	75 (0) [33,42]	75 (0) [35,42]	-	-
	11	75 (0) [27,48]	-	-	-
	20	75 (0) [34,41]	74 (0) [25,49]	74 (0) [25,49]	73 (0) [33,40]
	21	75 (0) [27,48]	74 (0) [33,41]	-	-

Generally speaking, the optimization of individual consumers is expected to provide better results. However, the aggregator is not able to implement one DR program tariff for each consumer. In the first schedule, the consumers are optimized as groups according to their initial tariffs that are defined according to their consumer's types, which is shown in Figure 4. With the reschedule made based on the consumers grouped according to the clustering input features, the groups are now optimized and the optimization results are better for the total number of groups equal to 3. The obtained actual amount of improvement can be seen as a small amount for a single event occurring in a short period of time. However, implementing several DR events during a year can have a great impact on the overall results for the aggregator.

Moving on to the results obtained for the scheduling of resources, a total number of clusters must be chosen and, therefore, the least expensive is picked in which the total number of groups is equal to 3. The initial scheduling results, before the resources are aggregated and a group tariff is assigned to the participant resources, are shown in Figure 5.

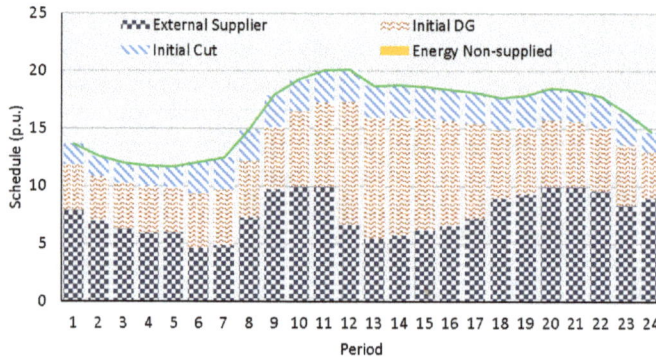

Figure 5. Scheduling before aggregation and remuneration processes.

The scheduling shows the contribution of all the resources considered including the external supplier, distributed generators, and consumers. Figure 6 presented the scheduling of the resources after the aggregating, i.e., stage 1 of phase two where the red outline demonstrates the periods and resource contributions that changed in relation to the initial scheduling. It is possible to see a reduction in terms of external supplier contributions and, consequently, a raise in the distributed generators participation. This variation is also related to the dynamic energy price offered by the external suppliers. Additionally, in both initial and final scheduling, the consumers were supplied without interruption where the "Energy Non-supplied" resource was not implemented. Moreover, this resource is considered the last option to be scheduled since it delimits energy interruptions in the consumers and affects their normal operation. In this way, the use for this is only justified in case of emergency situations where system reliability and security is at risk. In terms of demand response, we see a more or less constant behavior from the consumers with small quantities being used by the aggregator to obtain a valid scheduling of the resources.

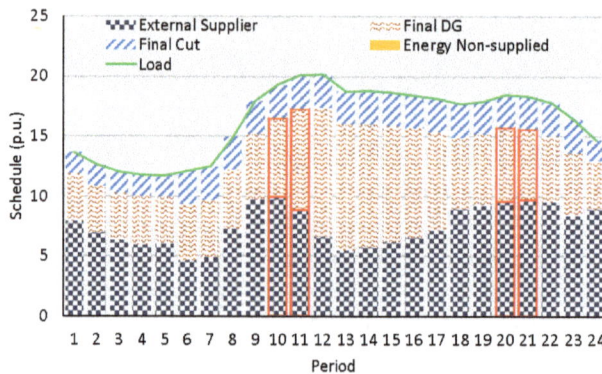

Figure 6. Scheduling after aggregation and remuneration processes.

Figure 6 shows that changes in the energy schedule were verified in periods number 10, 11, 20, and 21 when the total number of clusters is equal to 3.

Figure 7 shows the changes in prices of distributed energy resources between the initial prices and the ones resulting from phase one of the proposed methodology when the total number of groups is 3, which considered the evaluated periods shown in Table 2. For distributed generators (left hand-side graphs), the maximum raise noticed was around 0.019 m.u. (period 20 and 21, as seen in the top-right of each graph) in comparison with the initial price while around 0.042 m.u. was noted in period 10 as the maximum decreased. When considering demand response, the maximum increase was around 0.043 m.u. (periods 11 and 21) while around 0.023 m.u. was noticed in period 11 when the maximum decreased.

Table 3 presents the results obtained for the aggregation of resources, which was independently made for distributed generators and consumers participating in the demand response program for period 20. By matching Tables 2 and 3, it is possible to see that an additional generator was included in the aggregation when comparing with verified changes. The reason for this is that this resource was scheduled by the aggregator, but its price was not changed by the aggregation and remuneration processes performed before the second scheduling.

Table 3. Aggregation of resources—Period 20 | $K = 3$.

Resource	Number of Resources in Group			# Resources
	1	2	3	
Domestic	9	0	4	13
Large Industrial	3	3	0	6
Medium Commerce	12	0	1	13
Small Commerce	21	0	2	23
Large Commerce	9	10	1	20
Total	54	13	8	75
Wind	52	1	0	53
Biomass	1	0	0	1
Photovoltaic	0	0	0	0
Small Hydro	0	0	1	1
Co-generation	0	1	0	1
Total	53	2	1	56

Regarding the consumers, all of those who were scheduled by the aggregator and consequently participated in the aggregation and remuneration processes were also affected by price changes. The results for period 20 are presented for a total number of groups equal to 4 in Table 4 to compare group's number influence based on the tables mentioned before.

Similar results to the previous analysis of period 20 for a total number of groups equal to 3, which were obtained for a total number of groups equal to 4. Moreover, changes only occur at the resource's distribution amongst the groups, but patterns that were visible were the same number of resources assigned to a given group and, having one more group to fill in as showed by Table 4, group assignments of certain resources were changed. The choice of these two evaluations, period 20 for a total number of groups equal to 3 and 4, is based on the operation costs obtained (as shown in Table 4) by being the ones with lower costs when compared to the first scheduling.

Table 4. Aggregation of resources—Period 20 | $K = 4$.

Resource	Number of Resources in Group				# Resources
	1	2	3	4	
Domestic	0	4	0	9	13
Large Industrial	4	0	1	1	6
Medium Commerce	0	1	0	12	13
Small Commerce	0	2	0	21	23
Large Commerce	14	1	0	5	20
Total	18	8	1	48	75
Wind	52	0	1	0	53
Biomass	1	0	0	0	1
Photovoltaic	0	0	0	0	0
Small Hydro	0	1	0	0	1
Co-generation	0	0	0	1	1
Total	53	1	1	1	56

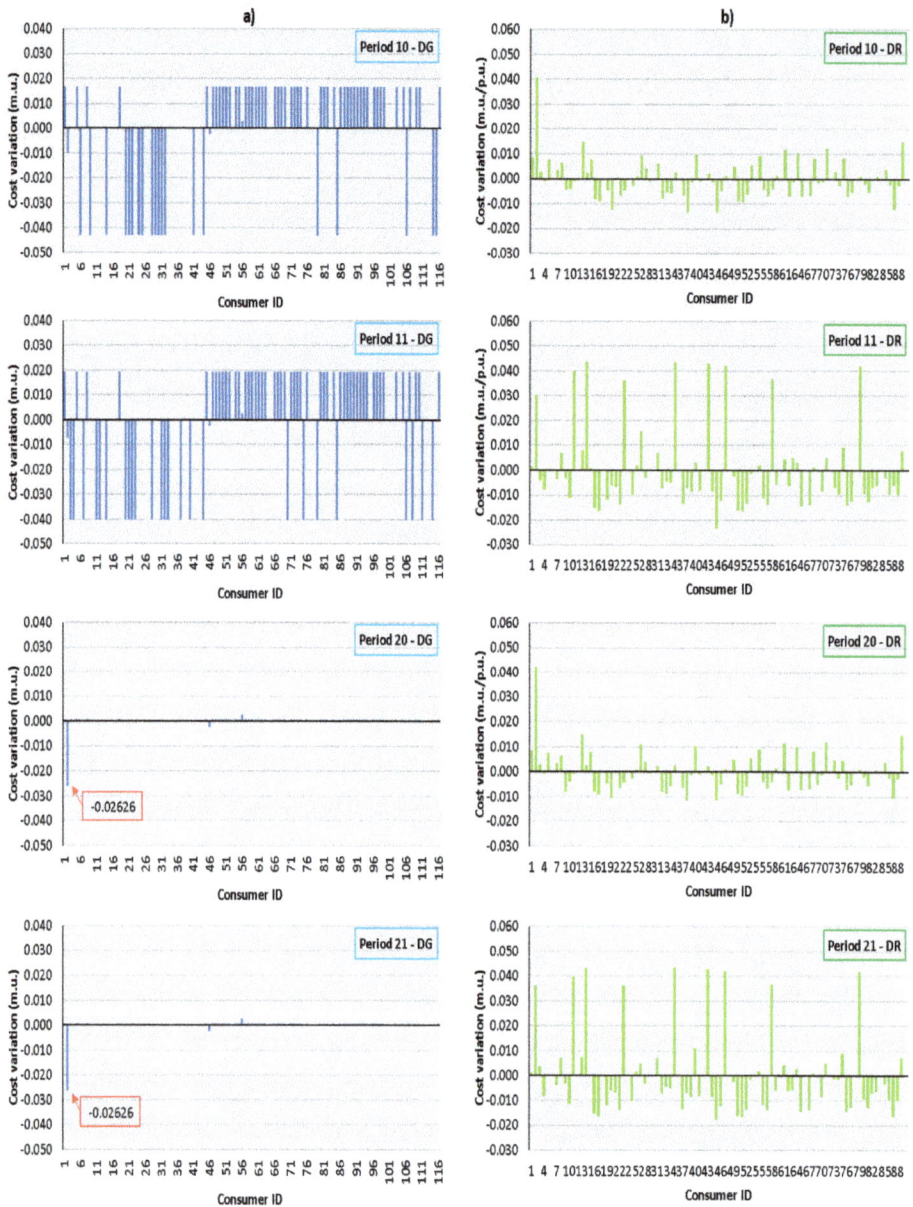

Figure 7. Changes in prices for (**a**) Distributed Generators and (**b**) Demand Response.

6. Conclusions

Aggregators in power systems and energy markets have become more often players in a deregulated environment provided by new legislation promoting the inclusion of distributed energy resources. Aggregators provide several solutions to the operation of power systems from easing complexity to fading energy transit throughout the network. This last feature is focused on integrating distributed energy resources that are capable of surgically injecting generation and/or load in certain points of the network to facilitate its operation.

The present paper proposed a methodology to support an aggregator in dealing with distributed energy resources with a focus on the rescheduling of resources following aggregation and remuneration processes. The aggregator, after an initial scheduling, aggregates the resources participating in the scheduling and computes a representative tariff for each group of distributed energy resources. The initial tariffs of the participating resources are updated to enter a new scheduling (rescheduling) of the aggregator. With the proposed methodology, the aggregator is able to have enriched information in order to have more balanced decisions regarding the consumer's participation and remuneration for DR programs implementation instead of providing a single final optimal decision.

With the results obtained from the first and second scheduling, the aggregator can compare operation costs and evaluate when is best to choose one or the other. Otherwise, the aggregator would not be aware of the impact of the actual scheduling after the new tariffs application resulting from the proposed methodology.

Acknowledgments: This work has received funding from the NETEFFICITY Project (ANI ｜ P2020—18015), from FEDER Funds through COMPETE program, and from National Funds through FCT under the project UID/EEA/00760/2013. This work was also supported by the European Union's Horizon 2020 Research and Innovation Programme under the Marie Sklodowska-Curie Grant Agreement 641794-DREAM-GO Project.

Author Contributions: Pedro Faria raised and developed the overall concept together with Zita Vale, organized the paper and discussed the work and the results with the rest of authors; João Spínola implemented the energy resources optimization model.

Conflicts of Interest: The authors declare no conflict of interest.

Nomenclature

Indexes

S	Total number of external suppliers
P	Total number of distributed generators
C	Total number of consumers
T	Total number of periods
K	Total number of clustering groups
N	Total number of clustering observations

Parameters

$C^{Sup}_{(s,t)}$	Energy tariff for external supplier s, in period t
$C^{DG}_{(p,t)}$	Energy tariff for the distributed generator p, in period t
$C^{cut}_{(c,t)}$	Energy tariff for the load curtailment of consumer c, in period t
$p^{Load}_{(c,t)}$	Energy consumption of consumer c, in period t
$P^{maxSup}_{(s,t)}$	Maximum energy that can be scheduled by the external supplier s, in period t
$P^{minSup}_{(s,t)}$	Minimum energy to be scheduled by the external supplier s, in period t
$p^{maxDG}_{(p,t)}$	Maximum energy that can be scheduled by the distributed generator p, in period t
$p^{minDG}_{(p,t)}$	Minimum energy to be scheduled by the distributed generator p, in period t
$p^{maxcut}_{(c,t)}$	Maximum curtailment that can scheduled by consumer c, in period t
$p^{mincut}_{(c,t)}$	Minimum curtailment that can scheduled by consumer c, in period t

Variables

$P^{Sup}_{(s,t)}$	Energy schedule for external supplier s, in period t
$p^{DG}_{(p,t)}$	Energy schedule for distributed generator p, in period t
$p^{cut}_{(c,t)}$	Energy schedule for load curtailment in consumer c, in period t
$X^{cut}_{(c,t)}$	Binary decision to apply curtailment in consumer c, in period t

References

1. Shen, B.; Ghatikar, G.; Lei, Z.; Li, J.; Wikler, G.; Martin, P. The role of regulatory reforms, market changes, and technology development to make demand response a viable resource in meeting energy challenges. *Appl. Energy* **2014**, *130*, 814–823. [CrossRef]
2. Carreiro, A.M.; Jorge, H.M.; Antunes, C.H. Energy management systems aggregators: A literature survey. *Renew. Sustain. Energy Rev.* **2017**, *73*, 1160–1172. [CrossRef]
3. Nosratabadi, S.M.; Hooshmand, R.-A.; Gholipour, E. A comprehensive review on microgrid and virtual power plant concepts employed for distributed energy resources scheduling in power systems. *Renew. Sustain. Energy Rev.* **2017**, *67*, 341–363. [CrossRef]
4. Li, J.; Wu, Z.; Zhou, S.; Fu, H.; Zhang, X.P. Aggregator service for PV and battery energy storage systems of residential building. *CSEE J. Power Energy Syst.* **2015**, *1*, 3–11. [CrossRef]
5. Li, B.; Shen, J.; Wang, X.; Jiang, C. From controllable loads to generalized demand-side resources: A review on developments of demand-side resources. *Renew. Sustain. Energy Rev.* **2016**, *53*, 936–944. [CrossRef]
6. Paterakis, N.G.; Erdinç, O.; Catalão, J.P.S. An overview of Demand Response: Key-elements and international experience. *Renew. Sustain. Energy Rev.* **2017**, *69*, 871–891. [CrossRef]
7. O'Connell, N.; Pinson, P.; Madsen, H.; O'Malley, M. Benefits and challenges of electrical demand response: A critical review. *Renew. Sustain. Energy Rev.* **2014**, *39*, 686–699. [CrossRef]
8. Walawalkar, R.; Fernands, S.; Thakur, N.; Chevva, K.R. Evolution and current status of demand response (DR) in electricity markets: Insights from PJM and NYISO. *Energy* **2010**, *35*, 1553–1560. [CrossRef]
9. Sorrell, S. Reducing energy demand: A review of issues, challenges and approaches. *Renew. Sustain. Energy Rev.* **2015**, *47*, 74–82. [CrossRef]
10. Fabi, V.; Spigliantini, G.; Corgnati, S.P. Insights on Smart Home Concept and Occupants' Interaction with Building Controls. *Energy Procedia* **2017**, *111*, 759–769. [CrossRef]
11. Theo, W.L.; Lim, J.S.; Ho, W.S.; Hashim, H.; Lee, C.T. Review of distributed generation (DG) system planning and optimisation techniques: Comparison of numerical and mathematical modelling methods. *Renew. Sustain. Energy Rev.* **2017**, *67*, 531–573. [CrossRef]
12. Picciariello, A.; Vergara, C.; Reneses, J.; Frías, P.; Söder, L. Electricity distribution tariffs and distributed generation: Quantifying cross-subsidies from consumers to prosumers. *Util. Policy* **2015**, *37*, 23–33. [CrossRef]
13. Ruggiero, S.; Varho, V.; Rikkonen, P. Transition to distributed energy generation in Finland: Prospects and barriers. *Energy Policy* **2015**, *86*, 433–443. [CrossRef]
14. Gkatzikis, L.; Koutsopoulos, I.; Salonidis, T. The Role of Aggregators in Smart Grid Demand Response Markets. *IEEE J. Sel. Areas Commun.* **2013**, *31*, 1247–1257. [CrossRef]
15. Ottesen, S.Ø.; Tomasgard, A.; Fleten, S.-E. Prosumer bidding and scheduling in electricity markets. *Energy* **2016**, *94*, 828–843. [CrossRef]
16. Mahmoudi, N.; Heydarian-Forushani, E.; Shafie-khah, M.; Saha, T.K.; Golshan, M.E.H.; Siano, P. A bottom-up approach for demand response aggregators' participation in electricity markets. *Electr. Power Syst. Res.* **2017**, *143*, 121–129. [CrossRef]
17. Goebel, C.; Jacobsen, H.A. Bringing Distributed Energy Storage to Market. *IEEE Trans. Power Syst.* **2016**, *31*, 173–186. [CrossRef]
18. Calvillo, C.F.; Sánchez-Miralles, A.; Villar, J.; Martín, F. Optimal planning and operation of aggregated distributed energy resources with market participation. *Appl. Energy* **2016**, *182*, 340–357. [CrossRef]
19. Honkapuro, S. Demand response in Finland-Potential obstacles in practical implementation. In Proceedings of the 11th Nordic Conference on Electricity Distribution System Management and Development, Stockholm, Sweden, 8–9 September 2014.
20. Tampere University of Technology, Tampere University of Applied Sciences, Lappeeranta University of Technology. *Demand Response—Practical Solutions and Impacts for DSOs in Finland.* 2015. Available online: http://dr.wordpress.tamk.fi/wp-content/uploads/sites/3/2015/01/DR-pooli_Summary_in-English.pdf (accessed on 5 January 2017).
21. Koponen, P. *DSM Situation in Finland in 2014*; 2014. Available online: http://sgemfinalreport.fi/files/DemandResponseInFinland2014final3December2014.pdf (accessed on 5 January 2017).

22. CAISO. Demand Response User Guide. CAISO, 2015. Available online: http://www.caiso.com/Documents/DemandResponseUserGuide.pdf (accessed on 5 January 2017).
23. Gils, H.C. Assessment of the theoretical demand response potential in Europe. *Energy* **2014**, *67*, 1–18. [CrossRef]
24. Zehir, M.A.; Batman, A.; Bagriyanik, M. Review and comparison of demand response options for more effective use of renewable energy at consumer level. *Renew. Sustain. Energy Rev.* **2016**, *56*, 631–642. [CrossRef]
25. Faria, P.; Spínola, J.; Vale, Z. Aggregation and Remuneration of Electricity Consumers and Producers for the Definition of Demand-Response Programs. *IEEE Trans. Ind. Informa.* **2016**, *12*, 952–961. [CrossRef]
26. Spínola, J.; Faria, P.; Vale, Z. Model for the integration of distributed energy resources in energy markets by an aggregator. In Proceedings of the 2017 IEEE PowerTech, Manchester, UK, 18–22 June 2017.
27. Eid, C.; Codani, P.; Perez, Y.; Reneses, J.; Hakvoort, R. Managing electric flexibility from Distributed Energy Resources: A review of incentives for market design. *Renew. Sustain. Energy Rev.* **2016**, *64*, 237–247. [CrossRef]
28. Eid, C.; Koliou, E.; Valles, M.; Reneses, J.; Hakvoort, R. Time-based pricing and electricity demand response: Existing barriers and next steps. *Util. Policy* **2016**, *40*, 15–25. [CrossRef]
29. Bhattarai, B.; Lévesque, M.; Bak-Jensen, B.; Pillai, J.R.; Maier, M.; Tipper, D.; Myers, K.S. Design and Cosimulation of Hierarchical Architecture for Demand Response Control and Coordination. *IEEE Trans. Ind. Inform.* **2017**, *13*, 1806–1816. [CrossRef]
30. Baker, P. Benefiting Customers While Compensating Suppliers: Getting Supplier Compensation Right. 2016. Available online: https://www.raponline.org/wp-content/uploads/2016/10/baker-benefiting-customers-compensating-suppliers-2016-oct.pdf (accessed on 15 February 2018).
31. Rious, V.; Perez, Y.; Roques, F. Which electricity market design to encourage the development of demand response? *Econ. Anal. Policy* **2015**, *48*, 128–138. [CrossRef]
32. Hurtado, L.; Mocanu, E.; Nguyen, P.; Gibescu, M.; Kamphuis, R. Enabling Cooperative Behavior for Building Demand Response Based on Extended Joint Action Learning. *IEEE Trans. Ind. Inform.* **2018**, *14*, 127–136. [CrossRef]
33. Soares, J.; Morais, H.; Vale, Z. Particle Swarm Optimization based approaches to vehicle-to-grid scheduling. In Proceedings of the 2012 IEEE Power and Energy Society General Meeting, San Diego, CA, USA, 22–26 July 2012.

MDPI

St. Alban-Anlage 66

4052 Basel

Switzerland

Tel. +41 61 683 77 34

Fax +41 61 302 89 18

www.mdpi.com

Energies Editorial Office

E-mail: energies@mdpi.com

www.mdpi.com/journal/energies

www.ingramcontent.com/pod-product-compliance
Lightning Source LLC
Chambersburg PA
CBHW051837210326
41597CB00033B/5688